ence in

ursing and
Health Care

PEARSON
Education

We work with leading authors to develop the
strongest educational materials in science,
bringing cutting edge thinking and best learning
practice to a global market.

Under a range of well-known imprints, including
Prentice Hall, we craft high quality print and
electronic publications which help readers to understand and apply their content, whether studying or
at work.

To find out about the complete range of our
publishing please visit us on the World Wide Web at:

www.pearsoned.co.uk

Trademarks

Many of the designations used by manufacturers and
sellers to distinguish their products are claimed as
trademarks. Pearson Education Limited has made every
attempt to supply trademark information about
manufacturers and their products mentioned in this
book. A list of the trademark designations and their
owners appears below:

Betadine® – Seton
Gelofusine® and Thermoscan – Braun
Haemaccel® – Hoechst Marion Roussel
Hespan® – Geistlich
Gentran® – Baxter
Sterets® – Seton
K-Y Jelly® – Johnson & Johnson
Konakion® – Roche

Science in
Nursing and
Health Care

Mark Foss and **Tony Farine**

Prentice
Hall

An imprint of **Pearson Education**

Harlow, England · London · New York · Reading, Massachusetts · San Francisco
Toronto · Don Mills, Ontario · Sydney · Tokyo · Singapore · Hong Kong · Seoul
Taipei · Cape Town · Madrid · Mexico City · Amsterdam · Munich · Paris · Milan

Pearson Education Limited
Edinburgh Gate
Harlow
Essex CM20 2JE
England

and Associated Companies throughout the world

Visit us on the World Wide Web at:
www.pearsoned.co.uk

First published 2000

ISBN 0-201-39846-X √S√b

British Library Cataloguing-in-Publication Data
A catalogue record for this book can be obtained from the British Library

Library of Congress Cataloging-in-Publication Data
Foss, Mark A.
    Science in nursing and health care / Mark Foss and Tony Farine.
        p. cm.
    Includes bibliographical references and index.
    ISBN 0-201-39846-X (alk. paper)
    1. Nursing – Study and teaching.   2. Medical sciences.   3. Science.
    4. Allied health personnel – Vocational guidance – Study and teaching.
    I. Farine, Tony.   II. Title.
        [DNLM: 1. Physiology Nurses' Instruction. 2. Biochemistry Nurses'
    Instruction. 3. Genetics Nurses' Instruction. 4. Health Physics
    Nurses' Instruction.   QT 104 F751s 2000]
    RT73.F57 2000
    612'.0024613–dc21
    DNLM/DLC
    for Library of Congress   99-16454
                                                                   CIP

10 9 8 7 6 5 4 3
05 04

Typeset in 10/12pt Sabon by 60
Printed and bound in China
GCC/03

Contents

Preface

Science in Nursing and Health Care has been written with the aim of introducing the reader to aspects of basic science which inform a number of health care disciplines The authors are involved in the teaching of science to nursing students at a British university and the text is primarily intended for use by those studying nursing at diploma and undergraduate levels. However, the general nature of the material means that other health care students may find the text of use, including those studying midwifery, physiotherapy, occupational therapy and dietetics. The expanding number of professionals involved in health care may mean that this book also finds its way onto the shelves of operating department assistants, perfusionists and paramedics.

Our approach

All of the topics covered in the text are also dealt with in a variety of other sources. For example, the concept of homeostasis is described widely in physiology textbooks, acids and bases are described in science texts and mechanical force in physics texts. The particular contribution of this book to the resources available to students is to bring apparently diverse topic areas together in one text. Furthermore, the content of each chapter is also given a context. For example, acids and bases are not described in isolation, but are discussed in the context of the acid–base balance of the body and electromagnetic radiation is placed in the context of diagnosis and treatment.

The aim of the authors is to produce a text which is accurate and thorough, while at the same time being one in which the science content is not dealt with separately from that of nursing. Instead the two are dealt with in an integrated way. The assumption which is made is that, as far as the physical care of individuals is concerned, the prescription of appropriate nursing interventions depends upon a thorough understanding of physical problems. This means that nurses caring for the physically ill must make constant reference to anatomy, physiology and pathology. A problem arises when students lack a knowledge of basic science, and this prevents them from developing a better understanding of the function of the body in health and illness and thus of nursing care too. It is intended that this text will help students to deal with this problem.

How this book is organised

The text is divided into 14 chapters. Although it is possible to read this book from cover to cover, it is expected that most students will instead dip into chapters as particular questions arise in their studies. Consequently, as far as possible, each chapter stands alone. However, information covered in one chapter is not repeated in another. For this reason it makes sense to read Chapter 2 (The physical world and basic chemistry) prior to Chapter 4 (Acids, bases and pH balance).

Each chapter begins with a short list of *learning outcomes*, and you may wish to think about these both before and after reading a chapter. Throughout the text questions appear in *italics*. These are intended to provoke thought as you read. When you come across such a question, pause for a moment and think about the answer. If you are unsure about the correct response, do not worry: the answer is given immediately below the question. However, do get into the habit of attempting to answer these questions before moving on. You will find that the text is a practical one in which science and health care are closely linked. However, some aspects are of a particularly practical nature and appear in boxed sections called *Practice Points*. In addition, bulleted *chapter summaries* will help you to reflect on the information covered, and these are followed by a number of *self-test questions*. The correct answers are also given, but it is clearly a good idea to attempt the questions before looking at the correct responses. Three types of question are used.

Simple multiple choice questions. Simply choose the correct response from a short list of possible answers. For example:

Which one of the following is considered to be a normal systolic blood pressure in mmHg?

a 12

b 120

c 8

d 80

The correct answer is b.

True/false questions. Identify which of a list of statements are true. There is no fixed number of true statements, but one statement at least will be true. For example:

Which of the following statements are true?

a Light is a transverse wave.

b Sound is a transverse wave.

c Sound is a mechanical wave.

d Light is an electromagnetic wave.

In this case a, c and d are true.

Matching pairs questions. These questions consist of two short lists. The items in each list have to be matched. For example:

Match the substances on the left with the appropriate descriptions on the right.

a Na | **i** An ion
b CO_2 | **ii** An atom
c NaCl | **iii** A covalent compound
d HCO_3^- | **iv** An ionic compound

The above match in the following way:

a ii

b iii

c iv

d i

At the end of each chapter there are also two further *study questions/exercises* with *references*. You may not always have the time to complete these at once, but keep it in mind to return to them when time allows. You will find that the questions asked are often of a practical nature and the references make interesting reading.

Summary of key features

- A practical text which integrates science and health care.
- Learning outcomes for each chapter.
- Key terms are highlighted.
- Friendly style with thought-provoking questions included in the text.
- Boxed practice points illustrate important practical applications.
- Bulleted chapter summaries.
- Self-test questions at the end of each chapter.
- Further study questions/exercises.
- Glossary.

All that now remains is for us to wish you well in your studies.

Mark Foss and Tony Farine
Nottingham June 1999

Acknowledgements

We are grateful to the following for permission to reproduce copyright material:

Figures 5.15, 5.17 and 5.18 from *Thoracic Surgery* (Foss, 1989), by permission of the publisher Mosby; Figure 6.2 from *The Addison-Wesley Science Handbook for Students, Writers and Science Buffs* (Coleman & Dewar, 1997), Addison Wesley, Canada; Figure 14.1 from *Longman Study Guides: A-Level and AS-Level Biology* (Cornwell & Miller, 1997), Longman, Harlow; Figures 14.2, 14.3 and 14.4 from *Longman Study Guides: GCSE Biology* (Millican & Barber, 1997), Longman, Harlow, all permission of Pearson Education Ltd; Table 7.4 from *Introduction to Nutrition and Metabolism* (Bender, 1997), 2nd edition, Taylor and Francis Ltd; Table 7.5 from *Beck's Nutrition and Dietetics for Nurses* (Barker, 1991), 8th edition, Churchill Livingstone; Figure 8.3 by permission of Braun, Kronberg; Figure 9.16 from The *Guide to the Handling of Patients: Introducing a Safer Handling Policy* (The National Back Pain Association, 1997), 4th edition, by permission of BackCare; Figure 14.6 from *Essentials of Human Physiology* (Sheeler, 1996), 2nd edition, William C. Brown.

1 Homeostasis – keeping the body in balance

Learning outcomes

After reading the following chapter and undertaking personal study, you should be able to:

1 Explain what is meant by the term the internal environment.
2 Explain what is meant by the term controlled condition and give examples.
3 Define the term *homeostasis*.
4 Identify the components of homeostatic control mechanisms and explain the principle of feedback.
5 Distinguish between negative feedback mechanisms and positive feedback mechanisms and give examples of each.

The internal environment

The smallest units of life are referred to as **cells**. Some organisms consist of just a single cell, whereas the body of an adult man or woman is comprised of billions of cells. Whether a cell forms the entire body of an organism or whether it is one of billions it must have access to the nutrients that it needs for its metabolism, and it must be able to eliminate its waste products. The similarity between single-celled organisms and the cells of our bodies probably ends here, as the following question demonstrates.

> *From where do cells acquire nutrients and where are waste products eliminated to?*

The answer to this question is the environment in which the cell lives. However, the environment of a single-celled organism and that of one of the cells of the human body are often very different. Let us consider a single-celled organism present in the ocean. Its body is so small that it can only be seen with the aid of a microscope, while in comparison its environment is enormous. This means that its metabolic activity (the consumption of nutrients and the production of waste products) has no impact on the environment in which it lives. Compare this condition with that of a cell that is found within the human body. Very few of the cells of a multicellular organism are actually in contact with the environment in which the organism lives.

What then is the environment of the cells of a multicellular organism?

The environment of a cell within the human body is best thought of as the conditions that prevail around it. Cells are in fact bathed in a fluid called **interstitial fluid**. It is from this fluid that cells obtain nutrients and into this fluid that they eliminate waste products. The environment of the cells of our bodies is not therefore the conditions that exist outside the body (*external environment*) but rather those that exist within it – the so-called *internal environment*. Interstitial fluid forms what might be regarded as a kind of 'internal sea'. However, there is a great difference between interstitial fluid and a real ocean, and that is its size. Even though interstitial fluid forms about 16% of the adult body weight, if it were stagnant it would quickly become depleted of nutrients and contaminated by waste products. Clearly there must be a turnover of interstitial fluid so that fresh nutrients replace those that are consumed by cells and so that waste products are eliminated. A number or organ systems contribute to maintaining the constancy of the interstitial fluid, although the cardiovascular system plays the principal role. The circulation of blood provides the means by which fresh materials are brought to cells and waste products are removed. Indeed, interstitial fluid is formed from the blood and is reabsorbed into it. Other systems also play a part by interacting with the environment outside the body. For example, the respiratory system replenishes the blood with oxygen and removes the waste gas carbon dioxide. The digestive system replenishes the blood with nutrients such as glucose, and the urinary system removes waste products, other than carbon dioxide, from the blood and eliminates them to the external environment. These relationships are illustrated in Figure 1.1.

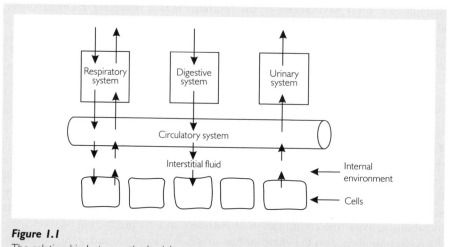

Figure 1.1
The relationship between the body's systems and the internal and external environments.

Homeostasis

An American physiologist of the early 20th century, Walter B. Cannon, noted that if the heat produced by exercising muscles were not dissipated then the ensuing rise in temperature would inactivate the body's proteins. In 1932, Cannon introduced the concept of **homeostasis** in order to explain the way in which our bodies strive to maintain the constancy of the 'internal environment'. Remember that the term internal environment primarily refers to the composition of the interstitial fluid. However, this is not regulated directly but indirectly, as a consequence of the regulation of the composition of blood. Any variable that the body regulates is referred to as a **controlled condition**. Examples of controlled conditions are given in Table 1.1. Any occurrence that provokes a change in controlled condition is referred to as a **stimulus**. How the body responds to stimuli is explained in the next section.

Table 1.1 **Some controlled conditions.**

Temperature	Blood pH
Water balance	Blood glucose level
Electrolyte balance	Blood pressure

Homeostatic control mechanisms

Regardless of the variable that is regulated, all homeostatic mechanisms possess the same three basic components – **receptors**, a **control centre** and **effectors**. Receptors are sensors that respond to a stimulus that brings about a change in a controlled condition. For example, *thermoreceptors* respond to changes in temperature and *osmoreceptors* to changes in blood osmolarity. Information about the controlled condition (*input*) is conveyed to the control centre, which then determines the **set-point** or range at which a controlled condition is to be maintained. It analyses the input from receptors and determines the appropriate *response* (*output*). Effectors are the means by which the control centre regulates the controlled condition. Effectors may be muscles or glands. For example, when we are cold, uncontrollable contraction and relaxation of our muscles (shivering) generates heat. In contrast, when we are hot the secretion of sweat by sweat glands cools down our bodies.

Receptors continually provide control centres with information about the effectiveness of the responses that they initiate. We might say that receptors provide control centres with 'feedback'. Indeed, this is the very term that is used. Homeostatic control mechanisms are described as **feedback mechanisms** or feedback loops. The principle of feedback is illustrated in Figure 1.2. As you consider this diagram, remember that feedback loops operate not intermittently but continually. There is a continuous input to control centres from receptors and a continuous output from control centres to effectors.

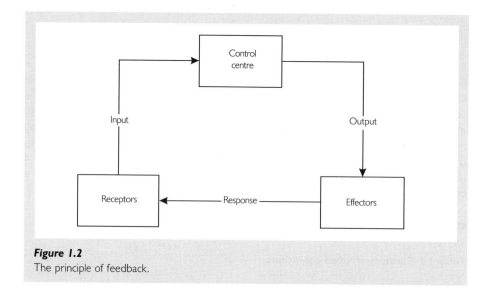

Figure 1.2
The principle of feedback.

Even the briefest description of homeostasis demonstrates that a key feature of control mechanisms is communication within the body (*internal communication*).

> *How is internal communication achieved?*

Internal communication is accomplished by the *nervous system* and the *endocrine system*. In the nervous system information is conveyed in the form of *electrical impulses* in nerves, whereas in the endocrine system ductless glands secrete chemicals, called **hormones**, directly into the bloodstream. A detailed consideration of these systems is beyond the scope of this book. However, they will form part of your study of physiology.

Negative feedback mechanisms and positive feedback mechanisms

Suppose that you were to step outside on a cold day. After some time your body temperature starts to fall.

> *Is the response of the body to initiate heat-generating mechanisms or cooling mechanisms?*

It is of course an obvious question – heat-generating mechanisms are indeed initiated. Now let us take an example that you might not be so familiar with. Suppose that it is some time since you ate last and that your blood glucose level has started to fall.

> *Is the response of the body to raise blood glucose or to lower it still further?*

Of course the body attempts to raise it once again. The purpose of these rather obvious questions is to illustrate a characteristic of most homeostatic control mechanisms – they operate in such a way that the response of the control centre is to reverse the change in the controlled condition. Such mechanisms are described as *negative* feedback mechanisms.

What then are positive feedback mechanisms?

Positive feedback mechanisms are those in which a change in a controlled condition is intensified rather than reversed. There are fewer examples of positive feedback mechanisms than negative feedback mechanisms because they do not result in the stability and control that are the essence of homeostasis. In contrast, they are self-perpetuating and appear to be 'out of control'.

What is their purpose?

Positive feedback mechanisms reinforce a stimulus until a key event is reached that terminates the feedback loop. The birth of a baby is a good example. As the uterus of the mother contracts during labour the baby is pushed against the cervix (muscular neck of the uterus). The stretching of the cervix is detected by pressure receptors that convey impulses (input) to the brain (control centre), which responds by causing the release of the hormone oxytocin (output) from the posterior lobe of the pituitary gland located beneath the brain.

What is the effect of oxytocin on the uterus?

Perhaps you have already been able to predict that the effect of oxytocin is to cause more vigorous uterine contractions that push the baby further towards the cervix, stretching it still further and leading to the release of more oxytocin. This sequence of events is only terminated when the baby is finally expelled from the uterus and is born. Incidentally, the above example illustrates the joint role played in internal communication by the nervous and endocrine systems. Nervous impulses from pressure receptors form the input to the control centre, while the output is in the form of the hormone oxytocin.

The purpose of this short introductory chapter has been to introduce the concept of homeostasis and to distinguish between negative and positive feedback mechanisms. As you work through this text you will encounter homeostasis in a number of different contexts. Perhaps the most notable controlled conditions that you will consider are water balance (Chapter 3), acid–base balance (Chapter 4), blood pressure (Chapter 5), blood glucose level (Chapter 6) and body temperature (Chapter 8). A detailed discussion of the homeostatic mechanisms involved in each case is beyond the scope of this book. However, you will look at these in more detail as you move on from basic science to study physiology as part of your course.

Summary

1 Interstitial fluid forms the internal environment.
2 Homeostasis refers to the regulation of the internal environment.

3 Any variable of the internal environment that is controlled is referred to as a controlled condition.

4 Any event that provokes a change in a controlled condition is referred to as a stimulus.

5 The basic components of all homeostatic control mechanisms are receptors, a control centre and effectors.

6 Control mechanisms are described as either negative feedback mechanisms or positive feedback mechanisms.

2 The physical world and basic chemistry

Learning outcomes

After reading the following chapter and undertaking personal study, you should be able to:

1 Identify three states of matter.
2 Describe the structure of the atom and differentiate between atoms and ions.
3 Define the terms *atomic number*, *mass number* and *relative atomic mass*.
4 Define the term *compound* and differentiate between ionic and covalent compounds.
5 Differentiate between polar and non-polar molecules and describe hydrogen bonds.
6 Define the terms *molecular mass* and *mole*.
7 Briefly explain the following types of chemical reaction: combination, decomposition, displacement, partner exchange, reversible and redox.
8 Explain the action of enzymes.
9 Differentiate between organic and inorganic chemistry.
10 Differentiate between aliphatic and aromatic compounds.
11 Outline the structure of the following groups of organic compounds: alkanes, alkenes, alkynes, alcohols, aldehydes, ketones, carboxylic acids and amines.

Introduction

To some students the prospect of studying chemistry is a daunting one. Nonetheless, a brief examination of the world around us reveals how important this subject is to nurses and to other health care workers too. Our bodies are made up of chemicals, the foods we eat and drink are chemicals, and we use some chemicals as drugs. Therefore a knowledge of chemistry becomes important when we take part in such activities as giving dietary advice and in the administration of medicines. Consequently, this chapter may very well become much more important to you than you might initially imagine! Despite its daunting appearance, the study of chemistry does not have to be difficult. Let us begin with a simple but important point.

Matter

Think about the universe for a moment.

What is it made up of?

At its most fundamental level we might say that the entire universe consists of **matter** and **energy**. The nature of energy is dealt with in Chapter 7 (Energy in the body), while in this chapter we shall think more about matter. Matter is anything that occupies space and it can exist in one of three *states*.

Do you know what the three states of matter are?

The states of matter

No doubt you will have noted that matter may exist in the **solid**, **liquid** or **gaseous** state. When we think of a substance such as oxygen, we tend to think of it in the one state in which we commonly encounter it – in this case the gaseous state. In other cases we may be familiar with a substance in all three states. Water is a case in point. Since its melting point (0 °C) and boiling point (100 °C) are within the range of our everyday experience of temperature, we are familiar with this substance in its solid state (ice), in its liquid state and as a gas (water vapour or steam). Obviously we also have some idea, from everyday experience, about the behaviour of matter in each of these states. For example, we know that solids have a shape and do not adapt to the shape of the container into which they are placed. If we take a rubber ball and place it in a square box it remains round and does not become square like the box! In contrast, liquids have a much less rigid structure and conform to the shape of the container into which they are poured. Nonetheless, their structure is still more rigid than that of a gas. It is possible, for example, to have a container half-filled with a liquid, but gases not only conform to the shape of the container into which they are placed but actually occupy the whole space of the container.

But are there not different kinds of matter?

Yes there are. Let us think about some of them.

Atoms and elements

Some time ago an elderly lady was stopped in the street by a television interviewer and asked about which political party she hoped would win the next general election. In response she said 'I don't give an atom!'. She may have known little else about chemistry but her reply indicated that she knew that an **atom** was something very small and that was how much she cared about politics! The term *atom* is derived from a Greek word meaning 'indivisible', and it was coined at a time when it was believed that the atom was the smallest particle of matter. We now know that atoms can be split into smaller particles, but the atom remains the smallest unit of matter that can enter a chemical reaction. Matter that consists of atoms of only one type is referred to as an **element**. There are 92 naturally occurring elements and a further 20 that are created artificially.

Table 2.1 **Some important elements in the body.**

Element	Importance
Calcium	Muscle contraction, nervous impulses, blood coagulation, structure of bones and teeth
Carbon	Principal element in organic compounds
Chlorine	Principal extracellular anion
Hydrogen	Component of all organic compounds
Iodine	Component of thyroxine
Iron	Component of haemoglobin
Nitrogen	Component of amines, amino acids and nucleic acids
Oxygen	Necessary for cellular respiration
Phosphorous	Component of adenosine triphosphate (ATP) – the body's chemical form of energy
Potassium	Nervous impulses, main intracellular cation
Sodium	Fluid balance, nervous impulses, main extracellular cation

> *Can you name any of the common naturally occurring elements?*

Your list might include hydrogen, oxygen, carbon, and nitrogen. In fact over 95% of a living organism is made up of these four elements alone, so they will often feature in your studies. Table 2.1 is a list of some of the elements that are important in the body.

Writing the names of elements is often a laborious process, so each is given a symbol (abbreviation) that usually consists of the first letter, or first and second letters, of its English or Latin name. If only one letter is used it is given as a capital. For example, the chemical symbol for oxygen is O. If two letters are used then the first is given as a capital and the second as a lower-case letter. Consequently, the symbol for calcium is Ca. The symbol Na for sodium looks confusing at first, but if you know that the Latin name for this element is natrium you will see that the same rules apply as before. However, when elements have similar names the first letter and another instead of the first and second letters are used. For example, the symbol for chlorine is Cl and for chromium it is Cr. Similarly, Mg is the symbol for magnesium and Mn is the symbol for manganese.

For convenience, the elements are grouped in a table referred to as the **periodic table** an example of which is given in Appendix B. Examine this table for a moment. You will see that, in addition to the symbol, the table gives a number of items of numerical data that we shall use later.

> *Now that you have seen the periodic table you are in a position to check the symbols of the elements in Table 2.1. Take a moment to do this now.*

The periodic table

Now that you have begun to use the periodic table we ought to consider it in a little more detail. Perhaps you have already begun to wonder why the elements are arranged in the way that they are and why the table of elements was given the name 'periodic' in the first place. You may be interested to learn that early scientists performed experiments which led them to conclude that atoms of different elements had different weights. The absolute weight of atoms was not being measured, but rather their relative weights. For example, it was shown that, in the reaction between hydrogen and chlorine, the weight of chlorine consumed was approximately 36 times that of hydrogen. It was then concluded that a chlorine atom weighs 36 times as much as a hydrogen atom. Such conclusions as this were based upon a number of assumptions, not all of which proved to be true. However, early scientists were able to ascribe a number, referred to as the **atomic mass**, which indicated the relative masses of atoms.

The Russian chemist Dmitri Mendeleyev noticed that when the known elements were listed in order of ascending atomic mass the properties of the elements, such as physical state and the ability to conduct heat and electricity, varied in a periodic manner. This led to the construction of a table in which elements with similar properties fell into the same vertical columns, and to this table the name 'periodic' was then applied. Incidentally, when using the periodic table it might be helpful to know that the vertical columns are referred to as *groups* and the horizontal rows as *periods*.

Elements to the left of the periodic table are nearly all solids, and they more readily conduct heat and electricity than elements to the right of the table.

> What name is given to these elements?

If you think that the answer is *metals* you are correct. In contrast non-metals are generally poor conductors of heat and electricity; they may be solids, but many are gases. Only the element bromine is liquid at room temperature.

> So where is the dividing line between metals and non-metals in the periodic table?

Unfortunately a rigid dividing line between metals and non-metals cannot be drawn. However, if you look at Appendix B once again you will see a step-like line drawn between the elements boron and aluminium, silicon and germanium and so on. Elements on either side of this line may have characteristics of both metals and non-metals. For example, carbon is a non-metal that exists in a number of forms, some of which do conduct electricity.

Compounds

Some substances consist of atoms of more than one type that are joined together by *chemical bonds*. Such substances are referred to as **compounds**. Compounds are named after the elements that comprise them. For example, sodium chloride is a compound that is formed from sodium and chlorine. In order to avoid the labour of writing the name of compounds the chemical **formula** is used. This is made up of the symbols of the elements that are contained in the compound.

What is the chemical formula for sodium chloride?

The formula is NaCl. Sometimes compounds have common names.

What is the common name for NaCl?

NaCl is known as common (table) salt. However, it is generally better to avoid the use of common names since they can be confusing. For example, there is a great deal of difference between bicarbonate of soda ($NaHCO_3$), caustic soda (NaOH) and washing soda ($NaCO_3$), yet each is referred to as soda. Note too that when different elements combine to form a compound they do not always do so in equal proportions, and in the chemical formula this is taken into account by the use of numerical subscripts. For example, one atom of carbon combines with two atoms of oxygen to form carbon dioxide (CO_2).

The structure of the atom

An atom is the smallest particle of an element that retains the characteristics of that element. Although atoms are very small they are made up of even smaller particles referred to as *sub-atomic particles* or *elementary particles*. These have been given the names **protons** (p), **neutrons** (n) and **electrons** (e^-). These particles have different characteristics. For example, although protons and neutrons have equal mass, that of an electron is negligible. In fact it is taken to be 1/1837 of the mass of a proton or a neutron. In addition, while the neutron carries no charge, the proton has a single positive charge and the electron a single negative charge.

But how are these particles distributed within the atom?

The atom contains a dense core, or **nucleus**, in which the protons and neutrons are located.

Does the nucleus have an overall charge?

You should have been able to work out that since the nucleus contains neutral neutrons and positively charged protons it has an overall positive charge.

So where are the electrons located?

Electrons rapidly orbit the central nucleus in what may be referred to as the **electron cloud**. This is illustrated in Figure 2.1.

Atomic number and mass number

The **atomic number** is simply the number of protons that an atom contains. All the atoms of a particular element have the same atomic number. For example, the atomic number of sodium is 11 and all the atoms of sodium have 11 protons in the nucleus. The atomic number is usually written as a subscript before or after the symbol for the element concerned. In the case of sodium it is $_{11}$Na.

> Go back to the periodic table in Appendix B once again and find out the atomic numbers of the elements in Table 2.1.

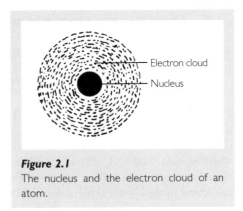

Figure 2.1
The nucleus and the electron cloud of an atom.

At this point it is also worth pointing out that atoms are electrically neutral.

> *This being so, what can you say about the number of electrons in an atom?*

You should have been able to work out that the number of electrons equals the number of protons. However, for the sake of definition, and for reasons that you will understand later, the atomic number is defined as the number of protons that an atom contains.

Atoms are also given a **mass number**, which is simply the sum of the number of protons and neutrons in the nucleus. The mass number is usually given as a superscript before or after the symbol for the element concerned. In the case of sodium it is written as ^{23}Na.

> *Why is the number of electrons not included in the mass number?*

To answer this question, remember that the mass of an electron is only 1/1837 of that of a proton or a neutron. Consequently, electrons make a negligible contribution to the mass of an atom. Furthermore, if both the mass number and the atomic number of an element are known the number of neutrons can be easily calculated.

> *How many neutrons are there in the case of $^{23}_{11}$Na?*

If you have written 12 you have the right answer (that is, $23 - 11$).

Isotopes

Isotopes are atoms that have the same atomic number but different mass numbers. Isotopes are therefore atoms of the same element, but they possess different numbers of neutrons. This is probably best illustrated by the case of hydrogen. All atoms of hydrogen have an atomic number of 1. The most common isotope of hydrogen also has a mass number of 1 ($^{1}_{1}$H), but there is an isotope of hydrogen with a mass number of 2 ($^{2}_{1}$H) and another with a mass number of 3 ($^{3}_{1}$H).

How many neutrons are there in ^1_1H, ^2_1H and ^3_1H?

There are 0, 1 and 2 neutrons, respectively.

The atoms of certain isotopes are unstable and decay with the emission of either high-energy particles or **electromagnetic radiation**. These emissions are described as **radioactivity**, and isotopes that decay in this fashion are described as **radio-isotopes**. This is explained more fully in Chapter 13 (Radiation in diagnosis and treatment).

Relative atomic mass

In order to explain the concept of relative atomic mass (sometimes referred to as average atomic mass or simply atomic mass) we shall consider the element chlorine. There are two isotopes of this element and they have the mass numbers 35 and 37. Of course, both of the isotopes have the atomic number 17.

So how many neutrons are there in the atoms of each of these isotopes of chlorine?

You should have been able to work out that the lighter isotope (^{35}Cl) has 18 and the heavier isotope (^{37}Cl) has 20.

If I had a sample of chlorine gas which isotope would be present?

The answer is both, but not in equal proportions. The lighter isotope is more abundant comprising 77.5% of the sample compared with the heavier isotope's 22.5%.

And how is the existence of two isotopes taken into account when we want to express the relative weight of chlorine atoms?

We clearly need to calculate an average of the mass numbers that takes into account the relative proportions of the two isotopes as shown below:

$$(35 \times 77.5\%) + (37 \times 22.5\%) = 27.125 + 8.325$$

$$= 35.45$$

It is this value that is referred to as the **relative atomic mass**, which we are now in a position to define as the average mass, in atomic mass units, of a single atom of an element that takes into account the relative proportions of the different isotopes of that element. The atomic mass unit (amu or u) is defined so that 12 u equals the mass of the most abundant isotope of carbon (^{12}C).

Energy levels

Electrons are not randomly distributed within the electron cloud of an atom but instead orbit the nucleus in orbitals of varying distance from the nucleus. This model of atomic structure may be described as an orbital model – electrons orbiting the nucleus of an atom may be likened to planets orbiting a star. The electron orbitals are, however, more usually referred to as **energy levels** or sometimes **energy shells**. Electrons with the least amount of energy are found in the lowest energy level (nearest the nucleus), while those with the greatest amount of energy are found in the highest energy level (furthest away from the nucleus). Each energy level is designated by a letter. The first energy level (the one nearest the nucleus) is designated the K shell; next comes the L shell, then the M shell and so on. Each energy level holds a specific number of electrons, and this can be calculated using a simple formula: $X = 2n^2$, where X is the maximum number of electrons in the energy level n. (For the K, L and M energy levels n equals 1, 2 and 3 respectively).

> *So how many electrons does the first energy level hold?*

Using the aforementioned formula you should be able to work out that the calculation is $2 \times 1^2 = 2$.

> *And what about the second and third energy levels?*

You should have worked out that the second energy level holds eight electrons (2×2^2) and the third holds 18 electrons (2×3^2). We should also note that the energy levels are filled in order, beginning with the lowest. Therefore K is filled before L, and L is filled before M. Let us now consider how the electrons are arranged in the atom of familiar elements beginning with sodium. Perhaps you remember that sodium has an atomic number of 11.

> *How are the 11 electrons of sodium arranged?*

You should not have too much difficulty in working out that the K level is filled with two electrons and the L level with eight; this leaves one electron left over in the M level. This electron configuration is usually written as 2.8.1, although it is often given in diagrammatic form, as Figure 2.2 shows.

We can consider chlorine in a similar manner. Chlorine has an atomic number of 17.

> *What is the electron configuration of chlorine?*

It is 2.8.7, and once again this arrangement can be given in diagrammatic form (Figure 2.3).

It should now be a simple matter to work out the electron configurations

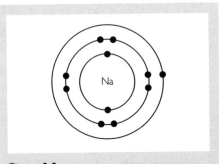

Figure 2.2
The arrangement of electrons in an atom of sodium.

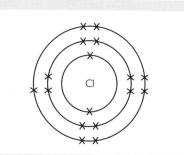

Figure 2.3

The arrangement of electrons in an atom of chlorine.

of the other elements listed in Table 2.1.

Take time to do this now.

Two observations about atomic structure and the position of elements within the periodic table are worth noting. Firstly, elements that have the same number of electrons in the outer energy level fall into the same *group*; secondly, elements that have the same number of energy levels occupied by electrons fall into the same *period*.

Chemical bonds

We have already noted that compounds are substances that consist of atoms of more than one kind joined by chemical bonds. We are now going to study compounds in a little more detail and examine the different kinds of chemical bond that can be formed. Before we do this it is helpful to look at a group of elements called *inert gases* (or noble gases). They can be found in Group VIII of the periodic table.

Take a moment to find them.

The first three members of this group are helium, neon and argon. Perhaps you are already familiar with some of the inert gases. Helium is lighter than air and is sometimes used to fill children's balloons, while neon is used in street lights and illuminated advertisements.

What are the atomic numbers of helium, neon and argon?

They are 2, 10 and 18, respectively.

How are the electrons of helium, neon and argon arranged?

You should have been able to work out that the electron configuration of helium is 2, of neon 2.8 and of argon 2.8.8.

Do you notice anything about the number of electrons in the outer energy levels of these elements?

No doubt you have noted that the outer energy level in each case is either full or contains eight electrons.

What is the significance of these observations?

The description of members of Group VIII as inert gases is made in reference to the fact that they do not react with other elements to form compounds. We now know that most atoms are in their most stable state when their outer energy levels possess eight electrons. In contrast to the inert gases, some elements are highly reactive – that is they readily form compounds with other elements. Sodium is one example of a reactive metal and chlorine is an example of a reactive non-metal. Examine the outer energy levels of sodium and chlorine for a moment.

How many electrons are there in the outer energy level of sodium and chlorine?

You should have noted that the outer energy level of sodium contains only one electron, while the outer energy level of chlorine possesses seven electrons. You will understand the significance of this as we move on to consider **ions** and **ionic bonding**.

Ions and ionic bonding

Remember that atoms contain equal numbers of protons and electrons, so they are electrically neutral.

But what would happen if an atom either lost or gained one or more electrons?

You should have been able to work out that it would no longer be electrically neutral but would possess a net charge. This charged structure is no longer called an atom, but is instead referred to as an **ion**. Let us think about the atomic structure of sodium and chlorine once again.

How could sodium achieve the stability of the noble gas neon?

It could do so by losing the single electron from its outer energy level.

What would the effect of this be on the sodium atom?

It would become a sodium ion with a single positive charge (Na^+).

What about chlorine? How could a chlorine atom attain the stability of the inert gas argon?

Perhaps you have worked out that, for chlorine, stability is achieved by gaining an additional electron in its outer energy level and thus becoming a chloride ion with a single negative charge (Cl^-). Incidentally, positively charged ions are referred to as **cations** and negatively charged ions are referred to as **anions**.

Good question – the answer is no. Generally, atoms with up to three electrons in the outer energy level achieve stability by losing electrons to form cations, while atoms with six or seven electrons in the outer energy level achieve stability by gaining electrons to form anions. Atoms with four or five electrons in the outer energy level do not form ions, although it is worth pointing out that there are exceptions to these general rules.

We are now in a position to explain the chemical reaction between sodium and chlorine that leads to the formation of the compound sodium chloride. Both atoms achieve their most stable state when sodium donates an electron to chlorine. This is represented in the form of an equation:

$$Na + Cl \longrightarrow Na^+ + Cl^-$$

Note that the sodium ion and the chloride ion carry opposite charges, and, as a consequence, they are strongly attracted to each other. We say that there is an **ionic bond** between them, and this results in a new substance – the compound sodium chloride (NaCl). Note that it is not usual to include the charges when writing the formula of ionic compounds. Ionic bonding can also be represented diagrammatically, as shown in Figure 2.4.

Before we leave ionic bonding let us consider a further example – calcium chloride.

What is the atomic number of calcium (Ca)?

It is 20.

And how are the electrons arranged in calcium?

You should have noted an electron configuration of 2.8.10.

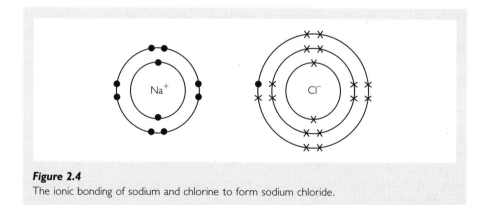

Figure 2.4
The ionic bonding of sodium and chlorine to form sodium chloride.

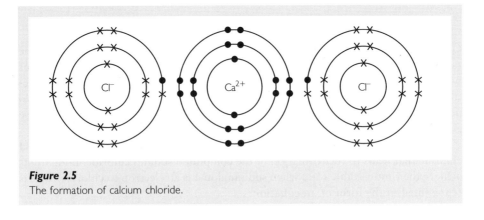

Figure 2.5
The formation of calcium chloride.

So how could calcium attain the stability of the inert gas argon?

It could do so by losing two electrons from its outer energy level to form a calcium ion.

And what charge does a calcium ion carry?

You should have been able to work out that, since two electrons have been lost, the calcium ion carries a double positive charge, which is written as Ca^{2+}. Let us think about this a little further. Suppose calcium were to react with chlorine to produce calcium chloride.

How many atoms of chlorine would react with each atom of calcium?

You should have worked out that the answer is 2. Consequently, the formula for calcium chloride is $CaCl_2$. An important point to note here is that not all atoms have the same combining power – a concept that is referred to as **valency**. Sodium and chlorine both have a valency of 1 and as a consequence they combine in an equal ratio. In contrast, calcium has a valency of 2 so when a calcium atom reacts with an element with a valency of 1, two atoms of the other element are involved in the reaction. Once again, this can be represented diagrammatically, as Figure 2.5 shows.

Since it is only the electrons in the outer-most energy level that are involved in chemical bonding, this energy level is referred to as the **valence shell**.

Polyatomic radicals

Some ions are actually made up of atoms of more than one element and they are referred to as **polyatomic ions** or **polyatomic radicals**. Table 2.2 contains a list of polyatomic radicals that are important within the body. At first sight they look a little complicated, but since polyatomic radicals act as though they are a single ion they are actually easy to deal with. For example, although the hydrogen carbonate

Table 2.2 **Important polyatomic radicals.**

Polyatomic radical	Importance
Ammonium (NH_4^+)	Regulation of urinary pH
Dihydrogen phosphate ($H_2PO_4^-$)	Regulation of urinary pH
Hydrogen carbonate (HCO_3^-)	Regulation of blood pH, transport of CO_2
Hydroxide (OH^-)	Produced by many bases
Monohydrogen phosphate (HPO_4^{2-})	Regulation of urinary pH
Phosphate (PO_4^-)	Component of adenosine triphosphate (ATP) – the body's chemical form of energy

ion (HCO_3^-) contains three different elements it behaves as an anion with a single negative charge. Consequently, it forms an ionic bond with a sodium ion to form sodium hydrogen carbonate ($NaHCO_3$).

Covalent bonding

We have already noted that certain atoms do not form ions.

Does this mean that they do not form chemical bonds with other atoms?

Good question. Atoms that do not form ions do form compounds with other atoms, but they do so through a process of electron sharing. The bonds that result from the sharing of electrons are called **covalent bonds,** and the structure that results from the covalent bonding of atoms is referred to as a **molecule**. Water (H_2O) is one example of a covalent compound. In the water molecule, an atom of oxygen shares two of its electrons with two atoms of hydrogen.

In order to understand this arrangement, make a note of the electron configurations of oxygen and hydrogen.

No doubt you have noted that, since oxygen has an atomic number of 8, its electron configuration is 2.6 and its valency is 2 since it requires two more electrons to achieve a stable valence shell. In contrast, hydrogen has an atomic number of 1 and possesses a single electron in its only energy level. It therefore has a valency of 1 since it requires one further electron in order to achieve a stable valence shell. In a water molecule the oxygen atom has six of its own electrons in its valence shell plus one shared with each of the two hydrogen atoms. Similarly, the valence shell of each of the two hydrogen atoms has one of its own electrons plus one shared with the oxygen atom. The shared electrons orbit the nuclei of both of the atoms involved in the bond, but they are most often in the region between the two nuclei. Covalent bonding can be illustrated diagrammatically, as Figure 2.6 shows.

At this point it is worth pointing out that covalent bonds can be formed between atoms of the same element. For example, two atoms of hydrogen share electrons to form a molecule of hydrogen – H_2. Similarly, oxygen and nitrogen do not exist as individual atoms but as molecules, O_2 and N_2 respectively. Consequently, these elements are referred to as **diatomic** gases (literally, two-atom gases). The structure of hydrogen, oxygen and nitrogen is shown in Figure 2.7.

Double and triple covalent bonds

Have a look at Figure 2.7 once again.

> *In the formation of molecules of hydrogen, oxygen and nitrogen, how many electrons are shared in each case?*

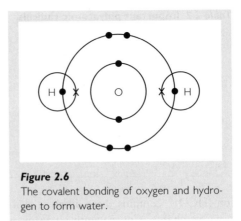

Figure 2.6
The covalent bonding of oxygen and hydrogen to form water.

You should have noted that in the case of hydrogen, each atom shares one of its electrons with another and a *single* covalent bond is formed. In the case of oxygen, each atom shares two of its electrons with another and this is referred to as a *double* covalent bond. Not surprisingly, when nitrogen atoms each share three electrons a *triple* covalent bond is formed.

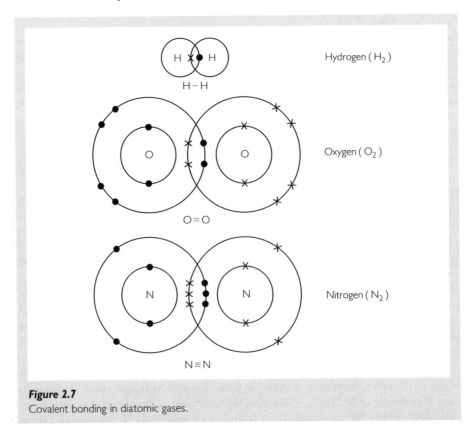

Figure 2.7
Covalent bonding in diatomic gases.

Non-polar and polar molecules

Covalent molecules may be described as either **polar** or *non-polar*. If the electrons that take part in a covalent bond are shared equally between the atoms of the molecule concerned the molecule is described as being non-polar. This means that the shared electrons spend equal amounts of time orbiting both nuclei and there is an even distribution of charge within the molecule. Non-polar molecules are the result of covalent bonding between identical atoms, as is the case in H_2, O_2 and N_2. Polar compounds result when the electrons that take part in a covalent bond are not shared equally between the atoms of the molecule concerned.

> *Why might the electrons in a covalent bond not be shared equally between the atoms of the molecule?*

Perhaps you have worked out that an atom with a greater number of protons will exert a greater attraction over shared electrons than an atom with fewer protons because of the greater positive charge carried by the larger nucleus. Such is the case in the water molecule, in which each oxygen atom shares electrons with two atoms of hydrogen.

> *How many protons do the nuclei of oxygen and hydrogen possess?*

Remember that this information is given by the atomic number. The oxygen nucleus is larger than the hydrogen nucleus; it possesses eight protons while hydrogen has only one. Consequently, the oxygen atom attracts shared electrons more strongly than the hydrogen atoms, and this leads to it possessing a slight negative charge, while the hydrogen atoms possess a slight positive charge. The polar nature of water is more fully explained in Chapter 3 (Water electrolytes and body fluids).

Table 2.3 shows the characteristics of ionic and covalent bonds.

Hydrogen bonds

When the covalent bonding of a hydrogen atom to an atom of another element results in the formation of a polar molecule, the positive (hydrogen) end of one molecule is attracted to the negative end of a second polar molecule and the resultant bond is called a hydrogen bond. Hydrogen bonds occur between water

Table 2.3 **Table showing the characteristics of ionic and covalent bonds.**

Ionic bonds	Ionic compounds
Formed by the transfer of electrons.	Tend to have high melting points.
Involve the formation of oppositely charged ions that attract each other.	Soluble in polar solvents such as water.
Covalent bonds	Covalent compounds
Formed by sharing electrons.	Tend to have low melting points.
	Dissolve in non-polar solvents such as alcohol.

molecules when the slightly positive hydrogen ends attract the partially negative oxygen ends of other molecules. Hydrogen bonds also occur between hydrogen atoms that are attached to nitrogen atoms and oxygen atoms that are part of a C=O group. Such hydrogen bonds are present in most proteins (see Chapter 6: Biological molecules and food) and in deoxyribonucleic acid (DNA). Hydrogen bonds are usually represented in diagrams as dotted lines in order to distinguish them from ionic and covalent bonds, which are between 10 and 20 times as strong.

Structural formulae

The type of chemical formula that we have used so far (for example, CO_2) is more correctly referred to as the **molecular formula**. It provides us with information about the atoms present in a compound and the ratios in which they occur. While this information is useful, a molecular formula does not show how the atoms of a compound are arranged. However, this information is shown by the **structural formula**, in which bonds between atoms are shown as solid lines. By way of example, the structural formulae of three compounds are given in Figure 2.8. It is also worth noting here that compounds that share the same molecular formulae but have different arrangements of atoms are referred to as **isomers**.

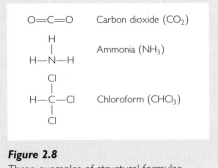

Figure 2.8
Three examples of structural formulae.

Molecular mass

The molecular mass of a compound is the sum of the masses of the elements that make up that compound. For example, since the atomic mass of carbon is 12.01 and that of oxygen is 16, the molecular mass of carbon dioxide (CO_2) is calculated as follows:

$$12.01 + (2 \times 16) = 44.01$$

Using data from the periodic table calculate the molecular mass of glucose ($C_6H_{12}O_6$).

You should have been able to calculate it in the following way:

$$(12.01 \times 6) + (1.01 \times 12) + (16 \times 6) = 72 + 12.12 + 96$$

$$= 180.18$$

The mole

Let us begin this section by asking a rather obvious question.

> *How many apples are there in a dozen?*

Yes: there are 12.

> *And how many oranges are there in a dozen?*

Once again the answer is 12 – a dozen always refers to 12, and it does not matter whether apples or oranges are being referred to. In a similar way a **mole** is an expression of the amount of a substance that contains a specific number of particles. They could be atoms, ions or molecules.

> *So what number of particles does a mole of a substance contain?*

Certainly not a dozen, but a very large number indeed – 6.022×10^{23} in fact. This number is sometimes referred to as Avogadro's number, which for the purpose of definition is the number of atoms in $12\,g$ of the ^{12}C isotope.

> *But what about other elements – what is the weight of one mole of sodium atoms, for example?*

This is actually very easy to work out. The weight of a mole of atoms is equal to the atomic mass of an element expressed in grams. Since the atomic mass of sodium is 23, one mole of sodium atoms weighs $23\,g$.

> *And what about the weight of one mole of NaCl?*

Once again this is easy to work out – one mole of a compound is equal to the molecular mass of that compound expressed in grams. In the case of NaCl it is $58.5\,g$.

> *Is there any point in learning about moles?*

Indeed there is. Moles are used by nurses and other health care workers as well as by scientists. One of the most common uses of the mole is in expressing concentration, whether it be of substances dissolved in blood, intravenous fluids or solutions of drugs. This is more fully explained in Chapter 3 (Water electrolytes and body fluids).

Chemical reactions

A chemical reaction is an interaction between chemical substances that results in a change of bonding between atoms. The substances that take part in a chemical reaction are referred to as **reactants** and the new substances produced as a consequence of a reaction are referred to as **products**. Chemical reactions are described in the form of an equation using chemical symbols. In such equations, reactants and products are separated by a horizontal arrow pointing away from the reactants on the left and towards the products on the right. When there is more than one reactant or more than one product the substances are separated by a + sign in the manner shown below:

$$A + B \longrightarrow AB$$

When using chemical equations it is important to remember the principle of the *conservation of matter*, which states that matter cannot be created or destroyed in a chemical reaction. That is, there must be the same number of atoms of each element on the left and right of the equation. When this is the case, the equation is said to be *balanced*.

There are many different types of chemical reaction, but most fall into one of several general types, some of which are shown below.

Combination reactions

Combination reactions involve two or more substances combining to form a single substance. Such reactions are sometimes referred to as *synthesis reactions*, and they conform to the pattern shown below:

$$A + B \longrightarrow AB$$

One simple example of a combination reaction is that between sodium and chlorine to form sodium chloride:

$$Na + Cl \longrightarrow NaCl$$

Note that the above equation is balanced – there is one atom of sodium and one atom of chlorine on the left and one ion of each on the right.

> *Are there examples of combination reactions in the body?*

There are, as the following practice point shows.

Practice point 2.1: combination reactions in the body

One example of a combination reaction in the body is that between carbon dioxide and water to produce carbonic acid according to the equation below:

$$CO_2 + H_2O \longrightarrow H_2CO_3$$

Carbon dioxide is a waste gas produced by the body's cells. You will encounter the above equation again when you study Chapter 4 (Acids, bases and pH balance).

The combination of oxygen with the red blood cell pigment haemoglobin (Hb) to form oxyhaemoglobin (HbO) is a further example of a combination reaction. Haemoglobin is a rather complex molecule, so no attempt is made here to give its chemical formula. However, we can illustrate its reaction with oxygen in the following way:

$$Hb + 4O_2 \longrightarrow HbO$$

A combination reaction is also involved in the formation of **adenosine triphosphate** (ATP – the body's chemical form of energy) from **adenosine diphosphate** (ADP) and a phosphate ion (P_i). Once again, no attempt is made to give the formulae of these rather complex molecules, but the reaction can be illustrated in the following way:

$$ADP + P_i \longrightarrow ATP$$

A combination reaction that involves a phosphate group is given the name **phosphorylation**. Such reactions are important in the body's energy-liberating processes. These are more fully explained in Chapter 7 (Energy in the body).

Finally, combination reactions provide the means by which the body grows and repairs itself. One example is the manufacture of proteins from amino acids. The structure of proteins is described in Chapter 6 (Biological molecules and food). Such combination reactions are often referred to as *anabolic* and the sum of these reactions as **anabolism**.

Decomposition reactions

Decomposition reactions involve the breakdown of one substance to produce two or more products according to the pattern shown below:

$$AB \longrightarrow A + B$$

You will no doubt realise that examples of decomposition reactions in the body include the reverse of those noted in the practice point above:

$$H_2CO_3 \longrightarrow CO_2 + H_2O$$
$$HbO \longrightarrow Hb + 4O_2$$
$$ATP \longrightarrow ADP + P_i$$

The digestion of food in the gut is also a form of decomposition reaction, and you can read more about this in Chapter 6 (Biological molecules and food). Decomposition reactions in the body are sometimes referred to as *catabolic* and the sum of these reactions as **catabolism**.

What word is used to collectively describe anabolism and catabolism?

The word is **metabolism**.

In addition, some compounds decompose (dissociate) into ions when dissolved in water. Not surprisingly, this form of decomposition is referred to as **ionisation**. It is illustrated below using the examples of hydrochloric acid, sodium hydroxide and sodium chloride:

$$HCl_{(aq)} \longrightarrow H^+ + Cl^-$$
hydrochloric acid   hydrogen ions   chloride ions

$$NaOH_{(aq)} \longrightarrow Na^+ + OH^-$$
sodium hydroxide   sodium ions   hydroxyl ions

$$NaCl_{(aq)} \longrightarrow Na^+ + Cl^-$$
sodium chloride   sodium ions   chloride ions

In the case of hydrochloric acid, ionisation produces hydrogen ions and chloride ions. **Acids** are defined as substances that liberate hydrogen ions. In the case of sodium hydroxide, ionisation results in sodium ions and hydroxyl ions. A substance that liberates hydroxyl ions is referred to as a **base**. Acids and bases are discussed in Chapter 4 (Acids, bases and pH balance). Finally, in the case of sodium chloride ionisation produces sodium ions and chloride ions. A substance that ionises to produce ions other than hydrogen ions or hydroxyl ions is referred to as a **salt**.

Displacement reactions

Displacement reactions involve the displacement of a less reactive element by a more reactive one, and they conform to the general pattern shown below:

$$A + BC \longrightarrow AC + B$$

This type of reaction is sometimes described as a *substitution* reaction, since one element is substituted for another.

Practice point 2.2: displacement reactions in the body

An abnormally elevated concentration of potassium ions in the blood is treated using a displacement reaction involving calcium polystyrene sulphate. Following the administration of this medicine, calcium is displaced by potassium in the following way:

$$K^+ + \text{calcium polystyrene sulphate} \longrightarrow Ca^{2+} + \text{K polystyrene sulphate}$$

By this means potassium ions are removed from the blood.

Partner exchange

Partner exchange reactions could be regarded as those in which there is a double displacement. That is, the substitution of elements between compounds results in the formation of more than one new compound. Such reactions conform to the pattern noted below:

$$AB + CD \longrightarrow AD + CB$$

An example of such a reaction is that between hydrochloric acid (HCl) and sodium hydrogen carbonate ($NaHCO_3$) according to the equation shown below:

$HCl_{(aq)}$	+	$NaHCO_3$	$\longrightarrow$	H_2CO_3	+	NaCl
hydrochloric acid		sodium hydrogen carbonate		carbonic acid		sodium chloride

---

Practice point 2.3: partner exchange reactions in the body

Hydrochloric acid is of course the acid produced by the stomach, while sodium hydrogen carbonate is a component of a number of antacid medicines which are used to treat excess stomach acidity. When such medicines are swallowed the above reaction takes place in the stomach. By this means a strong acid (HCl) is replaced by a weak one (H_2CO_3). In addition, you should also know from the section on decomposition reactions that H_2CO_3 decomposes into CO_2 and water.

What now happens to this CO_2?

No doubt you have just worked out why antacid medicines often make you burp!

Reversible reactions

In view of the convention of placing reactants on the left of an equation arrow and products on the right, we can think of reactions as having a direction. We might say that a reaction proceeds from left to right. However, in some cases the reverse reaction also takes place – that is, the reactants can be reformed from the products. Consequently, we might say that, in such cases, the reaction also proceeds from right to left. Such reactions are then said to be **reversible reactions**, and this is indicated by the presence of a double-arrow in the equation. An example of such a reaction is shown below:

H_2CO_3	$\rightleftharpoons$	HCO_3^-	+	H^+
carbonic acid		hydrogen carbonate ion		hydrogen ion

After a period of time the rate of the decomposition reaction (left to right in the above example) and the combination reaction (right to left) become equal and we say that a *chemical equilibrium* has been reached. For this reason reversible reactions are sometimes called *equilibrium reactions*.

Does this mean that there is an equal concentration of reactants and products?

No. When a reversible reaction reaches an equilibrium it simply means that the rate of the two reactions (left to right and right to left) are equal and there is no net increase in the concentration of the reactants and products. In the above reaction, the concentration of hydrogen carbonate ions in blood is 20 times that of carbonic acid molecules.

> *So could anything occur to change the rate of the reactions?*

Indeed it could. For example, if the concentration of carbonic acid were to become elevated the rate of the decomposition reaction (left to right) would increase until the carbonic acid:hydrogen carbonate ion concentration was restored to 1:20 once again. We might say that the reaction equilibrium had been *displaced* to the right. In fact, this is an important principle behind the behaviour of reversible reactions – any change in concentration of reactants or products displaces the equilibrium so as to restore these concentrations.

Oxidation–reduction reactions

Oxidation reactions are very important in the world around us. For example, the rusting of iron and the burning of natural gas are both examples.

> *What actually happens in an oxidation reaction?*

The original definition of oxidation was a reaction in which a substance gained an oxygen atom or lost a hydrogen atom. Consider the example below:

$$CH_3CHO \longrightarrow CH_3COOH$$
acetaldehyde   acetic acid

> *Why do we say that acetaldehyde has been oxidised to acetic acid?*

You should have been able to work out that it is because an oxygen atom has been acquired. You may not yet be familiar with these compounds, but simply count the number of oxygen atoms that each possesses. Acetaldehyde has one and acetic acid has two. Now consider another example involving nicotinamide adenine dinucleotide (NAD).

$$NADH \longrightarrow NAD^+$$

> *Why do we say that NADH has been oxidised?*

Because it has lost hydrogen. However, chemical reactions similar to these also occur in the absence of oxygen and hydrogen and so oxidation is currently defined as a chemical reaction in which a substance loses one or more electrons. In fact, in the case of NADH it actually loses two electrons as well as a hydrogen ion (H^+). Consequently, the product is NAD^+, not NAD.

And what about reduction; how is it defined?

Originally, **reduction** was defined as a reaction in which a substance lost an oxygen atom or gained a hydrogen atom. However, this definition has also been revised so that reduction is currently defined as a reaction in which a substance gains one or more electrons.

So reduction is in fact the opposite of oxidation?

Indeed it is. In fact, oxidation can never occur without reduction, as the electrons lost from one substance will be picked up by another. Consequently, it is more accurate to speak of *oxidation–reduction* reactions rather than one or the other. Chemists usually abridge this expression to *redox* reactions.

Chemical reactions and energy

Chemical reactions invariably involve energy. A certain amount of energy is required to bring about a chemical reaction in the first place, and this is referred to as the **activation energy**. It represents the energy required to break the chemical bonds of the reactants. If the products of a chemical reaction possess less energy than the reactants themselves then energy will be liberated during the reaction. Such a reaction is then described as **exergonic** (exothermic). For example, when ATP is broken down to ADP energy is liberated and used by the cell in order to sustain its energy-requiring processes. In contrast, if the products of a chemical reaction possess more energy than the reactants themselves then energy will have to be supplied in order for the reaction to proceed. Such chemical reactions are described as **endergonic** (endothermic). The formation of ATP from ADP and an inorganic phosphate molecule is obviously an example of an endergonic reaction.

But where does this energy come from in the first place?

It comes from the breakdown of the food that we eat.

Factors that affect the rate of chemical reactions

A number of factors affect the rate of chemical reactions, including the **concentration** of reactants and the temperature at which the reaction is taking place. In order to illustrate how these two factors affect reaction rate, consider the example of a simple combination reaction. A reaction will only take place between reactants when they collide with sufficient energy to break the bonds that initially exist.

How would an increase in the concentration of reactants or an increase in temperature affect the likelihood of a collision?

Both these changes would increase the likelihood of a collision and therefore the reaction rate would also increase. In addition, an increase of temperature causes reactants to collide with greater energy, and this makes a reaction upon collision more likely too. A reaction rate would also increase if the activation energy could be reduced. This can be achieved by the presence of a **catalyst**.

Catalysts

A **catalyst** is a substance that increases the rate of a chemical reaction without itself being changed by that reaction. Catalysts achieve their effect by lowering the activation energy. They are responsible for considerable improvements in the efficiency of industrial processes, but more important here are proteins, which fulfil the role of catalysts within the body.

What is the name by which biological catalysts are known?

They are called **enzymes**, one example of which is **carbonic anhydrase**. This enzyme is present in red blood cells, and it catalyses the reaction between carbon dioxide and water to form carbonic acid. By now you should have become quite familiar with this reaction! When an enzyme is involved in a reaction it is identified by placing its name above the equation arrow, as shown below:

$$CO_2 + H_2O \xrightarrow{\text{carbonic anhydrase}} H_2CO_3$$

Enzymes are highly effective catalysts and life would not be possible without them. They are also highly specific – each has an effect on one substance alone or on a number of closely related substances that are known as **substrates**.

But how do catalysts work?

The *lock and key hypothesis* is an attempt to explain enzyme action. It is illustrated in Figure 2.9 in the context of a decomposition reaction, but it should be remembered that enzymes also catalyse combination reactions, as the previous example has shown. The first step in enzyme action is the formation of a complex with the substrate. An *enzyme–substrate complex* forms because the unique three-dimensional structure of an enzyme leads to the existence of an *active site*. This is a small region which has a reciprocal shape to that of the substrate, into which the substrate fits, rather like a key fitting into a lock. The enzyme is believed to stretch, weaken and finally break chemical bonds within the substrate, and this

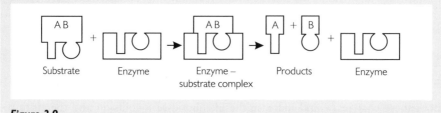

Figure 2.9
The lock and key hypothesis of enzyme action.

allows new products to form. These then move away from the active site, leaving the enzyme unchanged. Another related model of enzyme action is described as **induced fit**. In this explanation the active site is regarded as being flexible rather than rigid. It is suggested that the active site undergoes a change in shape during the formation of the enzyme–substrate complex so as to accommodate the substrate.

While some proteins are capable of functioning as enzymes alone, others need to be joined to other substances. In such a case the protein part of the enzyme is referred to as the **apoenzyme** and the non-protein part as a **co-factor**. Some co-factors are simple metal ions while others are quite complex organic molecules called **co-enzymes**. Many vitamins function as co-enzymes.

Sometimes enzymes lose their ability to function and they are then said to have been *denatured*. Heat, a change in pH and some chemical substances are commonly responsible.

> *What actually happens when an enzyme is denatured?*

Essentially, **denaturation** occurs when the active site is lost because of damage to the three dimensional structure of the enzyme. This occurs because of the disruption of the weak hydrogen bonds responsible for the unique shape of protein molecules.

Finally, you might find it helpful to know that enzymes are usually named by adding the suffix -ase to a prefix derived from the name of the substrate.

> *So what substrate does sucrase act upon?*

You should have been able to work out that it is sucrose – table sugar.

The carbon atom and organic chemistry

There are two major divisions in the study of chemistry – **organic** and **inorganic chemistry**. The word *organic* was originally used to refer to compounds found in organisms, but it then came to mean compounds containing carbon. Most compounds within the body do contain carbon, and often hydrogen and oxygen too. In current popular usage the word *organic* has connotations of naturalness and safety, but you may be surprised to learn that many of the products of the petrochemical industry, which certainly do not occur in nature, also contain carbon and are therefore organic. In this text the term *organic* is used in the way a chemist would use it – meaning a substance containing carbon. Inorganic chemistry, then, is the study of substances that do not contain carbon.

> *What is the electron configuration of carbon?*

By now you should have no difficulty in looking up the atomic number of carbon (6) and working out that its electron configuration is 2.4.

***Table 2.4* Table showing the carbon atom chain prefixes.**

1. Meth-	6. Hex-
2. Eth-	7. Hept-
3. Prop-	8. Oct-
4. But-	9. Non-
5. Pent-	10. Dec-

> *How many more electrons does carbon require in order to achieve a stable valence shell?*

The correct answer is 4, and these may of course be obtained by sharing. Were someone to attempt to count the number of organic molecules the figure would run into millions. One of the reasons for this abundance is the ability of carbon atoms to bond covalently with each other and thus form chains of varying lengths. Compounds in which the ends of these chains are not joined together are said to be **aliphatic**, while those in which the chains form a ring are said to be **cyclic** or **aromatic**. Some of the most important organic molecules within our bodies are carbohydrates, proteins, fats and vitamins. Indeed, they are so important that Chapter 6 has been given over to them along with other substances found in our food. In this section we shall concentrate on other important organic molecules. Let us first clarify how organic molecules are named.

The names of organic molecules usually begin with a prefix that indicates the length of the carbon atom chain. Table 2.4 identifies the first 10 carbon atom prefixes.

Chemical groups that are attached to a carbon atom chain are called *substituents*. The simplest organic molecules result when carbon forms compounds with the substituent hydrogen alone. The resultant compounds are collectively referred to as **hydrocarbons**.

Hydrocarbons

Organic molecules that contain only single bonds between carbon atoms are said to be *saturated*. The group of aliphatic hydrocarbons that are saturated are called **alkanes**, and they are named by adding the suffix -ane to the appropriate prefix. The simplest alkane is therefore methane.

> *How many carbon atoms does methane possess?*

Remember those prefixes? If you do you will have worked out that methane contains only one carbon atom. Its structural formula is given in Figure 2.10.

Perhaps you are already familiar with methane, since it is the main component of natural gas. You may also have used butane and propane as fuels in portable heaters and cookers.

> *Take a moment to draw the structures of butane and propane.*

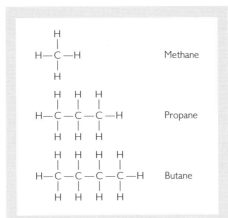

Figure 2.10
The structural formulae of methane, butane and propane.

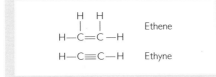

Figure 2.11
The structural formulae of ethene and ethyne.

Now check your diagrams against Figure 2.10.

Hydrocarbons that contain double or triple bonds are said to be *unsaturated*. This is a reference to the fact that the molecule contains less than the maximum number of substituents. The multiple bond could be chemically broken and further hydrogen atoms added; a process called *hydrogenation*. You will find reference is made to it on the back of margarine tubs! **Alkenes** contain a double bond, and the simplest member of this family is ethene (C_2H_4), the structure of which is shown in Figure 2.11. Note that alkenes are named using the suffix -ene. **Alkynes** contain a triple bond and the simplest member of this group is **ethyne** (C_2H_2), the structure of which is also given in Figure 2.11. Once again note that the family to which this molecule belongs is given by the use of a suffix – in this case it is -yne. You may be more familiar with ethene by its alternative name, acetylene.

Some of the chemical and physical properties of organic molecules are shared by many different compounds. For example, most are flammable. However, some properties are characteristic of certain groups only. Consider odour for a moment. Methane has no smell and its use as a fuel would be problematic, since gas leaks might go undetected were it not for the fact that gas companies add an odour to it. Other organic molecules have characteristic odours. For example, **ketones** (see later) smell like pear drops on the breath of patients who produce them in excess. The distinctive chemical or physical properties of organic compounds exist as a consequence of the possession of certain chemical groups called **functional groups**. These often contain elements such as oxygen and hydrogen, and sometimes nitrogen too. We shall continue our study of organic molecules by thinking about **alcohols**.

Alcohols

Alcohols are compounds that contain one or more hydroxyl groups (OH) and conform to the general formula RCH_2OH. (In organic chemistry R stands for a hydrogen atom or a carbon atom chain with hydrogen atoms attached). Alcohols are named by adding the suffix -ol to the relevant prefix. Consequently, the simplest alcohol is methanol (CH_3OH). You might buy methanol in the form of methylated spirits in order to fuel a camping stove or for use as a solvent. If ingested methanol may lead to blindness, so in addition to methanol, methylated spirit contains colouring in order to draw attention to its toxic nature. However, when the term *alcohol* is used in everyday language it is not methanol that is being referred to but ethanol (CH_3CH_2OH).

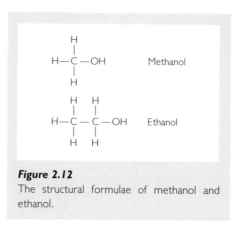

Figure 2.12
The structural formulae of methanol and ethanol.

> *Draw the structural formulae of methanol and ethanol.*

These are shown in Figure 2.12.

> *Now draw the structural formula of propanol.*

Remember that the prefix prop- indicates the presence of three carbon atoms.

> *Is there a problem in drawing the structural formula of propanol?*

No doubt you have realised that the hydroxyl group could be placed on an end carbon atom or on the middle one. We clearly need to be told which. This is done by adding a number to the name. For example, 2-propanol tells us to place the hydroxyl group on the second carbon atom as shown in Figure 2.13. Incidentally, 2-propanol, sometimes called isopropyl alcohol, is used in injection swabs and some preparations that are used to clean the skin prior to surgery.

Some alcohols actually possess more than one hydroxyl group. A good example is **glycerol** (glycerine), the structural formula of which is given in Figure 2.14. Glycerol is an important component of **glycerides**

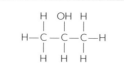

Figure 2.13
The structural formula of 2-propanol.

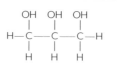

Figure 2.14
The structural formula of glycerol.

(a form of lipid), discussed more fully in Chapter 6 (Biological molecules and food). In addition, it is used as an *emollient* (a substance that softens the skin) and as a component of topical medications. Glycerol suppositories are sometimes used to soften the stool when constipation is present, and since it is sweet and has low toxicity it is also found as a dilutent in liquid oral medications. Note that in the name *glycerol* the aforementioned standard nomenclature has not been followed precisely. However, since this is the most commonly used name it is the one given here.

Aldehydes

Aldehydes are the first of three groups that possess a carbonyl group (C=O). In aldehydes a hydrogen atom is attached to the carbonyl group so that they conform to the general formula RCHO. They are named using the suffix -al and the simplest aldehyde is therefore methanal, the structural formula of which is shown in Figure 2.15.

Ketones

Like aldehydes, **ketones** are also characterised by the possession of a carbonyl group (C=O), but in this case it is located part-way along a carbon atom chain. Ketones conform to the general formula RCOR, and the simplest ketone is therefore propanone, a compound that is more usually known by its alternative name, acetone (Figure 2.16). Acetone is sometimes used by nurses as a solvent of sticking plaster adhesive, but more importantly it is one of the ketones that is sometimes produced by the body and excreted in the urine. This occurs in fasting and in patients with diabetes mellitus, when lipids, instead of glucose, are used as a source of energy. Some ketones also possess an acidic carboxyl group (see later) and the resultant accumulation of these leads to a condition referred to as ketoacidosis. Ketoacidosis may in turn lead to coma and death. Consequently, ketones will feature in your clinical studies.

Figure 2.15
The structural formula of methanal.

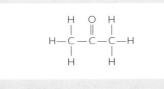

Figure 2.16
The structural formula of acetone.

Carboxylic acids

In **carboxylic acids** a hydroxyl group is attached to the carbonyl group so as to form a carboxyl group (COOH). Carboxylic acids therefore conform to the general formula RCOOH. The simplest member of this group is formic acid (HCOOH), but the most familiar is undoubtedly ethanoic acid (CH_3COOH) – sometimes called acetic acid or vinegar. The structural formula of both these compounds is shown in Figure 2.17. Carboxylic acids are sometimes referred to as *fatty acids*, since they are involved in the formation of lipids (fats).

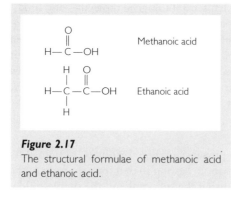

Figure 2.17
The structural formulae of methanoic acid and ethanoic acid.

Esters

Esters conform to the general formula RCOOR. They are formed from a reaction between an alcohol and a carboxylic acid such as is shown in Figure 2.18.

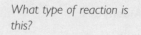

What type of reaction is this?

It is a combination reaction, but since it results in the elimination of water it is referred to as a **condensation reaction**. Of course, it is possible for an ester to undergo a decomposition reaction in order to reform a carboxylic acid and an alcohol, but in order for this to take place a water molecule would be required. A decomposition reaction of this type is referred to as **hydrolysis** – meaning to break with water. Since lipids (fats) are esters this group of organic compounds will be encountered frequently during your study of health and illness.

Amines

Amines are organic compounds that possess an amino group (NH_2) and conform to the general formula RCH_2NH_2. The simplest amine is therefore methyl amine, the structural formula of which is given in Figure 2.19. Amino acids, the molecules that join together to form proteins, possess both an amino group and a carboxyl group. The structure of the amino acid glycine is shown in Figure 2.20. Other important molecules that occur in the body and that contain an amino group include the **catecholamines** adrenaline (epinephrine), noradrenaline (norepinephrine) and dopamine. Adrenaline is a hormone and dopamine is a neurotransmitter, while noradrenaline functions in both these roles. The analgesic morphine and amphetamines are also amines, as their names imply.

Aromatic compounds

Aromatic hydrocarbons consist of unsaturated molecules in which the carbon atom chain is arranged in a ring, as the example of benzene (C_6H_6) shows in

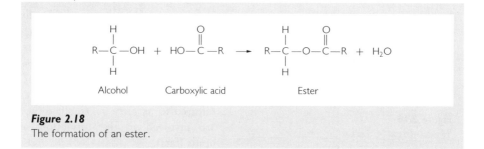

Figure 2.18
The formation of an ester.

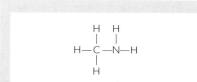

Figure 2.19
The structural formula of methyl amine.

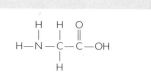

Figure 2.20
The structural formula of the amino-acid glycine.

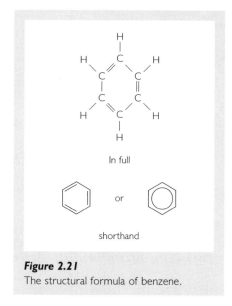

In full

or

shorthand

Figure 2.21
The structural formula of benzene.

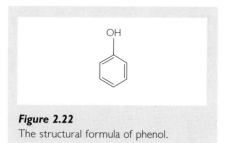

Figure 2.22
The structural formula of phenol.

Figure 2.21. It is not usual to draw the structural formula of benzene in full, and a simplified version is also shown. You may already have heard of benzene, since it is commonly used in industrial processes and is known to be *carcinogenic* (causes cancer). The benzene ring features in many aromatic compounds, including phenol, in which a hydroxyl group is attached directly to the benzene ring, as Figure 2.22 shows.

> *What kind of organic compound is phenol?*

It is an alcohol, since it possess a hydroxyl group. Phenol is a strong germicide (kills microorganisms) – in fact, it was originally used by Joseph Lister in 1867 as a surgical antiseptic, which he called carbolic acid.

Summary
1 Matter exists in three states: gas, liquid and solid.
2 An atom is the smallest particle of matter that can enter a chemical reaction.
3 A substance that consists of atoms of only one kind is referred to as an element.

4 A substance that consists of atoms of more than one kind chemically bonded together is referred to as a compound.

5 Atoms are made of protons, neutrons and electrons.

6 Ions result when atoms gain or lose electrons.

7 Ionic bonds are formed when electrons are exchanged between atoms.

8 Covalent bonds are formed when atoms share electrons.

9 When atoms share electrons equally a non-polar molecule is formed.

10 When atoms share electrons unequally a polar molecule is formed.

11 Hydrogen bonds form between polar molecules.

12 Chemical reactions may be described as combination, decomposition, displacement, partner exchange, reversible or redox.

13 Catalysts are substances that increase the rate of chemical reactions.

14 Enzymes are protein catalysts.

15 Organic chemistry is the study of compounds that contain carbon.

16 Organic molecules are either aliphatic or aromatic.

17 Important groups of organic compounds include alkanes, alkenes, alkynes, aldehydes, ketones, carboxylic acids and amines.

Self-test questions

2.1 Which of the following statements are true?

    a Matter exists in three states.

    b Elements consist of atoms of only one type.

    c Atoms are the smallest particles of matter.

    d Compounds consist of atoms of more than one type that are chemically bonded together.

2.2 Which of the following statements are true?

    a Protons are found in the nucleus.

    b Electrons carry a negative charge.

    c Neutrons orbit the nucleus.

    d Protons and neutrons have identical mass.

2.3 Which of the following statements are true?

    a The atomic number is the number of protons.

    b The mass number is the sum of the number of protons and electrons.

    c Atoms have no net charge.

    d In an ion there are the same number of protons as electrons.

2.4 Which of the following statements are true?

    a Ionic bonds result from the sharing of electrons.

    b Positively charged ions are called cations.

    c Negatively charged ions are called anions.

    d Sodium chloride is an example of an ionic compound.

2.5 Which of the following statements are true?

    a Covalent bonds result from the sharing of electrons.

    b Covalent bonds cannot form between identical atoms.

c Polar molecules result when electrons are shared equally between atoms.
d Hydrogen bonds are weak forces of attraction that exist between polar molecules.

2.6 Which of the following statements are true?
  a The second energy level contains a maximum of eight electrons.
  b The electrons in the outermost energy level are called valence electrons.
  c The first energy level contains a maximum of eight electrons.
  d An atom with electrons in the M energy level is most stable when this level contains eight electrons.

2.7 Match the equations of the left with the descriptions on the right.
  a $A + B \longrightarrow AB$ **i** Decomposition
  b $A + BC \longrightarrow AC + B$ **ii** Partner exchange
  c $AB + CD \longrightarrow AD + CB$ **iii** Displacement
  d $AB \longrightarrow A + B$ **iv** Combination

2.8 Match the general formulae on the left with the chemical groups on the right.
  a RCH_2NH_2 **i** Alcohol
  b RCH_2OH **ii** Amine
  c $RCOR$ **iii** Ketone
  d $RCOOH$ **iv** Carboxylic acid

2.9 Match the substances on the left with the chemical groups on the right.
  a Methane   **i** Alcohol
  b Acetone   **ii** Amine
  c Adrenaline (epinephrine)   **iii** Alkane
  d Ethanol   **iv** Ketone

2.10 Match the equations on the left with the descriptions on the right.
  a $CHOOH + CH_3OH \longrightarrow CHOOCH_3 + H_2O$ **i** Oxidation
  b $CHOOCH_3 + H_2O \longrightarrow CHOOH + CH_3OH$ **ii** Condensation
  c $NADH \longrightarrow NAD^+$ **iii** Reduction
  d $NAD^+ \longrightarrow NADH$ **iv** Hydrolysis

Answers to self-test questions

2.1 a, b and d
2.2 a, b and d
2.3 a and c
2.4 b, c and d
2.5 a and d
2.6 a, b and d
2.7 a iv
  b iii
  c ii
  d i

2.8 a ii
  b i
  c iii
  d iv
2.9 a iii
  b iv
  c ii
  d i

2.10 a ii
  b iv
  c i
  d iii

3 Water, electrolytes and body fluids

Learning outcomes

After reading the following chapter and undertaking personal study you should be able to:

1 Identify three states of matter and describe what happens when a substance undergoes a change of state.
2 Describe the structure of the water molecule and distinguish between polar and non-polar molecules.
3 Distinguish between forces of cohesion and adhesion and describe meniscus formation in water and mercury.
4 Describe what is meant by the term *surface tension* and give practical examples of surface tension effects.
5 Distinguish between different types of aqueous mixtures and give examples of each.
6 Distinguish between electrolytes and non-electrolytes and give examples of each.
7 Define the term *concentration* and explain the different units commonly used to express it, including percentage concentration and molarity.
8 Explain the physical processes diffusion, osmosis and filtration and give examples of each.
9 Define the terms *osmotic pressure*, *osmolarity*, *osmolality* and *tonicity*.
10 Define the term *oedema* and indicate possible causes.
11 Outline the means by which the body maintains sodium and water balance.
12 Briefly describe dialysis.

Introduction

In this chapter we are going to consider water, important substances dissolved in it, such as **electrolytes**, and body fluids such as **interstitial fluid**. In addition, we shall also look at fluid balance in the body and some of the different fluids used in **intravenous infusions** (drips). Before we do this, recall that all matter exists in one of three states – the **solid**, **liquid** and **gaseous** states. Since the melting point of water (0 °C) and its boiling point (100 °C) are within the range of our everyday experience of temperature, we are familiar with this compound in its solid state (ice), in its liquid state and as a gas (water vapour or steam).

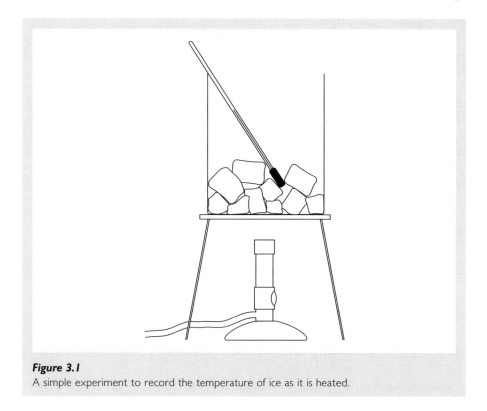

Figure 3.1
A simple experiment to record the temperature of ice as it is heated.

But what actually happens when matter undergoes a change of state?

To answer this question we shall describe a simple experiment which is illustrated below. In Figure 3.1, ice is taken from a deep freezer and is placed in a beaker at room temperature. A thermometer is also placed in the beaker. As the temperature of the ice slowly rises it undergoes a change in state from the solid to the liquid – in ordinary language we say that it melts.

Remind yourself at what temperature ice melts.

Ice melts at $0\,^\circ$C.

But what has actually happened to the ice?

The first thing to note here is that ice and liquid water are both water – no molecules have been broken. The change in state is a result of a change in the arrangement of the molecules. Perhaps we should say a little more about this. Solids consist of particles, atoms, molecules or ions, which are held together in a rigid arrangement by chemical bonds. Even though we regard solids as having a rigid structure, the particles of which they are comprised do have some energy of movement, which we call **kinetic energy**. In the solid state they are not moving freely, but instead they vibrate about a fixed position. Now let us consider what happens when the ice in our experiment is warmed by being allowed to stand at room temperature. When a solid is heated the kinetic energy of the particles begins to increase and they vibrate more rapidly. Eventually the particles have sufficient energy to escape the strong attraction which they have for each other and which is responsible for the rigidity of the solid state. At this point we observe a change in state (that is, the solid melts), and the temperature at which this occurs is called the melting point. In the liquid state the particles are not completely free of the attraction which they have for each other, but they are not held together in a rigid arrangement either. If we now light the Bunsen burner beneath the vessel, which contains liquid water, the kinetic energy of the water molecules is increased further and eventually some of the water molecules are able to break free from the attraction of others and they enter the gaseous state.

> *Have you noticed that when water is heated in this way bubbles form? Where do they come from?*

The bubbles are bubbles of gaseous water (steam).

> *Where in the container do they tend to form?*

If you look carefully they form first at the bottom, that is, at the hottest part of the container. As the temperature of the water rises, more and more of the liquid molecules enter the gaseous phase.

> *But does the temperature of the water continue to rise indefinitely?*

You will no doubt realise that the temperature of the water continues to rise until the boiling point is reached. Then, no matter how strongly the water is heated, the temperature remains stable.

> *Remind yourself what the boiling point of water is.*

It is 100 °C.

> *But why does the temperature remain constant at this point?*

The reason for this is that at the boiling point all the energy being delivered to the container is being used to convert water molecules from the liquid state to the gaseous state.

Thus melting and boiling are physical changes; that is, they involve a change in state. There has been no chemical change – no molecules have been broken and no new molecules created. The three states of matter are illustrated in Figure 3.2.

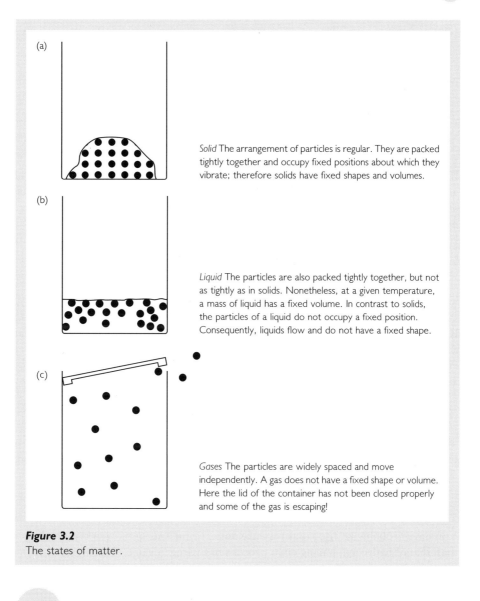

(a)

Solid The arrangement of particles is regular. They are packed tightly together and occupy fixed positions about which they vibrate; therefore solids have fixed shapes and volumes.

(b)

Liquid The particles are also packed tightly together, but not as tightly as in solids. Nonetheless, at a given temperature, a mass of liquid has a fixed volume. In contrast to solids, the particles of a liquid do not occupy a fixed position. Consequently, liquids flow and do not have a fixed shape.

(c)

Gases The particles are widely spaced and move independently. A gas does not have a fixed shape or volume. Here the lid of the container has not been closed properly and some of the gas is escaping!

Figure 3.2
The states of matter.

Water

Water is very abundant – two thirds of the Earth's surface is covered by it and it forms 60% of the weight of our bodies!

> *Do you remember which elements the water molecule consists of?*

We are sure that you remember that the water molecule consists of hydrogen and oxygen.

But how many atoms of these two elements does it take to form one water molecule?

No doubt you have noted that, in each water molecule, there are two atoms of hydrogen and one of oxygen – hence the molecular formula of water is H_2O. Water is a covalent compound.

Do you remember how covalent bonds are formed?

If not, it might be a good idea to go back and look at Chapter 2 (The physical world and basic chemistry). If you do remember that covalent bonds are formed when atoms share electrons you should have no difficulty in understanding the water molecule. Let us look at the oxygen and hydrogen atoms in turn.

Oxygen has an atomic number of eight and the arrangement of electrons in its electron shells is 2.6.

How many electrons are required to fill the outer shell of oxygen and so achieve stability?

If you have the answer eight in mind then you are quite right. This means that a single oxygen atom can achieve stability by sharing two electrons of another element. Now let us turn to hydrogen, which has an atomic number of only 1, and as a consequence it has a single electron in its first electron shell.

How many electrons does it take to fill this shell and so achieve stability?

Two electrons are required. This means that both elements could achieve stability if one atom of oxygen were to share one electron each with two atoms of hydrogen. This arrangement is illustrated in Figure 3.3.

Are the electrons shared equally?

You have probably realised that they are not. The nucleus of oxygen is much larger than that of hydrogen, having eight protons to hydrogen's one. This means that the oxygen nucleus exerts a greater pull on the electrons of the two hydrogen atoms than do the two hydrogen nuclei. This attraction is not sufficient to pull the electrons clean away from hydrogen, but it does result in an unequal sharing.

Does this unequal sharing have any important effects?

Yes, it does. Electrons do of course carry a negative charge, so the unequal sharing of them means that the oxygen end of the water molecule has a partial negative charge and the hydrogen end has a partial positive charge. Consequently, we describe water as a **polar molecule**. This is illustrated in Figure 3.4.

It is important to note that this redistribution of charge is small compared with the transfer of electrons in ionic bonding. In addition, covalent molecules have no net

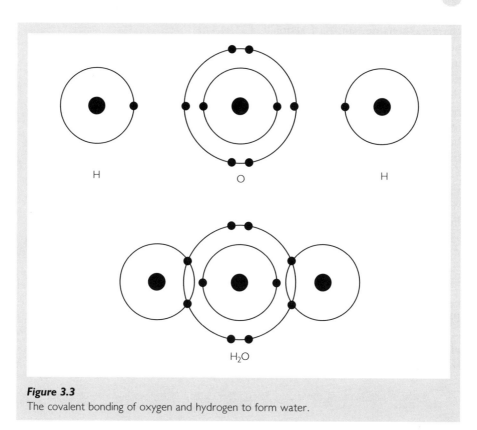

Figure 3.3
The covalent bonding of oxygen and hydrogen to form water.

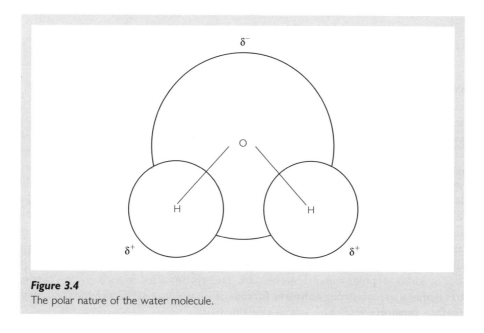

Figure 3.4
The polar nature of the water molecule.

charge as ions do, since a fractional negative charge in one part of the molecule is balanced by a fractional positive charge in another part of the molecule. Consequently, we do not use the symbols + and − to identify this redistribution of charge, but δ^+ (delta +) and δ^- (delta −) are used instead.

Incidentally you will also notice from Figure 3.4 that the atoms in a water molecule do not lie in a straight line – the molecule is bent.

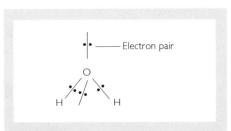
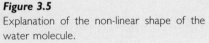

Figure 3.5
Explanation of the non-linear shape of the water molecule.

Why is this?

This happens because the four separate pairs of electrons in the outer shell repel each other and orientate themselves as far from each other as possible. This is illustrated in Figure 3.5.

In figuring out this arrangement you will need to remind yourself that only two electron pairs actually take part in bond formation – these are the bonding electrons. The other two pairs are non-bonding electrons.

What practical relevance does an understanding of the water molecule have?

It is important for several reasons. Firstly, water is an important solvent, and the polar nature of the water molecule determines which kinds of substance readily dissolve in it.

Can you work out which kinds of substance readily dissolve in water?

Polar water molecules are able to pull ions out of a solid crystalline lattice, so ionic compounds, such as sodium chloride, readily dissolve in water. In fact, in our bodies sodium ions are the most abundant extracellular cations (positively charged ions outside cells) and chloride ions are the most abundant extracellular anions (negatively charged ions outside cells). You will see the importance of this when you study aspects of physiology such as nerve and muscle function and when nursing patients with intravenous infusions (drips). More information about solutions is given a little later. In contrast, covalent compounds such as lipids (fats) are not readily soluble in water, and you might like to ask yourself the question: how does the body transport lipids around the body in the bloodstream? This is clearly important when looking at fats in the diet and their role in disease processes such as atherosclerosis. More information about this is given in Chapter 6 (Biological molecules and food).

In addition, polar water molecules are strongly attracted to each other – that is to say that there are strong **cohesive forces** within water, and this produces a strong **surface tension** effect. This is also explained later.

Finally, it is also worth pointing out that a very small number of water molecules dissociate to form ions according to the equation:

$$H_2O \longrightarrow H^+ + OH^-$$

The importance of this will be explained in Chapter 4 (Acids, bases and pH balance). For now, let us return to a consideration of the forces between water molecules and surface tension.

Cohesion and adhesion

We have already noted that the particles of a solid or liquid attract each other and that the strength of these forces of attraction determine the physical state of the substance. The force of attraction between *like* particles, such as molecules of water, are referred to as forces of *cohesion*. The effect of cohesive forces is quite simple to demonstrate. For example, suppose that you were to spill a little water on a flat surface.

> *Does the water spread out evenly?*

No – instead, it forms into droplets. Perhaps this is such a familiar phenomenon that we don't even think about it. Nonetheless, forces of cohesion do influence the behaviour of water, and one simple application of this can be seen in the practice point below. However, before this we should also note that there are also forces of attraction between *dissimilar* molecules – these are called forces of *adhesion*. One example is the attraction between molecules of a liquid and the glass of the container in which they are held.

Practice point 3.1: meniscus formation and reading manometers

Forces of cohesion and adhesion lead to the formation of a **meniscus** on the surface of a liquid. A meniscus is the bending of the surface, as illustrated by Figure 3.6. These show the menisci formed by water and mercury, such as might be seen in the case of the saline manometer or **sphygmomanometer**, respectively. (Manometers are explained more fully in Chapter 5: Pressure, fluids, gases and breathing).

> *Note that the shapes of the menisci formed by water and mercury differ.*

In the case of water the forces of adhesion are greater than the forces of cohesion, and this results in a bowl-shaped meniscus. The opposite is true of mercury, and this produces a dome-shaped meniscus.

> *Is this of immediate practical relevance?*

Let's ask another question.

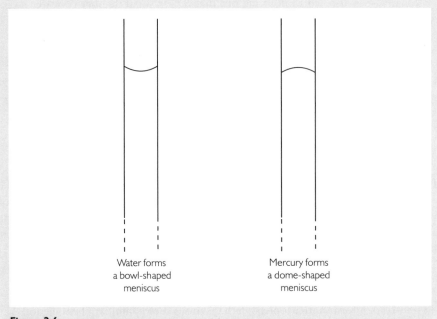

Figure 3.6
The shapes of the menisci formed by water and mercury.

If you had to read the level of fluid in a manometer would you take your reading from the top or the bottom of the meniscus?

Clearly this might be important when you are measuring blood pressure with a mercury sphygmomanometer or central venous pressure with a saline manometer.

An accurate reading will be obtained if measurements are taken from the *bottom* of the meniscus in the case of water and the *top* of the meniscus in the case of mercury.

Surface tension

Have you ever noticed how some insects, such as pond skaters, have the ability to glide across the surface of a pond? The water seems to behave as though it has a kind of skin! In a similar way, it is possible to take a pin, which is more dense than water, and with care place it on the water surface. If you can't find a pond with pond skaters you might like to have a go at this simple experiment instead! Clearly the water doesn't really have a skin, since if we agitate the surface a little the pin sinks. Perhaps it would be better to say that there appears to be a tension over the surface of the water – in fact, that is exactly the expression which is used – **surface tension**.

What could account for this surface tension?

Most of the molecules in a mass of water are surrounded by other molecules – in fact, all of them are except those on the surface of the water. This means that the surface molecules are pulled down by cohesive forces from below, but there is no corresponding force from above. Consequently, the surface of the water behaves like a skin, and we have already noted how some insects put this surface tension effect to good use. On the other hand you might have noticed what happens when an insect pierces the surface. Now the skin-like effect of surface tension acts to prevent the creature from escaping.

Surface tension is also responsible for the shape of liquid droplets.

> *Can you work out what shape a water droplet would be in space?*

You may have been able to figure out that in the absence of gravity the droplet would be spherical. On Earth the effect of gravity is to cause elongation of the droplet, and it becomes, well, droplet shaped! The ability of water to form droplets is put to good use in **nebulisers**, which are discussed in Chapter 5 (Pressure, fluids gases and breathing). In these devices a solution of a drug in water is formed into small droplets (nebulised) which can then be inhaled. This is just one medical use which we make of surface tension. Another is outlined in the practice point below.

Practice point 3.2: low surface tension and skin disinfection

It is worth noting that the surface tensions of different liquids are not the same. For example, the surface tension of the alcohol ethanol is less than a third that of water. Ethanol thus has a lower tendency to form droplets than does water and tends to spread out more when spilled; as a consequence it evaporates more readily. This property is also put to good use medically. For example, some antiseptics are presented as alcoholic solutions. These evaporate more readily than aqueous solutions and they may be preferred for the purpose of skin disinfection since they dry more quickly.

However, it is possible to ignite alcoholic preparations, for example with diathermy equipment, and this has been the cause of accidental burns in patients in operating theatres. As a consequence care should be taken to swab dry pools of alcoholic skin preparations, such as form in the umbilicus, when diathermy is to be used. Diathermy involves the use of high-frequency alternating current for the purpose of cutting tissue or coagulating bleeding vessels. It is explained in Chapter 10 (Electricity, magnetism and medical equipment).

In addition, the cleaning ability of a liquid is partially dependent upon its surface tension. Soaps and detergents have a lower surface tension than water, and as a consequence spread out and more effectively wet objects with which they make contact. They also have a reduced tendency to form droplets and a greater penetrating ability than does pure water, which tends to rest as droplets on the surface.

That tells us something about the science of washing clothes, but is there a practical point here?

You must have guessed that there is. Antiseptics and disinfectants also have a lower surface tension than water, and this means that they more thoroughly wet the surface of the wound with which they have contact – clearly an important property, since to be effective such a solution has to have contact with the wound!

Practice point 3.3: surface tension in the body – surfactant

We have already looked at surface tension in a number of practical situations, but now we turn to look at one important example of surface tension in the body. In the lungs there are thousands of microscopic sac-like structures called *alveoli*. It is across these that gaseous exchange takes place. A single alveolus and capillary are illustrated in Figure 3.7.

Each alveolus is lined with a film of fluid. We have already noted that water has a very high surface tension, and water is a major component of this fluid. The surface tension in alveolar fluid acts to reduce the alveoli to their smallest possible size. This is similar to the way in which the surface tension of a bubble tends to cause it to collapse. Indeed, if the alveolar fluid were pure water the alveoli would collapse (atelectasis). However, a **phospholipid surfactant**, which is secreted by the alveolar cells, is present in the fluid and this reduces the cohesive forces between water molecules and, as a consequence, reduces the surface tension. Incidentally, the term *surfactant* is a contraction of the expression *surface active agent*, which means a substance which reduces surface tension.

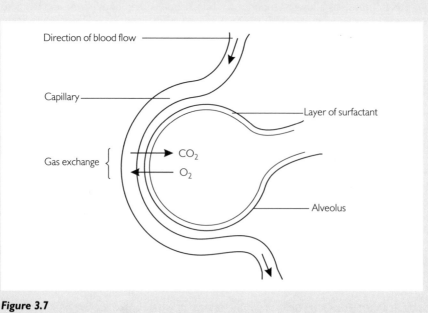

Figure 3.7
A single alveolus and capillary.

Surfactant is present in foetal lungs from about 23 weeks gestation, but before 28–32 weeks it rarely occurs in quantities sufficient to prevent alveolar collapse. This means that in babies born prematurely the alveoli may collapse and the condition neonatal respiratory distress syndrome (RDS) exists. In addition, some full-term infants also have a surfactant deficiency.

Aqueous mixtures

An aqueous mixture is quite simply a mixture of a substance in water. Nurses meet aqueous mixtures all the time. Cough linctus, antiseptic solutions and intravenous fluids are all obvious examples. However, there are a number of different kinds of mixture, as explained below.

Mechanical suspensions

This is the simplest type of aqueous mixture, which involves the dispersal of small particles of a solid in water. If a mechanical suspension is left to stand for any length of time the solid will separate out and the mixture will need to be re-created by shaking. The use of the word 'small' in connection with the particles of a mechanical suspension is relative. They are indeed small to the naked eye, but compared with water molecules they are very large indeed, and can easily be removed from the water by filtering through paper or cloth. Muddy water is one example of a mechanical suspension, and on first examination this appears to have little to do with nursing. However, you might be surprised to learn that one remedy for diarrhoea is a mechanical suspension of the clay kaolin, sometimes with the drug morphine added. Kaolin mixture is rather old-fashioned but nonetheless effective, and it is still available from pharmacies.

In addition, when presented as a paste (in other words the amount of water present is less), kaolin can be used to make a hot poultice which, when applied to painful joints, may provide relief. Kaolin poultices have also traditionally been used in the treatment of infected wounds where there is a collection of pus into which antibiotics do not readily penetrate. The application of the hot poultice to the wound causes blood vessels to dilate and may cause pus to discharge through the wound. Once again this is rather old-fashioned but still finds occasional use. *If you are involved in the application of hot poultices you should of course ensure that the patient is not in danger of receiving a burn.*

Perhaps a more familiar mechanical suspension is that of zinc oxide (ZnO) in water, otherwise known as calamine lotion. This is used in order to gain relief from sunburn. As is the case with muddy water the particles in kaolin mixture and calamine lotion separate out when left to stand. Consequently, it is important to shake the bottle before use.

Colloidal suspensions

The particles in a colloidal suspension are smaller than those in a mechanical suspension, being only 1–100 nm in diameter. Nonetheless, this is still larger

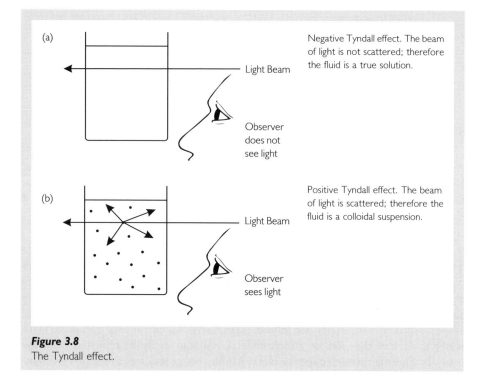

(a)

Light Beam

Negative Tyndall effect. The beam of light is not scattered; therefore the fluid is a true solution.

Observer does not see light

(b)

Light Beam

Positive Tyndall effect. The beam of light is scattered; therefore the fluid is a colloidal suspension.

Observer sees light

Figure 3.8
The Tyndall effect.

than atoms and ions and indeed larger than many molecules too. The particles in a colloidal suspension consist of groups of ions or atoms or of large molecules such as starch and proteins. If a colloidal suspension is allowed to stand the particles and water do not usually separate out, or if they do, they do so only very slowly.

The small size of colloidal particles means that they pass through ordinary filters but not through biological membranes such as the capillary endothelium or the **filtration** membrane of the nephron of the kidney.

Along with mechanical suspensions, colloids show a positive **Tyndall effect**; that is, they scatter a beam of light shone through them. As a consequence the beam can be seen from the sides of the container as well as from a position directly opposite the light source, as illustrated in Figure 3.8.

In addition, it is worth noting that there are a number of different types of colloidal suspension described below.

Foams: these are produced when a large volume of air is dispersed through a smaller volume of liquid or solid. When we whip cream we are in fact producing a foam, while other examples include shaving foam and the contents of foam fire extinguishers.

Emulsions: these are produced when one liquid is dispersed in another. No doubt you will have heard of emulsion paint; French dressing (oil and vinegar) is another example.

Practice point 3.4: emulsification of fats

We have already noted that a mixture of oil and vinegar is an example of an emulsion.

What happens when the mixture is left to stand?

They separate and have to be shaken together again.

Couldn't we add something to cause them to remain in an emulsified state?

Yes, we could: a pinch of mustard powder does the trick. Consequently, we might refer to the mustard as an emulsifying agent.

Are emulsifying agents important in the body?

You may have guessed that they are. When food is ingested, fats form large droplets which present a relatively small surface area to the fat-digesting enzyme lipase. Bile from the gall bladder passes into the duodenum, where bile salts emulsify the fats. That is, the fat is broken up into small droplets and an emulsion is formed. These emulsion droplets present a much larger surface area to lipase, so their digestion is facilitated. Note that bile salts are not enzymes: they do not digest the fat. Rather, their emulsifying action is required for the effective action of lipase. You will find more about fat digestion in Chapter 6 (Biological molecules and food).

Sols: this term is applied to many different colloidal mixtures in which a solid is dispersed through another solid, a liquid or a gas. The latter is referred to as an aerosol and these are of course encountered in everyday life as well as in nursing practice. For example, Betadine® dry powder spray is an aerosol of povidone–iodine used in the disinfection of minor skin wounds and infections.

Gels: these are colloidal mixtures of liquids dispersed in solids and they might be described as being on the borderline of solidity. They are not easily poured, but on the other hand they are only semi-rigid. Examples include gelatine, which is used in the manufacture of jelly desserts, and **agar**, which is manufactured from seaweed. In Victorian times agar was also used in making desserts, but it was subsequently chosen as a medium on which to grow bacteria and it is still employed for this purpose today. Perhaps agar was initially chosen for this purpose because an early scientist noticed colonies of organisms growing on a dessert?

Are gels encountered in nursing practice?

Indeed they are. Modern wound dressing products such as *hydrocolloids* and *hydrogels* are occlusive or semi-occlusive dressings which adhere to dry skin and interact with moisture in the wound to form a gel. They have a number of advantages over traditional dressings and you might like to find out what these are.

True solutions

Perhaps you have noticed that in discussing aqueous mixtures thus far we have used the term *suspension* and have avoided the term *solution* where possible. This is because the mixtures described up to this point have not been true solutions. (However, in clinical practice the word *solution* is commonly applied to aqueous mixtures, whether they are true solutions or not.) True solutions are characterised by having components which exist almost entirely as individual ions, atoms or molecules and not as groups of these. Consequently, the particles of a true solution are very small and do not settle when the solution is left to stand. In addition, they readily pass through filters and often through biological membranes too. They are also too small to show a positive Tyndall effect. The characteristics of different aqueous mixtures are summarised in Table 3.1.

In true solutions two or more substances are mixed so that their particles are uniformly distributed between each other. The most abundant substance is referred to as the **solvent** and the other substances are referred to as **solutes**. Furthermore, we do not simply describe solutes as being dispersed in the solvent, but instead we say that they are dissolved in it.

For example, suppose you are asked to get a bag of normal saline. Normal saline is a solution of 0.9 g of sodium chloride (NaCl) dissolved in 100 ml of water.

> Which is the solute and which is the solvent?

Sodium chloride is the solute and water is the solvent.

We can represent the formation of a true solution in the manner shown in Figure 3.9. Note that the solute particles are evenly distributed throughout the solvent, but because the proportion of solute particles is small there are many solvent particles which have no immediate contact with the solute.

Table 3.1 Summary of characteristics of aqueous mixtures.

Characteristic	Mechanical suspension	Colloidal suspension	True solution
Particles	Visible lumps	Collections of ions, atoms or molecules or single large molecules	Single ions atoms or molecules
Particle size	>100 nm	1 – 100 nm	<1 nm
Settling	Particles quickly settle out	Particles do not settle out or do so only slowly	Particles never settle out
Separation by simple filters	Yes	No	No
Separation by semi-permeable membranes	Yes	Yes	No
Tyndall effect	Yes	Yes	No

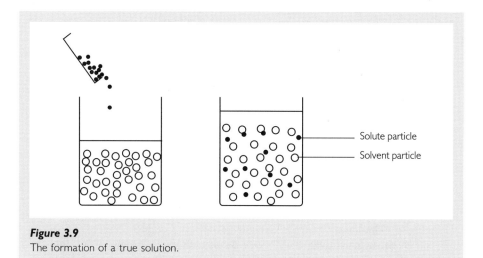

Figure 3.9
The formation of a true solution.

Practice point 3.5: crystalloid and colloid intravenous fluids

Many patients in hospital have to be given fluids directly into a vein. This procedure is referred to as an intravenous infusion. There are many reasons for administering an intravenous infusion, and perhaps you might like to find out what some of these are.

Some intravenous fluids are colloidal suspensions; these include blood, **albumin solution** (at least 95% of the protein present is albumin), and **plasma protein solution** (at least 85% of the protein present is albumin). The latter two are both derived from human plasma. In addition there are a number of plasma protein substitutes such as suspensions of gelatin (for example, Gelofusine® and Haemaccel®), hetastarch (for example, Hespan®) and dextran (for example, Gentran®).

Other intravenous fluids, for example, sodium chloride and dextrose (glucose), are true solutions, although in clinical practice they are often referred to as *crystalloids* because of the ability of the solute to form crystals. This can be achieved by gently heating the solution so that the water is evaporated away to leave the solute crystals behind.

Are there specific reasons for choosing one fluid for infusion rather than another?

Yes there are, and you might like to find out more about this subject by consulting a nursing text. Briefly, crystalloid solutions would be used to rehydrate a patient who has experienced prolonged vomiting and diarrhoea. On the other hand, the patient who has suffered severe bleeding will require a blood transfusion, since it is necessary to replace the lost red blood cells as well as the water and solutes which blood contains. Sometimes blood may not be immediately available – perhaps you are part of a team which has just arrived at the scene of a road traffic accident. In this case plasma substitutes may initially be given until blood becomes available.

Plasma substitutes, such as those previously mentioned, have advantages over crystalloid solutions since they do not readily cross biological membranes and thus do not 'leak' out of the vascular compartment. Instead, they remain within the blood vessels and continue to exert an influence on blood volume. Consequently, they may be referred to as **plasma expanders**. There are of course many other uses of intravenous infusions but greater detail is reserved for your further study.

Whichever type of fluid is chosen there are certain observations which you should make before connecting a bag of fluid to an infusion-giving set. You should of course note that it is the correct fluid for the patient concerned and that the concentration chosen is the one prescribed. You should note the expiry date of the fluid and the batch number. The hospital pharmacy and manufacturer will need to know these in the event of an untoward reaction to any fluid. *You should not, of course, administer any fluid once its expiry date has passed.*

Next you should check that the bag and its packaging are intact. If not, the contents will no longer be sterile and they should be discarded. You should also check the fluid itself. Hold the bag up to the light. Remember that only colloidal suspensions should show a positive Tyndall effect. True solutions should not. In addition, colloidal suspensions often have a pale amber or straw colour, whereas true solutions are uncoloured. Finally, although true solutions are often referred to as crystalloids, there should be no particles of solid material in any type of intravenous fluid.

Electrolytes and non-electrolytes

Before defining these terms we should look at solution formation in more detail, beginning with sodium chloride as an example. Sodium chloride is an ionic compound and in the solid state it exists as a crystal lattice. This is more fully explained in Chapter 2 (The physical world and basic chemistry). The ionic bonds between the positively charged sodium ions (Na^+) and the negatively charged chloride ions (Cl^-) are very strong. In order to break the ionic lattice and melt sodium chloride the temperature would have to be raised to $804\,°C$. At this point the ions would no longer be held in a fixed position, but would instead become free to move. Consequently, whereas solid sodium chloride does not conduct electricity molten sodium chloride does.

In view of this you may now be surprised to learn that when ionic compounds are mixed with water their component ions are pulled out of the lattice and become dissolved.

> *How is this so easily achieved by water when we have just noted that it requires a high temperature to break the ionic lattice of the solid compound?*

First of all you need to recall that water is a polar molecule and that water molecules are attracted to each other. In a similar way, the hydrogen end of the water molecule (slightly positive) is attracted to negatively charged chloride ions (Cl⁻) and the oxygen end of the water molecule (slightly negative) is attracted to positively charged sodium ions (Na^+).

> *Surely these forces of attraction are relatively weak compared with the forces of attraction between ions in a crystal lattice?*

Yes they are, but there are two more points which have to be considered here. Firstly, ions located at the outside of a crystal lattice are not completely surrounded by other ions and they are therefore held less securely. As a consequence, they are also more readily pulled out of the lattice by polar water molecules. Secondly, once an ion has been pulled free from the lattice it immediately becomes surrounded by water molecules, each of which exerts forces of attraction upon it. Although each of these forces is relatively weak, the combined effect of the numerically superior water molecules is considerable. This is illustrated in Figure 3.10.

The water molecules have, in effect, 'come between' the ions, and as a consequence the forces of attraction between the ions are very much reduced.

> *What has this to do with the term* electrolyte?

The point is that when an ionic compound is dissolved in water the result is a solution of free ions. Such solutions are good conductors of electricity, and any substance which dissociates into ions when dissolved in water is referred to as an electrolyte.

By now you may have sufficient time to work out that while ionic compounds are soluble in water they do not dissolve in non-polar solvents such as hexane. This is because such non-polar liquids do not exert forces of attraction upon ions in a crystal lattice. Perhaps you have also worked out that many covalent compounds do not dissolve in water.

> *Why is this?*

It is because these uncharged molecules cannot disrupt the forces of attraction between polar water molecules. In contrast, covalent compounds do readily dissolve in non-polar solvents, since the forces of attraction between solvent and solute molecules are similar to those between solute molecules themselves.

However, some covalent compounds, such as glucose, **amino acids** and urea (the waste product of amino acid metabolism), do dissolve in water.

> *How can this be explained?*

Covalent compounds which do dissolve in water are always those with a polar nature, like water itself. Consequently, the forces of attraction between the different molecules in the mixture (solvent–solvent, solute–solute and solvent–solute) are of similar strength, and this allows polar covalent compounds to dissolve in water. However, since polar covalent compounds do not dissociate into ions upon dissolving they are described as non-electrolytes.

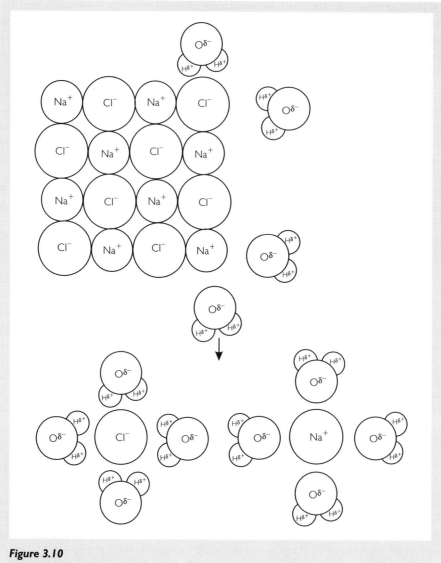

Figure 3.10
The effect of water molecules on an ice lattice.

The concept of concentration

Thus far we have looked at different kinds of aqueous mixture, including true solutions. When we deal with such fluids it may be important to know what kind of mixture it is. It will certainly be important to know what the components of a mixture are and probably the concentration too. There are a number of ways of describing concentration, but what this measurement tells us is the relative proportions of solute to solvent. It would of course be convenient if one unit of concentration could be used all the time, but unfortunately this is not the case and as nurses we have to be familiar with a number of different units. Some of these are described below.

Percentage concentration

You will encounter this expression of concentration in a number of contexts, including the concentration of intravenous fluids. The most commonly used concentration of sodium chloride for intravenous infusion is 0.9% (w/v) and for dextrose (glucose) it is 5% (w/v).

> *But what do these figures mean?*

The expression 'w/v' means weight per unit volume, and the percentage concentration (w/v) is defined as the mass of a solute dissolved in 100 ml of solution. It is given by the formula below:

$$\text{percentage concentration (w/v)} = \frac{\text{mass of solute (g)}}{\text{volume of solution (ml)}} \times 100$$

Thus a 0.9% solution contains 0.9 g of solute in every 100 ml of solution while a 5% solution contains 5 g of solute in every 100 ml.

Percentage concentrations can also be expressed as the weight of solute per unit weight of solvent (w/w) or volume of solute per unit volume of solvent (v/v) if both solute and solvent are liquids. These two expressions are much less common. Nonetheless two examples are given below:

Aluminium hydroxide mixture (non-proprietary) 4% (w/w) Al_2O_3 in water for use as an antacid (reduces stomach acidity).

Sterets® pre-injection swabs are saturated with 70% (v/v) isopropyl alcohol.

Other expressions of concentration as a weight per unit volume

Instead of expressing concentration as a percentage we could give the weight of solute in a different volume of solution – say in 5 ml instead of 100 ml.

> *Can you think of occasions when this might be more convenient?*

One example would be liquid medicines – we doubt that 100 ml of a medicine would ever be required! The concentration of liquid medicines is usually given in terms of the number of milligrams (mg) per 5 millilitres (ml) of mixture. Take the example of Pholcodine linctus, which is used to suppress a dry cough. It is commonly presented at a concentration of 5 mg/ml.

> *Why is the 5 ml measure so important?*

Have you realised that this is the volume of a standard medicine spoon? It is also the volume of a culinary teaspoon, but of course cutlery does not conform to a standard volume, so *special medicine spoons or graduated medicine pots should always be used to measure the volume of preparations containing drugs*. The 5 ml measure is not only a matter of accuracy but of convenience too. A typical dose of the concentration of Pholcodine mentioned is 5–10 ml (1–2 spoons) 3–4 times each day.

This way of expressing concentration is also important when administering continuous infusions of drugs – perhaps solutions of analgesics in the management of postoperative pain.

Molarity

Another way of expressing concentration is in terms of **molarity** – that is, in terms of the number of *moles of solute per litre of solvent* (mol/l).

> *Do you remember what the term* mole *means?*

If you are unsure you should go back to Chapter 2 (The physical world and basic chemistry). However, you might have remembered that a mole of a substance is its molecular mass in grams. For example, the molecular mass of glucose ($C_6H_{12}O_6$) is 180, so one mole of glucose weighs 180 g. The formula for this expression of concentration is given below:

$$\text{concentration (mol/l)} = \frac{\text{number of moles}}{\text{solution volume in litres (1)}}$$

However, even a concentration of 1 mol/l is quite high for most medical and nursing applications, so concentrations are often measured in terms of *millimoles per litre* (mmol/l) instead. A solution of 1 mmol/l contains one thousandth of a mole per litre of solution.

A number of applications of this measurement could be given. One such example is that of sodium bicarbonate (sodium hydrogen carbonate) 1.26% intravenous fluid, which contains 150 mmol/l each of sodium ions (Na^+) and bicarbonate ions (hydrogen carbonate ions/HCO_3^-). In addition, the concentration of substances dissolved in blood is usually given in terms of mmol/l.

Practice point 3.6: plasma concentrations

When caring for physically ill patients we often need to make a note of the concentration of various electrolytes and non-electrolytes. Important electrolytes include sodium ions (Na^+), chloride ions (Cl^-), potassium ions (K^+), bicarbonate ions (hydrogen carbonate ions/HCO_3^-) and calcium ions (Ca^{2+}). Important non-electrolytes include glucose and urea.

You might like to consider the circumstances in which the concentration of these substances in blood are measured. Two examples include the plasma glucose concentration in diabetes mellitus or electrolyte concentrations during a period of prolonged vomiting and diarrhoea.

When completing laboratory request cards, doctors often use abbreviations for the tests they require. It is not always a good idea but perhaps you will already be able to recognise that 'u's & e's & glc' is a request for the measurement of the concentrations of urea, electrolytes and glucose. The numerical values of some important solute concentrations are given below.

Na^+	135–145 mmol/l
K^+	3.5–5.0 mmol/l
Cl^-	100–106 mmol/l
Ca^{2+}	2.1–2.6 mmol/l
HCO_3^-	19–29 mmol/l
Glucose	3.9–5.6 mmol/l
Urea	2.9–8.9 mmol/l

Finally for this section, you should note that some blood values are not given in terms of mmol/l. The concentration of hydrogen ions (H^+) is described in terms of the pH scale (see Chapter 4: Acids, bases and pH balance) and haemoglobin concentration is given in terms of grams per decilitre (g/dl; that is, g/100 ml).

Other units – the milliequivalent per litre (meq/l)

This unit is uncommon but it is used in the USA and is described here for the convenience of any readers who might be employed there. The *milliequivalent* takes into account the charges carried by particles. The formula for milliequivalents per litre (meq/l) is given below:

$$\text{number of milliequivalents per litre (meq/l)} = \text{molarity (mmol/l)} \times \frac{\text{number of charges}}{\text{carried by particle}}$$

For example, sodium ions (Na^+) carry a single charge and the numerical value of their concentration in blood expressed in meq/l is therefore the same as that expressed in mmol/l (within the range 135–145). The same is of course true of all ions which carry a single charge, such as potassium ions (K^+) and chloride ions (Cl^-). However, magnesium ions (Mg^{2+}) carry a double charge, so the numerical value of their concentration in blood expressed in meq/l (2 meq/l) is twice that of the value expressed in mmol/l (1 mmol/l).

Diffusion and osmosis

Diffusion and **osmosis** are important physical processes which you may have experienced in everyday life without knowing it. If you have studied biological science before you will certainly have encountered these processes and performed experiments to demonstrate their effects. Before we proceed, however, it might be worthwhile describing a very simple experiment to show the random movement of particles of a gas.

Imagine that you have a small glass container into which some smoke is blown. The container is then sealed. Smoke consists in part of small particles which result from the burning of a solid, and we could examine these by placing the small glass container under the light microscope.

> *If we were to look down the microscope at the particles of smoke what would we see? Would the particles be stationary?*

If you have decided that they would not, but would be seen to be moving randomly, you are right.

> *Why is this?*

The movement of the smoke particles is caused by the collisions which they experience with the molecules of the gases which form the air. We cannot see these molecules with the ordinary light microscope but we can see their effects on the larger smoke particles. What this simple experiment demonstrates is that the molecules of the gases which form the air atmosphere are in constant random movement, and this is referred to as **Brownian motion**.

Diffusion

Now let us apply this understanding to a real-life example. Suppose someone were to allow a volume of a gas to escape into the corner of a room. Imagine that the door and windows are closed and that you are sitting in the opposite corner.

> *Would the gas remain confined in the corner into which it was released?*

You obviously know from experience that it would not, but our observation of Brownian motion now enables us to explain why not. The collisions which our mysterious gas molecules have with the molecules of the gases which form the atmosphere mean that they eventually become dispersed throughout the room. After some time you might even detect them from your seat in the opposite corner – perhaps the gas has an unpleasant smell. Whether you realise it or not you have just experienced diffusion – that is, the movement of a gas from an area of high concentration, down a concentration gradient, to an area of low concentration. The expression 'down a concentration gradient' refers to the fact that the concentration of gas particles is greater at the moment of their release into the room than after their dispersal.

Diffusion is not of course confined to gases – it is a process which takes place in liquids too. Imagine that we take a small volume of ink and drop it into a glass of water.

> *Would it remain confined as a small drop?*

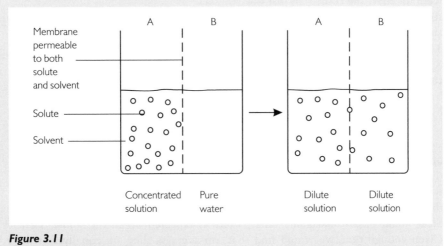

Figure 3.11
An experiment to demonstrate diffusion.

No: eventually it would become dispersed in the water and we would have a dilute ink solution. Once again the cause of this dispersal of the ink particles is collisions, this time with randomly moving water molecules.

How could we speed up the process of the dispersal of the ink?

We could agitate the glass or stir the water. This is an obvious but important point. Although the process of diffusion is quite capable of leading to a uniform dispersal of ink throughout the water, it does take some time. You should note then that diffusion is much more rapid in the gaseous phase.

When you study the human body you will encounter diffusion once again. It is one of the important processes by which substances move within the body. However, here diffusion often involves movement through a membrane such as the plasma (cell) membrane or the respiratory membrane of the lungs. You might wish to study these in more detail later, but we can simulate diffusion across a membrane in the experiment illustrated in Figure 3.11.

In this experiment a vessel is divided equally into two compartments (A and B) by a membrane. An equal volume of fluid is poured into each compartment, but in A the fluid is a concentrated solution of some substance, let's say glucose or sodium chloride, while in B there is pure water only. The membrane separating the compartments has pores and the point you need to bear in mind is that the diameter of these pores is greater than that of the solute particles.

So what happens next?

Let us concentrate on just one solute particle. It suffers many collisions with solvent particles (water molecules) and with other solute molecules too. Consequently, it experiences the kind of random motion which we have already described. If we were to 'chart the course' of just one solute particle it would consist of a zigzag line as represented in Figure 3.12.

Each change of direction is the result of a collision – lesser changes by glancing blows and greater ones by near 'head-ons'. Sometimes the solute particles will bounce off the sides or bottom of the container and off the membrane too. However, if the course of a particle coincides with a pore it will pass straight through into compartment B. Later on it might pass back again. This pattern is repeated by all the solute particles.

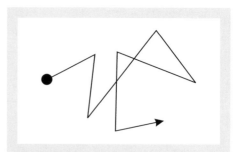

Figure 3.12
Representation of the movement of a single solute particle.

Given enough time we would find that, on average, they are equally distributed between the two compartments, although at any one moment in time there might be more particles in one compartment. So instead of A containing a concentrated solution and B pure water, both now contain dilute solutions. This is illustrated in Figure 3.11.

Now we are in a position to give a fuller definition of diffusion which takes into account the presence of a membrane:

Diffusion is the movement of solute particles from an area of high solute concentration, through a semi-permeable membrane, to an area of low solute concentration, until an equilibrium is reached.

Here the phrase 'until equilibrium is reached' refers to the fact that after a sufficient period of time, the number of solute particles moving from A to B equals the number moving in the opposite direction. It would also be true to say that the average number of solute particles on either side of the membrane is the same, or that the concentrations of the solutions on either side are equal.

In addition, we should perhaps clarify the term *semi-permeable*. This is applied to a membrane which has pores of a certain size and which therefore allows particles of a smaller size to diffuse through it. Not surprisingly such particles are then referred to as *diffusible* particles and larger particles as *non-diffusible*. In the case of artificial membranes the permeability of the membrane is fixed by the pore size and the term 'semi-permeable' is applied. However, in the body, the permeability of biological membranes can often be varied and the term 'selectively permeable' is applied.

> *Are there factors which influence the rate of diffusion?*

Perhaps you already know that there are. A number are given below.

Surface area: the greater the surface area of the membrane the greater the number of particles that come into contact with it and the faster the rate of diffusion. This principle is employed in the body. For example, the alveoli of the lungs provide a surface area for gas exchange which is one thousand times the area of the surface of the body! If this sounds a lot you should perhaps also know that the absorption of nutrients by the ileum of the gut is enhanced by villi and microvilli which provide a surface area one million times that of the surface of the body!

Particle size: more energy is required to move larger particles than is required to move smaller ones. Consequently, the rate of diffusion of individual ions (such as electrolytes in plasma) is much greater than that of large molecules (such as plasma proteins).

Concentration: the greater the difference in concentration of two solutions the greater the rate of diffusion between them. We refer to a difference in the concentration of two solutions as a **concentration gradient**.

Temperature: higher temperatures cause faster movement of particles and a greater diffusion rate.

Charge: some diffusible particles (ions) are charged and you may already know that like charges repel each other while unlike charges attract. These electrostatic forces affect the diffusion rate. For example, in the resting state the concentration of sodium ions inside nerve cells is much less than the concentration of the same ions outside the cell. In addition, the inside of the cell is charged negatively compared with the outside. We refer to this difference in charge as an electrical **potential difference**. Thus there is not only a concentration gradient favouring the movement of sodium ions into nerve cells but also an *electrical gradient* too. That is, the positively charged sodium ions (Na^+) are attracted to the negatively charged interior of the cell. However, when the nerve cell is in the resting state sodium ions do not readily pass through the cell membrane.

> *Do you remember that we said that biological membranes, such as the cell membrane, are selectively permeable?*

Perhaps you recall that this means that their permeability to a particular solute can be altered. When the nerve cell is stimulated, sodium channels through the cell membrane are opened and now the rate of diffusion of the sodium ions into the cell is so great that textbooks often describe them as 'rushing' in. This influx of sodium ions is the first phase in the production of a nervous impulse, and is dealt with in more detail in Chapter 10 (Electricity, magnetism and medical equipment).

Pressure: an increase in pressure on one side of the membrane results in a greater rate of diffusion.

Osmosis

Now let us consider another experiment very similar to the diffusion experiment described previously. Our new experiment is illustrated in Figure 3.13. Here, as before, there is a vessel which is divided into two equal compartments (A and B), and these are filled with the same liquids as before. The difference between this experiment and the previous one is that the membrane has very small pores – too small in fact for the solute particles to pass through, but still larger than water molecules.

> *If the solute molecules cannot pass from A to B is there any change in the contents of the two compartments?*

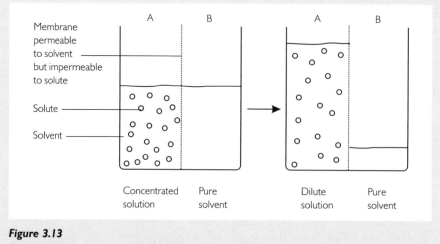

Figure 3.13
An experiment to demonstrate osmosis.

To answer this question we need to focus on the movement of solvent particles through the membrane, through which the solute particles can no longer pass.

In which compartment is the solvent concentration the greatest?

Note that we said *solvent* concentration – this is greatest in compartment B. The concentration of solvent molecules in compartment A is in fact 'diluted' by the presence of solute. We do not of course normally talk this way, but it is perfectly correct to do so. Since solvent molecules move randomly and pass down concentration gradients, just as solute molecules do, a sufficient lapse of time will reveal that there has been a net movement of water from compartment B to compartment A. This can be seen in the raised fluid level in A, as shown in Figure 3.13.

This movement of water is referred to as **osmosis** and we are now in a position to define it:

Osmosis is the movement of solvent molecules from an area of high solvent concentration, across a semi-permeable membrane, to an area of low solvent concentration.

Note that in this definition the phrase 'until equilibrium is reached' does not appear. This is because an equal concentration either side of the membrane could never be achieved. No matter how much water were to move from compartment B to compartment A, the solute concentration in compartment A will always be greater than in compartment B. The factor which limits the net movement of water from B to A is the pressure developed by the expanded volume of fluid in A. This pressure serves to oppose the movement of water by osmosis. This does not mean that there is now no movement of water molecules at all. Indeed they are still moving randomly and experiencing collisions as they did before. Rather it is that the rate of movement of water from A to B now equals the rate of movement in the opposite direction. Consequently, the volume of liquid in each compartment is stable, but A contains a greater volume than does B. This discussion leads us on to consider the concept of **osmotic pressure**.

Osmotic pressure

Osmotic pressure may be considered to be the pressure required to prevent the movement of solvent from a region of pure solvent and into a solution. The SI (Système Internationale) unit of pressure is the **pascal** (Pa), but sometimes the unit millimetres of mercury (mmHg) is also used. You will find out more about measuring pressure in Chapter 5 (Pressure, fluids, gases and breathing).

For now, the point to note is that the greater the concentration of solutes in solution the greater the pressure required to stop the net movement of solvent into that solution by osmosis. However, when we consider osmotic pressure we often think of it as a kind of water pulling power. It is easy to see how this comes about, since water will move by osmosis from one solution and into another which has a greater concentration of solutes, and therefore a higher osmotic pressure. It is worth bearing this in mind, since in practice you may hear the expression 'osmotic pull' being used as well as osmotic pressure.

Finally, the osmotic pressure of a solution depends upon the number of particles dissolved in solution. In the blood, sodium ions (Na^+) are the most abundant solute particles and as a consequence they exert the greatest influence upon osmotic pressure. In addition, there are many other ionic and non-ionic solutes as well as larger suspended molecules such as plasma proteins. Sometimes a distinction is made between the total osmotic pressure and that fraction of it for which the plasma proteins are responsible. When the latter is being referred to the term **oncotic pressure** is used.

Osmolarity and osmolality

We have already noted that the osmotic pressure is dependent upon the number of particles in solution. Keep in mind that a particle could be an ion, an atom or a molecule. No matter how large or small, each counts as one particle as far as osmolarity is concerned.

> *Perhaps we need a new unit to express concentration of solute in terms of the number of particles.*

What do you think? Can't we stick with using moles? After all, life is complicated enough already. On first examination it may appear that there is no need to consider an additional unit. For example, suppose we dissolve 1 mole of glucose in water.

> *How many particles do we have?*

To answer this question you will need to remember that a mole is the molecular mass of a substance in grams (180 g in the case of glucose) and that one mole of a substance has the same number of particles as one mole of any other substance.

Do you remember how many particles a mole of a substance contains?

Perhaps you remember that a mole of any substance contains 6×10^{23} particles of that substance (Avogadro's number). If you are not sure about this you may need to go back and look at Chapter 2 (The physical world and basic chemistry).

Thus there seems little need to consider additional units further. To work out the number of particles we just need to know how many moles of a solute have been dissolved. It may look that way so far, but if you think carefully about electrolytes and non-electrolytes you might realise that there is a problem.

Is glucose an electrolyte or a non-electrolyte? Does it dissociate into ions in solution?

We are sure you remember that glucose is a non-electrolyte and that it does not dissociate into ions in solution.

But what about sodium chloride (NaCl)?

Yes indeed, sodium chloride is an electrolyte and it does dissociate into ions in solution. Consequently, if we take one mole of sodium chloride and dissolve it in water we produce one mole of sodium ions (Na^+) and one mole of chloride ions (Cl^-) – that is, twice as many particles as one mole of glucose produces. Perhaps we had better consider another unit after all! Meet the **osmole**.

Osmolarity

To work out the number of osmoles in a solution, simply multiply the number of moles by the number of particles which a solute particle produces in solution. The mathematical formula for the osmole is given below:

$$\text{number of osmoles} = \text{number of moles} \times \frac{\text{number of particles which a}}{\text{solute particle dissociates into}}$$

Go back to glucose as an example and you should see that, since glucose does not dissociate in solution, one osmole of glucose has the same numerical value as one mole. However, in the case of sodium chloride (NaCl) one mole produces two osmoles in solution.

In addition, just as we often talk about molarity (that is the concentration of a solution in mol/l or mmol/l) we also refer to **osmolarity** – that is, the number of osmoles per litre (osmol/l) or milliosmoles per litre (mosmol/l). The formula for osmolarity is given below:

$$\text{osmolarity} = \text{molarity} \times \frac{\text{number of particles which a}}{\text{solute particle dissociates into}}$$

Thus far we have imagined aqueous solutions which consist of only one solute.

Table 3.2 **Composition of intracellular and extracellular fluid.**

Substance	Concentration in mosmol/l		
	Plasma	Interstitial fluid	Intracellular fluid
Na^+	142	139	14
K^+	4.2	4.0	140
Ca^{2+}	1.3	1.2	0
Cl^-	108	108	4.0
HCO_3^-	24	28.3	10
Protein	1.2	0.2	4.0

But what about solutions with more than one solute? How would we work out their osmolarity?

Let us imagine that we have 1 l of a solution in which there is dissolved 1 mole of glucose and 1 mole of sodium chloride (NaCl).

What would the osmolarity of this solution be?

From what has gone before you should be able to work out that since 1 mole of glucose generates 1 mole of glucose in solution and 1 mole of NaCl generates 1 mole of sodium ions (Na^+) and 1 mole of chloride ions (Cl^-) the total number of particles generated is 3 osmol and the osmolarity of the solution is 3 osmol/l.

At this point you should also recognise that a solution with a greater osmolarity than another is referred to as being **hyperosmotic** and one with a lower osmolarity as being **hypo-osmotic**. Two solutions which have the same osmolarity are of course described as being **iso-osmotic**.

However, it should be noted that although iso-osmotic solutions have the same concentration of particles they do not necessarily have the same particles.

Remember that a particle could be an ion, an atom or a molecule. Each counts as one particle as far as osmolarity is concerned.

Let us take a real-life example. When we compare intracellular fluid (inside cells) and extracellular fluid (outside cells) we discover that both have an osmolarity of approximately 280 mosmol/l (that is, they are iso-osmotic). However, they each have very different solute concentrations (Table 3.2).

Osmolality

It is also worth pointing out that we sometimes talk about **osmolality** instead of osmolarity. The formula for osmolality is given below:

$$\text{osmolality} = \text{molality} \times \frac{\text{number of particles which a}}{\text{solute particle dissociates into}}$$

Molality differs from molarity in that it is the number of moles of solute per kilogram (kg) of solution rather than the number of moles of solute per litre (l) of solution. Consequently, osmolality gives the number of particles of solute per kilogram of solution rather than per litre. However, since one litre of water weighs very nearly one kilogram the units osmolarity and osmolality are virtually identical.

Tonicity

In this section we are going to consider the effect of different fluids on the movement of water in and out of cells. The erythrocyte (red blood cell) is commonly chosen as an example, since blood samples are easy to obtain. In addition, an effect upon blood cells *in vitro* (that is outside the body) may be reproduced *in vivo* (that is within the body) by the administration of fluid intravenously. This is clearly of interest to nurses who regulate the administration of infusions.

First of all, we need to remind ourselves that, although intracellular fluid and extracellular fluid have different concentrations of solutes, they have the same osmolarity. In addition, the cell membrane is selectively permeable and regulates the movement of solutes across it. However, water is able to move through the cell membrane freely.

> *In the circumstances described above is there a net movement of water in or out of the cell?*

If you think not, then you are right. The intracellular fluid and extracellular fluid are iso-osmotic and there is no net movement of water. Since this is the case we also say that intracellular fluid and extracellular fluid are **isotonic**. This literally means that they have the same strength. The intracellular fluid cannot pull water into the cell from the extracellular fluid and the extracellular fluid cannot pull water out of the intracellular fluid.

Now let us perform an experiment in order to determine the effect of hyper-osmotic and hypo-osmotic solutions upon our red blood cells by placing some cells in each of the solutions and then examining them under the microscope. In each case you should assume that the solute is *non-diffusible* – that is, it does not readily cross the cell membrane and enter the cell. Sodium chloride (NaCl) and glucose are examples of non-diffusible solutes.

> *What would be the effect of placing the cells in a hyperosmotic solution?*

You should have worked out that water would move out of the cell by osmosis and the cell would shrink. This is illustrated in Figure 3.14 and it is referred to as **crenation**. When a solution has the ability to draw water out of a cell in this way it is described as being **hypertonic**.

> *In contrast, what would happen to the cells placed in a hypo-osmotic solution?*

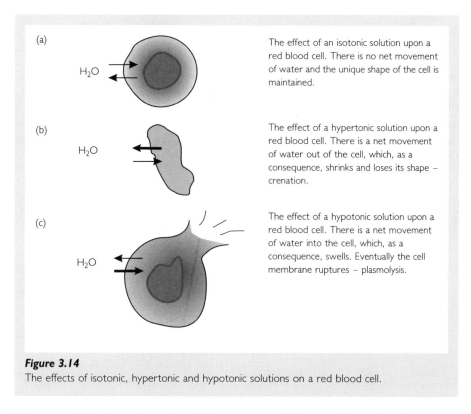

Figure 3.14
The effects of isotonic, hypertonic and hypotonic solutions on a red blood cell.

Water would be drawn into the cells by osmosis and they would swell and rupture – a process referred to as **plasmolysis**. This is also illustrated in Figure 3.14. Solutions which cause plasmolysis are referred to as **hypotonic**.

So, isotonic solutions cause no net movement of water in or out of cells, hypertonic solutions cause crenation and hypotonic solutions cause plasmolysis. Note that, in our example, the isotonic solution is also iso-osmotic, the hypertonic solution is hyperosmotic and the hypotonic solution is hypo-osmotic.

> *So then isn't tonicity the same as osmolarity?*

You could be forgiven for thinking so, but in reality they are different. Osmolarity is concerned with the concentration of solute particles in solution, while tonicity deals with the tendency of a solution to cause crenation or plasmolysis. While it is true that isotonic solutions are usually iso-osmotic, hypertonic solutions are usually hyperosmotic and hypotonic solutions are usually hypo-osmotic, it is not always the case.

> *But how could it be otherwise?*

First of all, note that in our red blood cell experiment we made a point of saying that the solute was non-diffusible.

What would happen if we were to use a solution with a diffusible solute, such as urea?

Suppose we place a sample of our red blood cells in an iso-osmotic solution of urea. Remember that iso-osmotic means the same concentration of particles, not necessarily the same particles. Urea molecules will diffuse into the cell, the contents of which now become hyperosmotic. Consequently, water is drawn into the cell by osmosis and it swells and plasmolyses. Thus osmolarity and tonicity are not the same thing at all.

Are there any practical applications of this?

We are sure that you will not be surprised that there are – have a look at the following practice point.

Practice point 3.7: tonicity, osmolarity and intravenous solutions

Suppose that we need to administer fluids to a patient intravenously.

If crenation or plasmolysis of red blood cells is to be avoided what type of fluid is usually administered?

Isotonic solutions are chosen. The most commonly administered isotonic crystalloid solutions are sodium chloride 0.9% (w/v) and dextrose (glucose) 5% (w/v). (Incidentally, in some parts of the world the abbreviation 'D5W' is used for dextrose 5% in water.) Both these solutions are iso-osmotic with blood plasma. Remember that this does *not* mean that plasma is a 0.9% solution of sodium chloride and a 5% solution of glucose. Rather, it means that these solutions have the same concentration of particles as does plasma. It is just that plasma has very many different kinds of particle, whereas the infusion fluids mentioned are pure solutions of just one particle.

Hypertonic and hypotonic solutions are occasionally administered intravenously, but the reasons for this are beyond the scope of this book. However, if you see them being used in clinical practice you might like to have staff explain their use to you.

Filtration

At this point it is worthwhile distinguishing diffusion and **filtration**. Filtration is the process whereby a liquid is forced through a porous membrane by a pressure difference on either side of the membrane. What filtration has in common with diffusion is that what prevents the movement of some particles through the membrane is their size. However, in reality the two processes are quite different.

In the case of diffusion the net movement of particles of a substance is from an area of high concentration to an area of low concentration as a consequence of their random motion. Particles do actually move in both directions, as we have discussed previously; it is their net movement which is down a concentration gradient. In filtration, particles are forced in one direction through a membrane by a pressure gradient.

In the body the formation of interstitial fluid and its reabsorption involves filtration and the blood is filtered by the kidneys – a process which leads to the production of urine.

The distribution of water within the body

We have already mentioned intracellular fluid and extracellular fluid, but in fact there are three fluid compartments within the body. The extracellular compartment can be further divided into the *intravascular compartment* (plasma) and the *interstitial compartment*. **Interstitial fluid** is the fluid found both outside cells and outside the circulation. It bathes cells and is the medium through which substances diffuse between cells and the blood. Interstitial fluid is formed from blood, but contains only some of its constituents. For example, there are no cells or large molecules, such as plasma proteins, in interstitial fluid. These are simply too large to escape through blood vessels in normal circumstances. The electrolyte concentrations of plasma and interstitial fluid are very similar, since water and ions move freely across blood capillary walls. Table 3.3 shows, in adults, the distribution of fluid between the three compartments as a percentage of body weight.

When we think of fluid in the body our minds often turn first to blood. However, when looking at Table 3.3 you should note that while 60% of the body weight is water, most of this is inside cells rather than outside them. Even when we consider extracellular fluid alone we find that interstitial fluid accounts for the greatest proportion, and that blood is only 4% of body weight.

This pattern is repeated for adolescents, but it is not the case for infants and children, as Table 3.4 illustrates. In the case of the neonate 80%, rather than 60%, of the total body weight is water, and the greater proportion of it is extracellular rather than intracellular.

Table 3.3 **The distribution of water within the body as a percentage of the total body weight.**

	Percentage of body weight (%)
Total body fluid	60
intracellular fluid	40
extracellular fluid	20
intravascular (plasma)	4
interstitial	16

Table 3.4 **The distribution of fluid within the body for different age groups.**

Age group	Percentage of body weight (%)	Extracellular (%)	Intracellular (%)
Puberty	60	20	40
I year	70	25	45
Neonate	80	45	35

The formation of interstitial fluid

Interstitial fluid is formed at the arterial end of *capillaries* and reabsorbed at their venous ends. The process of the formation of interstitial fluid is illustrated in Figure 3.15.

Capillaries connect *arterioles* and *venules*. The driving force of the blood through the circulation is referred to as the **hydrostatic pressure** (blood pressure). Note that, at the arterial end of a capillary, the hydrostatic pressure (blood pressure) is approximately 32 mmHg, while at the venous end it is only 12 mmHg.

Why does the hydrostatic pressure fall along the length of the capillary?

It falls because fluid is lost from the capillary as interstitial fluid is formed.

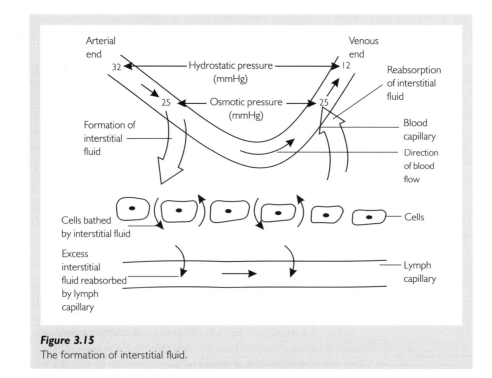

Figure 3.15
The formation of interstitial fluid.

Hydrostatic pressure tends to force small molecules, such as water, through the capillary wall, but it is opposed by **osmotic pressure** which is approximately 25 mmHg.

> *At the arterial end of the capillary, which is greatest: hydrostatic pressure or osmotic pressure?*

Hydrostatic pressure is greater than osmotic pressure, and there is a net pressure of 7 mmHg (that is, 32–25) in favour of the formation of interstitial fluid.

> *But what about the venous end? Which is the greater pressure?*

At the venous end the situation is quite different and hydrostatic pressure is less than osmotic pressure. Here a net pressure of −13 mmHg (12–25) favours the reabsorption of water.

In summary we can say that interstitial fluid bathes the cells of the body and that it is formed at the arterial end of capillaries but reabsorbed at their venous ends. Remember that, in normal circumstances, only small molecules can escape through capillary walls, so interstitial fluid contains no cells or large molecules, such as plasma proteins.

However, it is worth noting here that not all the interstitial fluid formed at the arterial end of capillaries is in fact reabsorbed at their venous ends.

> *What happens to the interstitial fluid which is not reabsorbed?*

If you have studied biology before you will no doubt remember that interstitial fluid which is not returned immediately to the circulation is collected by *lymph capillaries* and this too is illustrated in Figure 3.15. The lymph system is a very important system of vessels within the body and has roles in normal health and also in disease. When you have a moment you might like to find out more about it.

Practice point 3.8: oedema

Let us begin this section by imagining that the capillaries of a patient have become excessively permeable or 'leaky', and as a consequence molecules which normally remain in the circulation, such as the protein albumin, are able to pass into the interstitium.

> *What effect do you think this loss of albumin will have on capillary osmotic pressure and interstitial osmotic pressure?*

If you have decided that capillary osmotic pressure will fall and interstitial osmotic pressure rise then you are right.

> *What will the effect of these changes be upon the formation of interstitial fluid?*

Yes, there will be an increase in the production of interstitial fluid. An excess of interstitial fluid is called **oedema**. Oedema may be a generalised or localised condition which is common cause of swelling of the tissues. For example, oedematous ankles show an increase in girth, and gentle compression with the pads of the fingers will produce indentations referred to as pitting.

Increased capillary permeability may be caused by a lack of oxygen or by the release of so-called *vasoactive substances* during an inflammatory reaction. Think about the swelling which occurs at the site of a wasp sting. The term *vasoactive* literally means 'having effects on blood vessels', and one effect is increased capillary permeability.

Do you know the names of any vasoactive substances?

There are many, although you may not yet have encountered them. Probably the most familiar is *histamine*, which is important in local injuries, such as our wasp sting; in allergic reactions, such as hay fever; and in asthma too. Some drugs which help to reduce swelling do so by blocking histamine and are, not surprisingly, referred to as *antihistamines*. One such drug is *terfenadine*, which is used in the treatment of hay fever and allergic skin conditions.

At this point it is worth mentioning that, in some patients, the symptoms of an allergy are widespread and include cardiovascular collapse and oedema of the airway which impairs their breathing. The condition is called *anaphylactic shock* (allergic shock), and once again a fuller discussion is beyond the scope of this book. However, you may wish to find out more about it for yourself.

Increased capillary permeability is not the only cause of oedema. Raised hydrostatic (blood) pressure, reduced plasma osmotic pressure and lymphatic obstruction are other possible causes.

Raised hydrostatic pressure: hydrostatic pressure may be elevated following the intravenous administration of an excessive volume of fluid. Therefore care should be taken to regulate infusion rates accurately. The most serious effects are seen in the pulmonary circulation, where pulmonary oedema is responsible for impaired gas exchange and *decreased lung compliance* (that is, the lungs are 'stiff' to inflate). Both contribute to *dyspnoea* (the experience of difficulty in breathing). *Tachypnoea* (fast pulse rate), cough and the production of clear sputum, which may be blood streaked, are other manifestations. Pulmonary oedema is managed by the intravenous administration of the *diuretic* drug (increases urine output) *frusemide* and in addition the patient should be sat upright and given oxygen.

Pulmonary oedema may also occur when blood volume is normal, but where the pumping ability of the heart is impaired, such as is the case following a heart attack (myocardial infarction). Once again the mechanism is raised pulmonary hydrostatic pressure, but the cause is a failure of the damaged left ventricle to 'clear' blood returning to it from the lungs – you might say that there is a backlog! Backlogging of blood in the pulmonary circulation increases the workload of the right side of the heart and eventually right ventricular failure may occur, with subsequent development of systemic oedema. This is often seen in the elderly as swollen ankles.

Much less serious is lower limb oedema, which occurs when standing for prolonged periods. Standing causes venous congestion, since the compression of the veins of the leg by muscle contraction when walking is essential for the promotion of venous return. It may be prevented, in those whose occupations require prolonged standing, by periodic contraction of the leg muscles and by wearing support stockings which compress the limb with uniform pressure and so aid venous return. Obviously resting with the affected limb on a stool above the level of the hips will relieve lower limb oedema when it does occur, but this should be avoided when pulmonary oedema is present since it will be exacerbated by the increased venous return.

Reduced plasma osmotic pressure: the importance of plasma proteins in the maintenance of blood osmotic pressure has already been noted. Consequently, *hypoproteinaemia* (low blood protein), such as occurs in starvation and liver disease (plasma proteins are manufactured in the liver), is another cause of oedema.

Lymphatic obstruction: since interstitial fluid which is not reabsorbed by blood capillaries is drained by lymphatic capillaries, obstruction of these also causes oedema. Lymphatic obstruction may be caused by the growth of a tumour or, in tropical countries, by the filarial worm *Wuchereria bancrofti*. The resultant gross swelling of the lower limbs gives rise to the name *elephantiasis*.

Specialised body fluids

It is worth noting that a very small proportion of the fluid present in the body is localised in spaces where it performs specialised functions. For example, *cerebrospinal fluid* (CSF) surrounds and cushions the brain, *synovial fluid* is a lubricant in joints and *pleural fluid* allows the two pleural membranes to slide over each other during breathing. You will find a little more information on specialised fluids in Chapter 5 (Pressure, fluids, gases and breathing), but for the fullest discussion you should refer to a physiology text.

The regulation of fluid balance within the body

The term *fluid balance* implies the state in which fluid intake and output are equal. This is the normal state of the body over a period of time, as the figures in Table 3.5 show.

Table 3.5 **Balance of water intake and output.**

Fluid intake (ml)		Fluid output (ml)	
Drinking	1250	Insensible losses (such as sweat, faeces, breath)	1100
Water in food	1000	Urine	1500
Water produced by metabolic reactions	350		
Total	2600	Total	2600

How is this balance achieved?

When we consider fluid balance in the body we are primarily concerned with plasma, since this is the only fluid which can be directly acted upon to control its volume and composition. However, since water and ions move freely across capillary walls, when plasma is regulated the volume and composition of interstitial fluid are also regulated. The intracellular fluid is, in turn, influenced by changes in extracellular fluid, but this is limited by the selective permeability of the cell membrane.

Control of sodium balance is important in regulating extracellular fluid volume.

We have previously noted that sodium ions are the most abundant particles in ECF, and so they exert the principal influence upon osmotic pressure. We also noted that osmotic pressure is sometimes referred to as a 'pull' on water and can be thought of as 'water-holding ability'. Consequently, the mass of sodium ions in extracellular fluid determines the extracellular fluid volume, and regulation of extracellular fluid volume depends upon controlling sodium balance.

How is sodium balance regulated?

An obvious point to note here is that, in order to maintain sodium balance at a certain level, ingestion must equal excretion.

So do we regulate sodium intake?

The answer to this question is that our sodium intake may vary, but we do not regulate it specifically to meet our needs at a particular point in time. We may add salt (sodium chloride) to our meals because we like the taste of it or we may be unaware of the salt content of particular foods. Alternatively, we may avoid salt because we have heard that an excess intake of it may be associated with **hypertension** (high blood pressure). The point is that sodium intake is not regulated as such, and a typical daily intake of 10–15 g is very much in excess of the required 1 g. So, if we do not regulate sodium intake we must regulate its excretion.

How is sodium lost from the body?

Sodium (in sodium chloride) is of course lost in sweat, but sweating is part of the *thermoregulatory* mechanism. We sweat because we need to cool down, not because we need to control sodium balance. Similarly, the sodium lost in faeces is not controlled. However, the amount of sodium lost in urine is regulated. The mechanisms are not dealt with here – you can find out more about them in a physiology text. The important thing to note is that our kidneys regulate salt balance and maintain the volume of extracellular fluid.

Are there mechanisms which act upon water balance too?

Yes there are, as you will see from the example below.

Regulation of blood osmolarity by antidiuretic hormone (ADH) and thirst

In the case of water we do regulate our intake, driven by thirst, and our excretion in urine too. Of course there are water losses which we cannot control – the so-called insensible losses. These losses include water in sweat and faeces and water vapour in breath.

> *So how does the body determine how much water to lose?*

A part of the brain called the *hypothalamus* monitors the osmolarity of blood passing through it. The cells which achieve this monitoring function are, not surprisingly, called *osmoreceptors*. Let us take the case of an elevated osmolarity as a means of illustrating what happens next. First of all, an elevated osmolarity stimulates thirst and the individual will seek out a drink. The thirst centre of the brain is also located in the hypothalamus.

> *Do you see how increasing fluid intake will tend to reduce osmolarity to normal again?*

But suppose that a drink is not immediately available – perhaps you are in the middle of a game of squash! In this case water conservation is the priority. The hypothalamus is linked to the *posterior lobe* of the *pituitary gland*, which is located beneath the brain. From this gland a hormone called *antidiuretic hormone* (ADH/vasopressin) is released.

> *What does ADH do?*

Its name tells all – antidiuretic hormone reduces diuresis (urine output). So increasing water intake and reducing water loss work together to restore blood osmolarity. These mechanisms are illustrated in the diagram in Figure 3.16.

> *Do you recognise the above as an example of a negative feedback mechanism?*

> *Now attempt to draw a similar diagram for reduced blood osmolarity.*

So the body regulates the volume and composition of extracellular fluid primarily by regulating sodium and water excretion and water intake.

> *Could something go wrong?*

It could. But this is not a pathology text, so we will consider just one example – that of dehydration.

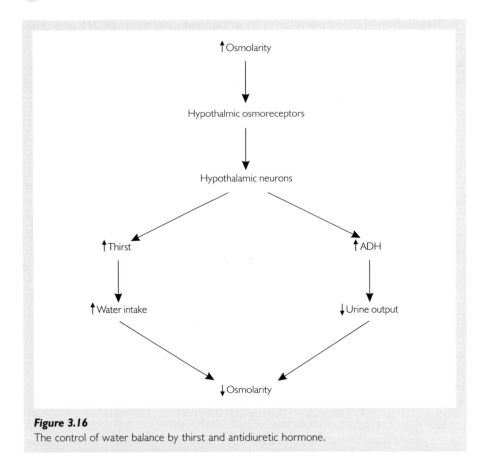

Figure 3.16
The control of water balance by thirst and antidiuretic hormone.

Practice point 3.9: dehydration

Dehydration may be described as a negative water balance – that is, the state in which losses exceed gains. Reduced water intake may accompany problems such as a swallowing defect, confusion in the elderly or perhaps the water jug simply being placed too far away from a frail patient. When you next visit a placement where the frail, elderly, confused or those with learning difficulties are cared for, think about basic physiological needs, such as that for water.

List some possible causes of excessive water loss.

Your list may include vomiting, diarrhoea or excessive urine production. The latter occurs in the conditions *diabetes insipidus* (a lack of ADH) and *diabetes mellitus*. While it is true that many causes of dehydration involve water and electrolyte loss, water is often lost in a greater proportion.

What effect would you expect this to have on blood osmolarity?

You should be able to work out that blood osmolarity will rise.

What will the effect of this be upon intracellular fluid?

Water will be drawn out of the cells by osmosis. Consequently, we refer to this state as extracellular fluid hypertonicity. We might initially detect dehydration from the appearance of the tissues when water has been lost from the cells. For example, the eyeballs appear sunken, the tongue and mucus membranes are dry and the skin is inelastic. More serious effects are related to the shrinkage of brain neurons and range from confusion and irrational behaviour to convulsions and death. Incidentally, castaways on desert islands and shipwrecked mariners in open boats are sometimes portrayed as confused and irrational – at least in some rather bad films! No doubt this image is partly based on the real-life manifestations of severe dehydration.

Specific gravity

We have seen that, in order to maintain fluid balance, the body is able to regulate sodium and water excretion. Consequently, samples of urine obtained from the same individual, but at different times of the day, differ in composition. Sometimes it is useful to determine whether there has been any gross change in the urine composition before resorting to expensive laboratory tests. One easily taken measurement is that of **specific gravity** or relative density. This is the ratio of the density of a substance to that of water, and the formula for it is given below:

$$\text{specific gravity} = \frac{\text{density of a substance}}{\text{density of water}}$$

Density is the mass of a substance per unit volume and the SI (Système Internationale) unit of density is the kilogram per cubic metre (kg/m^3). If you are unsure about this, see Appendix A: Weighing and Measuring – SI units. The density of water is $1000 \, kg/m^3$ and its specific gravity is of course 1.

What about urine – what is its specific gravity?

The first thing to note here is that urine is an aqueous solution and the presence of solutes means that it has a greater density than that of pure water. Consequently, its specific gravity will be above 1. Secondly, since urine composition changes with the body's fluid balance status its normal specific gravity exists within a range, between 1.001 and 1.035.

In the modern health care setting the specific gravity of urine is easily measured using a reagent strip, and this simple test is regularly performed as part of a general health check – perhaps when you register with a doctor for the first time or visit a hospital for almost any reason. A specific gravity outside the normal range may indicate the need for additional tests. The point to remember is that specific gravity may show that urine composition is abnormal, but it is a non-specific test – it does not indicate what the abnormality is. Furthermore, some changes in urine composition are not reflected in changes in specific gravity – it is a crude test.

Dialysis

In the presence of partial or complete kidney failure the role of the kidneys in the maintenance of water, electrolyte and acid–base balance and in the elimination of waste may be undertaken by an 'artificial kidney' in a process referred to as **dialysis**. There are two forms of dialysis – **haemodialysis** and **peritoneal dialysis**.

Haemodialysis

In haemodialysis, blood is diverted from the circulation to an 'artificial kidney' (the dialyser) and then pumped back to the patient again. A number of different designs of artificial kidney exist, but the principle of haemodialysis is the same in each case. Blood is passed over one side of a semi-permeable membrane, while a special fluid (**dialysate**) flows over the other side. A pressure difference is maintained on either side of the membrane so that both water and solutes are filtered from the blood. In addition, the composition of the dialysate is such that a concentration gradient is also maintained across the membrane and solutes such as urea diffuse from the

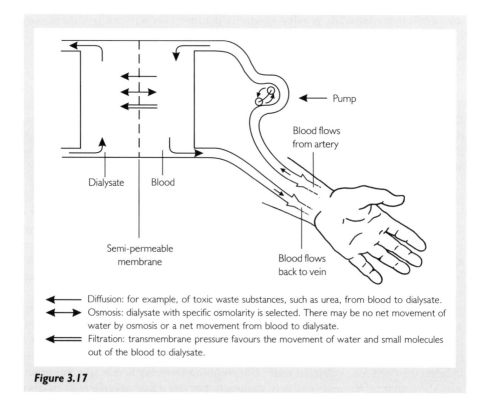

Figure 3.17

blood and into the dialysate. Finally, the composition of the dialysate determines its osmolarity, which in turn regulates the movement of water from the blood to the dialysate by osmosis. Despite great advances in the technology of dialysis it is only effective during the time that it is performed, and the patient has to return for treatment frequently.

Peritoneal dialysis

In this procedure the patient's own peritoneum serves as a membrane and dialysate is instilled into the peritoneal cavity, where it remains for a period of time before being drained out. One form of peritoneal dialysis may be performed by patients in their own homes, and consequently a good deal of independence is achieved.

Summary

1 Water is a polar molecule.
2 The shape of a meniscus is determined by the relative strengths of the forces of adhesion and cohesion.
3 Cohesive forces are responsible for surface tension.
4 Aqueous mixtures include mechanical suspensions, colloidal suspensions and true solutions.
5 True solutions consist of a solute and a solvent.
6 Solutes which dissociate into ions in solution are referred to as electrolytes.
7 Ionic compounds and polar molecules dissolve in polar solvents.
8 Non-polar covalent molecules dissolve in non-polar solvents.
9 The concentration of a solution identifies the relative proportions of solute and solvent.
10 Concentration may be expressed as a percentage or in terms of molarity.
11 Diffusion is an important physical process that accounts for the movement of particles of a solute or of a gas.
12 Osmosis is an important physical process that accounts for the movement of solvent particles.
13 The number of particles of a solute is expressed in terms of osmolarity or osmolality.
14 The tendency of a solution to produce plasmolysis or crenation is described in terms of tonicity.
15 The control of sodium balance is important in the regulation of extracellular fluid volume.
16 Blood volume is regulated by antidiuretic hormone (ADH) and thirst.
17 Specific gravity is the ratio of the density of a liquid to that of water.
18 The process of dialysis involves filtration as well as diffusion and osmosis.

Self-test questions

3.1 Which one of the following statements is true of water? It is
   a An ionic compound

   b A polar molecule
   c An element
   d A non-polar molecule
3.2 Match the substance on the left with the appropriate description on the right.
   a Hexane   **i**   non-polar solvent
   b NaCl   **ii** ionic compound
   c Glucose   **iii** polar solvent
   d Water   **iv** polar solute
3.3 Which one of the following is an electrolyte?
   a Urea
   b Glucose
   c Sodium chloride
   d Acetone
3.4 Which one of the following saline concentrations (w/v) is described as isotonic?
   a 0.09%
   b 0.9%
   c 0.5%
   d 5%
3.5 Which one of the following glucose concentrations (w/v) is described as isotonic?
   a 0.09%
   b 0.9%
   c 0.5%
   d 5%
3.6 If a solution of sodium chloride has a concentration of 1 mol/l, what will its osmolarity (in osmol/l) be?
   a 0.1
   b 1.0
   c 2.0
   d 10.0
3.7 Which one of the following represents a normal specific gravity for urine?
   a 1.020
   b 10.20
   c 1.050
   d 10.50
3.8 Which one of the following is *not* a possible cause of oedema?
   a Elevated blood osmotic pressure
   b Increased venous hydrostatic pressure
   c Increased capillary permeability
   d Lymphatic obstruction
3.9 When a concentration is expressed as 5% (w/v), what volume of solution (in ml) contains 5 g?
   a 1
   b 10
   c 100
   d 1000

3.10 If blood osmotic pressure is 25 mmHg and hydrostatic pressure at the arterial end of a capillary is 32 mmHg what is the net pressure leading to the formation of interstitial fluid (in mmHg)?

a 7

b 25

c 32

d 57

Answers to self-test questions

3.1 b

3.2 a i

   b ii

   c iv

   d iii

3.3 c

3.4 b

3.5 d

3.6 c

3.7 a

3.8 a

3.9 c

3.10 a

Further study/exercises

3.1 In giving practical examples relevant to this Chapter we have mentioned intravenous infusions. What care is required by the patient who has an intravenous infusion?

Bohony J. (1993). 9 common IV complications and what to do about them. *American Journal of Nursing*, **93**(10), 45–49

Goodinson S. (1990). The risks of IV therapy. *Professional Nurse*, **5**(5), 235–236, 238

Goodinson S. (1990). Good practice ensures minimum risk factors. Complications of peripheral venous cannulation. *Professional Nurse*, **6**(3), 175–177

Mathewson M. (1989). Intravenous therapy. *Critical Care Nurse*, **9**(2), 21–23, 26–28, 30–36

Wilkinson R. (1991). The challenge of intravenous therapy. *Nursing Standard*, **5**(28), 24–27

3.2 Mention has been made of hydrocolloid, hydrogel and alginate wound dressings. Identify examples of each and find out how they are used.

Cullum N. (1995). Hydrocolloid dressings. *Elderly Care*, **7**(2), 19–21, 23

Dealey C. (1993). Role of hydrocolloids in wound management. *British Journal of Nursing*, **2**(7), 358, 360, 362

Erwin-Toth P. and Hocevar B. J. (1995). Wound care: selecting the right dressing. *American Journal of Nursing*, **95**(2), 46–51

Fowler E., Cuzzell J. Z. and Papen J. L. (1991). Healing with hydrocolloids. *American Journal of Nursing*, **91**(2), 63–64

Milward P. (1991). Examining hydrocolloids. *Nursing Times*, **87**(36), 70, 72–74

Seymour J. (1997). Alginate dressings in wound care management. *Nursing Times*, **93**(44), 49–50, 52

Thomas S. (1990). Making sense of . . . hydrocolloid dressings. *Nursing Times*, **86**(45), 36–38

Further reading

Fischer R. G., Morgan S. R. and Parks B. (1985). What is the role of oral electrolyte solutions in diarrhoeal dehydration in children? *Pediatric Nursing*, **11**(3), 215, 227

Metheney N. M. (1996). *Fluid and Electrolyte Balance: Nursing Considerations*. Philadelphia: Lippincott

Miller J. A. (1989). Intravenous therapy in fluid and electrolyte imbalance. *Professional Nurse*, **4**(5), 237–241

Rutherford C. (1989). Fluid and electrolyte therapy: considerations for patient care. *Journal of Intravenous Nursing*, **12**(3), 173–183

Terry J. (1994). The major electrolytes: sodium, potassium and chloride. *Journal of Intravenous Nursing*, **17**(5), 240–247.

4 Acids, bases and pH balance

Learning outcomes

After reading the following chapter and undertaking personal study you should be able to:

1 Identify examples of acids and bases which are either common or which have special biological significance.
2 Define the terms *acid* and *base*.
3 List some of the properties of acids and bases.
4 Outline the first-aid measures to be taken in chemical burns.
5 Describe the pH scale.
6 Identify the products of a reaction between an acid and a base.
7 Distinguish between the meaning of the terms *strength* and *concentration* when used in connection with acids and bases.
8 Define the term *buffer*.
9 Outline the regulation of acid–base balance in the body by reference to buffer systems, respiratory regulation and renal regulation.
10 Using the results of laboratory tests, identify an acid–base imbalance as either acidosis or alkalosis and distinguish the cause as either respiratory or metabolic.

Introduction

Some students who have never studied science before may not be familiar with the terms **base** or *alkali*. (In this text the terms *base* and *alkali* are taken to be synonymous.) However, few will not have heard of **acids**, although the term may conjure images of dangerous and caustic substances which are to be avoided at all cost! Some might be surprised to learn that we encounter acids and bases all the time in our everyday lives. They are important substances in the environment, we use them as drugs and they have particular relevance to us in health or disease.

Can you identify some of the acids or bases which are important within the body or which may be found in the home?

Tables 4.1 and 4.2 identify some of the acids and bases which we regularly meet.

Table 4.1 Some common acids.

Acid	Common name/function
Acetylsalicylic acid	Aspirin, a common analgesic and antipyretic
Amino acid	The group of acids which make up proteins
Ascorbic acid	Vitamin C
Carbonic acid	Formed from the reaction between CO_2 and water
Citric acid	Found in citrus fruits and important in the energy-producing reactions of the body
Deoxyribonucleic acid	Carries genetic information
Fatty acid	Important molecules in fats
Folic acid	One of the B group vitamins
Hydrochloric acid	Important in digestion in the stomach

Table 4.2 Some common bases.

Base	Common name/function
Aluminium hydroxide	Gastric antacid
Magnesium hydroxide	Gastric antacid
Magnesium sulphate	Epsom salts, a laxative
Magnesium trisilicate	Gastric antacid
Sodium hydrogen carbonate	Baking soda, a gastric antacid also used to correct acidosis
Sodium sulphate	Glauber's salt, a laxative

Acids

An acid is a substance (a molecule or an ion) which donates a hydrogen ion (H^+) to another substance during a chemical reaction. A hydrogen ion results when an atom of hydrogen loses an electron:

$$H \longrightarrow H^+ \quad + \quad e^-$$

hydrogen atom   hydrogen ion   electron

Which sub-atomic particles does a hydrogen atom consist of?

Since a hydrogen atom normally consists of only one **electron** and one **proton**, it is not difficult to see that a hydrogen ion is simply a proton. For this reason, an acid may sometimes be defined as a *proton donor*. Having noted this, it is worth pointing out that hydrogen ions probably do not exist in a free state in solution, that is as H^+, but rather they react with water (H_2O) to form hydroxonium ions (H_3O^+). You may find some texts which use H^+ and others which use H_3O^+, so both are mentioned here. However, throughout this text H^+ is used. You will see this as we move on to consider some examples of acids in a little more detail.

Hydrochloric acid

This is an important acid in the body.

> Do you know where it is found?

Perhaps you remember that hydrochloric acid is found in the stomach, where it is responsible for turning pepsinogen into the protein-digesting enzyme pepsin and also for conferring some protection against microorganisms.

> Clearly hydrochloric acid is an important substance in health, but what happens if it is produced in excess or when the stomach is not adequately protected against its action?

You may know that in these circumstances a gastric ulcer may result.

Hydrochloric acid only exists in the presence of a solvent such as water. In the absence of water the substance is more correctly referred to as hydrogen chloride (HCl) – a gas formed from the covalent bonding of hydrogen and chlorine. The dry gas hydrogen chloride does not produce any protons, so it is not an acid. However, hydrogen chloride is extremely soluble, and in the presence of water the dissolved molecules dissociate (break up) into hydrogen ions (H^+) and chloride ions (Cl^-) as indicated below:

$$HCl \xrightarrow{\;H_2O\;} H^+_{(aq)} \quad + \quad Cl^-_{(aq)}$$

$$\text{hydrogen chloride} \qquad \text{hydrogen ion} \qquad \text{chloride ion}$$

The subscript (aq) means *aquated* and refers to the fact that ions in solution (dissolved in water) have a number of water molecules attached. For the reasons described above you should use the symbols $HCl_{(aq)}$ when referring to hydrochloric acid and HCl when referring to the dry hydrogen chloride gas.

You should note that in the equation showing the dissociation of hydrochloric acid into hydrogen ions and chloride ions the arrow indicating the direction of the reaction is from left to right. This is to show that in water hydrochloric acid dissociates completely into ions. It is therefore referred to as a strong acid.

Ethanoic acid

Ethanoic acid is sometimes referred to as acetic acid, and the reader will certainly have encountered it in dilute solution – vinegar! Ethanoic acid belongs to a group of acids called **carboxylic acids**, and these may be represented by the general formula below, where R represents a hydrogen atom or a chain of carbon atoms:

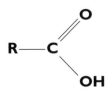

Carboxylic acids are also important in health and illness, since they form important molecules when combined with the alcohol **glycerol**.

Do you know what these molecules are?

Carboxylic acids combine with glycerol to form **glycerides** (fats) and for this reason carboxylic acids are also referred to as *fatty acids*. In your studies of health and diet you will encounter fatty acids once again, but for now let us return to ethanoic acid. Unlike hydrochloric acid, ethanoic acid is a weak acid, which means that it is not completely dissociated in water. Its dissociation is represented in the following way:

$$CH_3COOH \rightleftharpoons CH_3COO^- + H^+$$

ethanoic acid   ethanoate ion   hydrogen ion

In this case the reaction can proceed in either direction:

From left to right – ethanoic acid dissociates into ethanoate ions and hydrogen ions.

From right to left – ethanoate ions and hydrogen ions combine to form ethanoic acid.

For this reason two arrows are used to represent the reaction, but since, at room temperature, only 7% of the ethanoic acid molecules dissociate into ions, the arrow pointing from right to left is emphasised.

Hydrogen carbonate (bicarbonate) ion

The two previous examples of acids were covalently bonded molecules which dissociated in solution. Now we consider an ion which can act as an acid:

$$HCO_3^- \rightleftharpoons H^+ + CO_3^{2-}$$

hydrogen carbonate ion   hydrogen ion   carbonate ion

You will note that the reaction direction arrows are equally emphasised. In this case the equilibrium of the reaction may be tipped to the left or to the right depending upon the concentration of hydrogen ions. If in low concentration, the reaction equilibrium is tipped to the right and more hydrogen ions are generated. If hydrogen ions are in excess then the reaction equilibrium will be tipped to the left and hydrogen ions will be removed by combination with carbonate ions.

Bases

A base (alkali) is a molecule or an ion which accepts a hydrogen ion during a chemical reaction. An older definition of a base is a substance which yields hydroxide ions (OH^-) in solution, but in reality the two definitions are saying the same thing. For example, since a small number of water molecules dissociate into hydrogen ions and hydroxide ions thus:

$$H_2O \rightleftharpoons H^+ + OH^-$$

the addition of any substance which removes hydrogen ions leads to a relative excess of hydroxide ions.

Sodium hydroxide

Perhaps sodium hydroxide (NaOH) is unfamiliar to you but you may have heard it referred to by its common name *caustic soda*. Indeed, you may have even bought products which contain it in order to unblock a drain or clean the oven! The use of sodium hydroxide in oven cleaners is related to its ability to dissolve fat, and since cell membranes contain a high proportion of fat it is potentially dangerous to body tissue, especially if splashed into the eyes.

> In this event what would be an appropriate first-aid measure?

In order to check what you have decided you should refer to the section on first aid later in this chapter and consult a first-aid manual.

Sodium hydroxide is an ionic compound. However, even in the solid state it does not exist as molecules of NaOH but rather as sodium ions (Na^+) and hydroxide ions (OH^-). Since it is the hydroxide ions which accept protons we can for the moment ignore the sodium ions and demonstrate the action of sodium hydroxide as a base in the following way:

$$OH^- \quad + \quad H^+ \quad \rightleftharpoons \quad H_2O$$

hydroxide ion   hydrogen ion   water

Sodium hydroxide is described as a strong base because it is completely dissociated into sodium ions and hydroxide ions.

Liquid ammonia

Ammonia is actually a gas (NH_3). However, it does dissolve in water to produce an opaque solution (cloudy ammonia) which may be used as a cleansing agent. You may have bought a product which contained ammonia for cleaning the toilet! Aqueous ammonia is a weak base; that is, only a small proportion of the ammonia molecules are ionised by the acceptance of hydrogen ions:

$$NH_{3(aq)} \quad + \quad H^+ \quad \rightleftharpoons \quad NH_4^+$$

aqueous ammonia   hydrogen ion   ammonium ion

Hydrogen carbonate (bicarbonate) ion

The behaviour of the hydrogen carbonate ion as an acid has already been discussed, but it can also behave as a base according to the following equation:

$$HCO_3^- \quad + \quad H^+ \quad \rightleftharpoons \quad H_2CO_3$$

hydrogen carbonate   hydrogen ion   carbonic acid

Note once more that the reaction direction arrows are equally emphasised, which indicates an equilibrium reaction in which the balance may be tipped to the left or the right depending upon the concentration of reactants and products.

Properties of acids and bases

Those acids which are safe to taste are sour – think about citrus fruits such as lemons (citric acid) and vinegar (ethanoic acid). They also turn blue litmus paper red (litmus is a vegetable dye). Bases which are safe to taste have a bitter or metallic taste – think about the taste of soap. They also turn red litmus paper blue. Both acids and bases may be caustic; that is, they can cause burns and destroy body tissue. Acids and bases react together to form a **salt** and *water*. For example:

$$HCl_{(aq)} \quad + \quad NaOH \quad \longrightarrow \quad NaCl \quad + \quad H_2O$$

hydrochloric   sodium   sodium chloride   water
acid   hydroxide

In the above example the salt which is formed is indeed common salt (sodium chloride). However, the term *salt* specifically refers to an ionic compound that dissociates into ions other than hydrogen ions or hydroxyl ions.

Practice point 4.1: first aid

The caustic (corrosive) nature of many acids and bases has already been noted, and the burns which result from contact with them can be severe. For this reason you should wear safety glasses (or shields if you already use spectacles) all the time you are in a science laboratory. When part of the body comes into contact with an acid or a base the best first-aid measure is to wash the body part with plenty of clean water. If you are in a laboratory the first-aid box will contain an eyewash bottle and you should know where this is located before beginning any experiment involving acids. Alternatively there may be taps set aside for this purpose. Never attempt to neutralise an acid with a base or a base with an acid. You may do more harm than good. Once first-aid measures have been carried out, medical advice should be sought.

Measuring acidity

Before considering how acidity is measured it is important to clarify some terms. The terms *weak* and *strong* have already been used in connection with acids and bases. Here the tendency for the acid or base to dissociate is being referred to. This should not be confused with the **concentration**. For example, hydrochloric acid is referred to as a strong acid, because in solution it completely dissociates into hydrogen ions and chloride ions. Conversely ethanoic acid is described as a weak acid because only a small proportion of the molecules of ethanoic acid dissociate in solution to yield hydrogen ions and ethanoate ions. In a similar way, sodium hydroxide is described as a strong base because it dissociates completely in solution to sodium ions and hydroxide ions, while ammonia is incompletely ionised in water and is therefore a weak base.

In contrast, concentration has to do with the amount of a substance dissolved in a solvent. For example, a 1 molar (1 M) solution contains 1 mole of the solute dissolved in 1 litre of solvent (1 mol/l), a 0.01 molar solution contains 0.01 moles of a solute dissolved in 1 litre of solvent (0.01 mol/l) and so on.

You should now be able to work out that it is possible to have the following combinations:

1 A concentrated solution of a strong acid, such as 10 M hydrochloric acid.
2 A dilute solution of a strong acid, such as 0.05 M hydrochloric acid.
3 A concentrated solution of a weak acid, such as 10 M ethanoic acid.
4 A dilute solution of a weak acid, such as 0.05 M ethanoic acid.

Of course, it is also possible to have a range of concentrations for any named acid, and in a similar way we may talk of concentrated and dilute solutions of strong and weak bases.

> It can appear a little confusing at first and you might need to read the last section again.

Once you have grasped the idea that, as far as acids and bases go, concentration and strength are different things you will probably realise that in describing acidity what we really need to know is the concentration of hydrogen ions ($[H^+]$). In chemistry the use of square brackets refers to concentration, so that $[H^+]$ means the concentration of hydrogen ions.

The hydrogen ion concentrations which we will consider are in the range 1 – 0.000 000 000 000 01 mol/l. Clearly we are looking at some very low concentrations and the use of mol/l is somewhat clumsy with all those noughts. We might of course use a logarithmic scale; that is, we could speak of the concentration in terms of the powers of 10. The range of concentrations would then be $10^0 - 10^{-14}$ mol/l. This is somewhat easier, but it can be made simpler still.

Whenever acidity needs to be described, whether it be by scientists, nurses or even brewers, the **pH** scale is used (the letters pH stand for power of hydrogen). pH is the negative log of the hydrogen ion concentration and it enables us to describe acidity in terms of simple numbers. That is, a logarithmic scale gets rid of the noughts, and by defining the scale as a negative log the negative powers $10^0 - 10^{-14}$ are turned into positive figures. Table 4.3 is a comparison of $[H^+]$ and pH, while Table 4.4 gives the pH of various body fluids.

Table 4.3 Comparison of [H⁺] and pH.

$[H^+]$(mol/l)	pH
10^0	0
$10^{-3.5}$	3.5
10^{-7}	7
10^{-14}	14

Table 4.4 The pH of various body fluids.

Fluid	pH
Bile	7.6–8.6
Blood	7.35–7.45
Cerebrospinal fluid	7.35–7.45
Gastric secretions	1–3
Pancreatic secretions	8–8.3
Saliva	6–7
Semen	7.2–8
Urine	4.6–8

More about the pH scale

When using the pH scale a number of points must be considered. Firstly, since it is a logarithmic scale every change of one unit in pH represents a *tenfold* change in hydrogen ion concentration, a change of two units in pH a *hundredfold* change in hydrogen ion concentration and so on. For this reason, the apparently narrow normal range of serum pH (7.35–7.45) is not as narrow as it first appears. In addition, apparently small changes in serum pH represent large changes in hydrogen ion concentration, which you may need to report to a doctor. Secondly, the pH scale is a negative scale. That is a falling pH represents a rise in hydrogen ion concentration and a rising pH a falling hydrogen ion concentration.

Pure water has a pH of 7 and an identical concentration of hydrogen ions and hydroxide ions, and is therefore referred to as **neutral**. If hydrogen ions are added then $[H^+]$ rises and pH falls – that is, acids have a pH of less than 7. In contrast, if hydrogen ions are removed then $[H^+]$ falls and pH rises – that is, bases have a pH of greater than 7.

Salts as acids and bases

When acids and bases react together the salt formed may be neutral, acidic or basic, depending on the strengths of the acid and base used in the reaction. If a strong acid is added to a strong base, or a weak acid is added to a weak base, the resultant salt is neutral. In contrast, the reaction between a strong acid and a weak base results in the formation of an acidic salt, while the reaction between a weak acid and a strong base produces a basic salt.

Acid–base balance

We have already looked at the concept of **homeostasis** in Chapter 1. We noted that the cells of the body require a stable environment in which to function and that this environment is comprised of such things as temperature and the concentration of various electrolytes, including hydrogen ions. The pH of intracellular fluid is not easily measured, but blood pH is often measured in clinical practice and the normal value is found to be within the range 7.35–7.45.

> *Look at these figures again. Have you realised that blood is normally slightly alkaline?*

If the pH rises above 7.45 the condition **alkalosis** exists and we would describe the patient as being *alkalotic*. On the other hand, if the pH were to fall below 7.35 the condition of **acidosis** exists and we would describe the patient as *acidotic*. Note that since neutrality is 7, but the lower normal limit of blood pH is 7.35, the terms *acidosis* and *acidotic* are used even when the blood remains slightly alkaline but less alkaline than usual! In fact, it is not usual for patients to survive if the pH is less than 7. Similarly a pH of 8 is also usually fatal.

> *How does the body regulate acid–base balance?*

Three mechanisms exist:
1 Buffer systems
2 Respiratory regulation
3 Renal regulation

Buffer systems

A buffer system is a solution which resists a change in pH. That is, if an acid or base is added the change in pH which occurs is much less than that which might otherwise be expected. Buffer solutions consist of a mixture of a weak acid and its basic salt. The weak acid partially dissociates to release hydrogen ions and tends to cause the pH to fall, while the basic salt removes hydrogen ions and tends to cause the pH to rise. Of course, the actual pH of the solution depends upon the ratio of weak acid to its basic salt in solution.

Following the addition of hydrogen ions to a buffer solution a fall in pH is resisted by the presence of the basic salt. Conversely, following the addition of a base a rise in pH is resisted by the presence of the weak acid. As far as the body is concerned a change in blood pH is resisted by a number of buffers.

The hydrogen carbonate system

If a hydrogen carbonate buffer system were being prepared in the laboratory, a solution of carbonic acid (the weak acid) would be added to a solution of sodium hydrogen carbonate (the basic salt). In the blood, carbonic acid is formed from the reaction of carbon dioxide (produced by cellular metabolism) with water. In addition, hydrogen carbonate ions are also present from the dissociation of this acid and we can illustrate the reactions involved in the following way:

$$CO_2 \quad + \quad H_2O \quad \rightleftharpoons \quad H_2CO_3 \quad \rightleftharpoons$$

carbon   water   carbonic
dioxide   acid

$$H^+ \quad + \quad HCO_3^- \quad \rightleftharpoons \quad H^+ \quad + \quad CO_3^{2-}$$

hydrogen   hydrogen   hydrogen   carbonate
ion   carbonate ion   ion   ion

> *Do you remember that the hydrogen carbonate ion can act as an acid or a base? Can you see how the presence of hydrogen carbonate ions in blood buffers a change in pH?*

Suppose for a moment that an acid were added to the blood and the concentration of hydrogen ions began to increase. These would be removed by combination with hydrogen carbonate ions to form carbonic acid. That is, the reaction equilibrium is tipped to the left. On the other hand, if a base were added to the blood and hydrogen ions removed their concentration would be restored by the further dissociation of carbonic acid (which yields hydrogen ions and hydrogen carbonate ions). That is, the reaction equilibrium would be tipped to the right.

Later we shall see the close relationship of this buffer system to the renal and respiratory systems. For example, the concentration of hydrogen carbonate ions is regulated by the kidneys, while in the lungs carbonic acid dissociates into water and carbon dioxide, which are eliminated through breathing. In this way the body avoids a build-up of carbon dioxide which would otherwise push the reaction equilibrium to the right.

Phosphate buffer system

This system is similar to the hydrogen carbonate system. Two ions are important – the dihydrogen phosphate ion ($H_2PO_4^-$), and the monohydrogen phosphate ion (HPO_4^{2-}). Monohydrogen phosphate ions are able to combine with excess hydrogen ions to form dihydrogen phosphate ions:

$$HPO_4^{2-} \quad + \quad H^+ \quad \longrightarrow \quad H_2PO_4^-$$

monohydrogen   hydrogen   dihydrogen
phosphate ions   ions   phosphate ions

while the addition of hydroxide ions to dihydrogen phosphate ions results in the formation of monohydrogen phosphate ions and water:

$$H_2PO_4^- \quad + \quad OH^- \quad \longrightarrow \quad HPO_4^{2-} \quad + \quad H_2O$$

dihydrogen   hydroxide   monohydrogen   water
phosphate ions   ions   phosphate ions

The concentration of dihydrogen phosphate ions and monohydrogen phosphate ions is low in extracellular fluid but high in intracellular fluid and in the fluid in renal tubules. Therefore the phosphate buffer system has little effect in blood, but it is important within cells and in renal tubules.

Protein buffer system

Proteins consist of chains of **amino acids**. An amino acid consists of a weak acid group (the carboxyl group – COOH) and a weak basic group (the amino group – NH_2) as illustrated below:

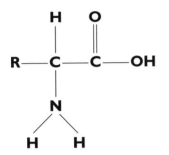

The carboxyl group may donate a hydrogen ion to the amino group to form a **zwitterion** (dipolar ion). This is a covalent compound containing charged ionic regions:

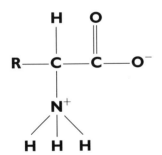

Zwitterions can function as buffers in the following way. A fall in pH is resisted when an excess of hydrogen ions are removed by combination with the COO⁻ group:

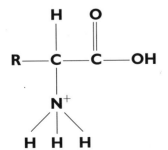

Conversely, an excess of hydroxide ions results in the release of hydrogen ions from the amine group to form water:

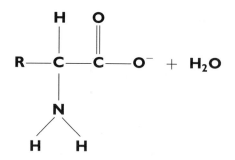

The haemoglobin buffer system

Do you remember where haemoglobin is found?

Perhaps you can recall that haemoglobin is the protein found in erythrocytes (red blood cells) which is responsible for the transportation of oxygen. However, erythrocytes, and in particular haemoglobin, also play an important role in buffering pH change. In addition to the possession of haemoglobin, erythrocytes also have the enzyme **carbonic anhydrase**, which catalyses the reaction between carbon dioxide and water to yield carbonic acid. We have previously noted the dissociation of this acid into hydrogen ions and hydrogen carbonate ions, and it is the fate of these ions which concerns us here. Hydrogen carbonate ions formed within erythrocytes diffuse into the plasma and their role has already been discussed. The retention of hydrogen ions by erythrocytes might be expected to cause a fall in intracellular pH but such a change is buffered as a consequence of the combination of hydrogen ions with haemoglobin. In fact, *reduced haemoglobin* (haemoglobin not combined with oxygen) has a much greater affinity for hydrogen ions than does *oxyhaemoglobin* (haemoglobin combined with oxygen). We might therefore regard the function of haemoglobin once it has delivered oxygen to the cells as transporting hydrogen ions from the cells to the lungs. Here reduced haemoglobin combines with oxygen to form oxyhaemoglobin, from which hydrogen ions are released. In the plasma these combine with hydrogen carbonate ions and the carbonic acid which results dissociates into carbon dioxide and water, which are of course eliminated in the breath.

It should be noted that the electrical imbalance that might be expected as a result of the efflux of (negatively charged) hydrogen carbonate ions from erythrocytes is prevented by the simultaneous influx of (negatively charged) chloride ions. Similarly, the efflux of (positively charged) hydrogen ions from erythrocytes in the capillaries of the lung is accompanied by the simultaneous efflux of chloride ions. Not surprisingly, the movement of chloride ions in and out of erythrocytes is referred to as the *chloride shift*.

In summary, haemoglobin serves as a buffer by chemically combining with hydrogen ions and so temporarily removing them from the plasma.

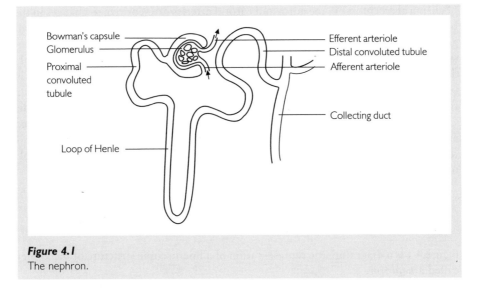

Figure 4.1
The nephron.

The ammonia buffer system

This system operates in the renal tubules (see Figure 4.1). Amino acids are *deaminated* (broken down) in tubular cells to form ammonia (NH_3) which diffuses into the renal filtrate. Ammonia molecules are then able to combine with free hydrogen ions to form ammonium ions (NH_4^+) thus:

$$NH_{3(aq)} \quad + \quad H^+ \quad \longrightarrow \quad NH_4^+$$

aqueous ammonia   hydrogen ions   ammonium ions

Ammonium ions combine with chloride ions (Cl^-) and the resultant salt, ammonium chloride (NH_4Cl), is eliminated in the urine. The presence of the phosphate and ammonia buffer systems means that excess hydrogen ions can be eliminated in the urine without high urine acidity.

Respiratory regulation

It is important to note that buffer systems do not actually remove hydrogen ions from the body – this role is performed by the respiratory and renal systems. Furthermore, the supply of buffers is not without limit and it is possible to reach the maximum buffering capacity of the body. Consequently, other mechanisms are required to ensure acid–base balance.

We have already noted that carbon dioxide has a great influence on blood pH due to its reaction with water to form carbonic acid. Therefore the extent to which carbon dioxide is eliminated by the lungs is a major factor in acid–base balance. It has been shown that structures called *chemoreceptors* (a contraction of chemical receptors) in the aorta and carotid arteries are sensitive to changes in the blood pH and that chemoreceptors in the respiratory centre of the medulla oblongata (part of the brain stem) are highly sensitive to changes in the pH of cerebrospinal fluid (CSF). Perhaps it is not surprising to discover that a falling pH detected by these chemoreceptors is responsible for stimulating an increase in both rate and depth of breathing (hyperventilation) and that this tends towards a restoration of acid–base balance by increasing the elimination of carbon dioxide. Conversely, when blood pH returns to normal the respiratory stimulus is removed and the individual ceases to hyperventilate. If you have already read Chapter 1 (Homeostasis – keeping the body in balance) you will recognise this as an example of a negative feedback mechanism.

Renal regulation

Figure 4.1 is a diagrammatic representation of a microscopic structure of the kidney called a *nephron*.

> *Look at it carefully for a moment.*

There are about 1.3 million nephrons in each kidney and they are sometimes referred to as the functional units of the kidney. The term *functional unit* implies that the nephron is the smallest part of the kidney which has the same functions as the kidney as a whole. The *glomerulus* is a capillary network responsible for the process of filtration. This is the process whereby a fluid (filtrate) is formed from the blood, but this fluid does not retain all the components of blood. Cells and proteins are not filtered from the blood, but water and small molecules are. The filtrate passes down the *tubule* of the nephron, where some molecules are reabsorbed from it and others are secreted into it. Eventually the modified filtrate is excreted as urine. For further detail on renal function you should consult a physiology text, but for now it is sufficient to know that the secretion of hydrogen ions into the filtrate and the reabsorption of hydrogen carbonate ions from it are important processes in the maintenance of pH balance.

Cells in the distal part of the renal tubule are sensitive to changes in blood pH. Carbonic anhydrase within these cells catalyses the reaction between carbon dioxide from blood and water to form carbonic acid, which dissociates into hydrogen ions and carbonate ions. The hydrogen ions are actively transported into the filtrate in exchange for sodium ions, which, along with hydrogen carbonate ions, pass into the blood. The secretion of hydrogen ions is accelerated by a falling pH, and this, together with the release of hydrogen carbonate ions into the blood, accounts for one way in which the kidney regulates acid–base balance.

It should be noted that an increase in hydrogen ion secretion into tubular fluid does not cause the pH of filtrate to fall as much as might be expected. This is due to the effect of the phosphate and ammonia buffer systems previously described.

In addition to the secretion of hydrogen ions into filtrate, hydrogen carbonate ions can be reabsorbed from filtrate when required to maintain acid–base balance.

However, the size and negative charge of hydrogen carbonate ions makes this difficult. Instead of being reabsorbed directly they are converted to carbon dioxide in the following way. Hydrogen carbonate ions in renal filtrate react with hydrogen ions to form carbonic acid, the dissociation of which yields carbon dioxide and water. The water is excreted in the urine – thus removing hydrogen ions from the body. The carbon dioxide diffuses into tubular cells where it forms carbonic acid which in turn dissociates into hydrogen ions and hydrogen carbonate ions, the fate of which has already been described.

In summary, the kidney plays an important role in acid–base balance, firstly through the secretion of excess hydrogen ions into renal filtrate and secondly through the selective reabsorption of hydrogen carbonate ions, in accordance with the body's need.

It is also worth pointing out that the renal regulation of acid–base balance may also be affected by the endocrine glands. For example, the hormone aldosterone increases the rate of sodium reabsorption and potassium secretion in the distal convoluted tubule. However, when aldosterone levels are elevated hydrogen ion secretion is also stimulated. Consequently, patients who produce aldosterone in excess (aldosteronism) may experience an elevated blood pH (alkalosis). Conversely, abnormally low levels of aldosterone (Addison's disease) may lead to acidosis. Having noted this, the renal mechanisms for maintaining acid–base balance are primarily influenced by pH of body fluids and not by the endocrine system.

Acid–base imbalance

An acid–base imbalance may either be acidosis or alkalosis.

> *Do you remember that acidosis is a fall in pH below 7.35 while alkalosis is a rise above 7.45?*

An acid–base imbalance may be further described as either *respiratory* or *metabolic*. Respiratory disturbances involve carbon dioxide elimination, while metabolic disturbances involve fixed (non-volatile) acids normally excreted by the kidneys. However, it must be admitted that this terminology is not ideal, since carbon dioxide is of course produced as a consequence of the metabolic activity of the cells. Carbonic acid is therefore as much of a metabolic acid as are fixed acids.

Respiratory acidosis

This results whenever carbon dioxide retention occurs. The accumulation of carbon dioxide within the blood pushes the carbonic acid equilibrium to the right, resulting in an excess of hydrogen ions. The cause may be hypoventilation following head injury or excessive use of sedating drugs; an excess of pulmonary secretions; a pulmonary infection; asthma or a chronic obstructive pulmonary disease such as chronic bronchitis. Other causes include collapsed alveoli (atelectasis) and pulmonary embolism.

Respiratory alkalosis

Respiratory alkalosis occurs when there is an increased elimination of carbon dioxide as a consequence of hyperventilation. The cause may be hysteria or poorly adjusted mechanical ventilation.

Metabolic acidosis

Metabolic acidosis results when the ability of the kidney to secrete hydrogen ions into filtrate is impaired, as is the case in chronic renal failure and Addison's disease. Another important cause is the production of acidic ketone molecules, which may occur in diabetes mellitus. Whenever a lack of insulin results in a cellular deficiency of glucose, fats are used as a fuel instead. This process is more fully explained in Chapter 7 (Energy in the body), but for now it is sufficient to know that the metabolic pathway involved in fat metabolism results in an over-production of ketones (**ketoacidosis**). Finally, acidosis which occurs following excessive ingestion of acidic drugs such as aspirin (acetylsalicylic acid) and that which results from the loss of alkali from the body, such as diarrhoea, are also included here.

Metabolic alkalosis

Metabolic alkalosis may result from excessive loss of hydrogen ions from the body and may follow vomiting or administration of diuretics (drugs which increase urine output). Alkalosis in aldosteronism has already been mentioned, and one other cause may be the excessive ingestion of alkaline indigestion medication.

Practice point 4.2: recognising acid–base imbalance

Help in recognising acid–base imbalance is often required by students and qualified staff alike. This involves comparing laboratory data with normal acid–base laboratory values, such as those given in Table 4.5.

First of all, look at the patient's most recent blood laboratory results and check the pH. Now describe an imbalance as acidosis or alkalosis. Next, identify the cause as a respiratory or metabolic one. For respiratory imbalances check the partial pressure of carbon dioxide in arterial blood ($PaCO_2$). The normal range is 4.7–6.0 kPa; a value above this is respiratory acidosis and a value below this is respiratory alkalosis. For metabolic imbalances check the hydrogen carbonate ion concentration ($[HCO_3^-]$). The normal is 19–29 mmol/l; a value above this is metabolic alkalosis and a value below this is metabolic acidosis.

However, someone may point out that the $[HCO_3^-]$ is not only affected by metabolic disturbances but also respiratory ones. For example, in respiratory acidosis $[HCO_3^-]$ rises along with $PaCO_2$ according to the by now

Table 4.5 Normal acid–base laboratory values.

pH	7.35–7.45
$PaCO_2$	4.7–6.0 kPa
$[HCO_3^-]$	19–29 mmol/l
base excess	$^-2 - ^+2$

familiar equation below:

$$CO_2 + H_2O \rightleftharpoons H_2CO_3 \rightleftharpoons H^+ + HCO_3^-$$

The answer to this problem is in reality a simple one. Prior to the measurement of $[HCO_3^-]$ in the laboratory the blood sample is artificially brought under standard conditions of $PaCO_2$ and temperature. For this reason the measurement of $[HCO_3^-]$ is referred to as the *standard* hydrogen carbonate ion concentration.

Perhaps you would like to attempt some examples. Consider the following:

Patient 1

$pH = 7.33$, $PaCO_2 = 5.6\,kPa$, $[HCO_3^-] = 18\,mmol/l$
1 pH is low – acidosis
2 $PaCO_2$ is normal
3 $[HCO_3^-]$ is low
Conclusion: metabolic acidosis

Patient 2

$pH = 7.47$, $PaCO_2 = 5.5\,kPa$, $[HCO_3^-] = 35\,mmol/l$
1 pH is high – alkalosis
2 $PaCO_2$ is normal
3 $[HCO_3^-]$ is elevated
Conclusion: metabolic alkalosis

Patient 3.

$pH = 7.31$, $PaCO_2 = 7.5\,kPa$, $[HCO_3^-] = 22\,mmol/l$
1 pH is low – acidosis
2 $PaCO_2$ is high
3 $[HCO_3^-]$ is normal
Conclusion: respiratory acidosis

Patient 4

$pH = 7.51$, $PaCO_2 = 3.2\,kPa$, $[HCO_3^-] = 22\,mmol/l$
1 pH is high – alkalosis
2 $PaCO_2$ low
3 $[HCO_3^-]$ is normal
Conclusion: respiratory alkalosis

When you next have sight of laboratory reports you can work through them in a similar way. Perhaps you have seen such reports already. If so, you will no doubt have noticed results of another test used to assess the severity of a patient's metabolic disturbance – the base excess. This is a measure of the amount of acid which would need to be added to the sample of patient's blood to restore acid–base balance to normal. The normal value of base excess is therefore zero; in metabolic alkalosis it is positive: in metabolic acidosis it is negative. For the sake of simplicity, values of base excess have not been included in the examples given here.

So far we have considered acid–base disturbances which affect only one system – the respiratory system or the kidneys. It is worth remembering that the body attempts to compensate for a disturbance in one system by a change in the other. Look at a further three examples:

Patient 5

$pH = 7.32$, $PaCO_2 = 7.2\,kPa$, $[HCO_3^-] = 32\,mmol/l$
1 pH is low – acidosis
2 $PaCO_2$ is high
3 $[HCO_3^-]$ is high
Conclusion: compensated respiratory acidosis

Patient 6

$pH = 7.35$, $PaCO_2 = 4.3\,kPa$, $[HCO_3^-] = 18\,mmol/l$
1 pH is on the low side of normal – acidosis
2 $PaCO_2$ is low
3 $[HCO_3^-]$ is low
Conclusion: compensated metabolic acidosis

Table 4.6 **Management of acid–base imbalance.**

Disturbance	Aetiology	Intervention
Respiratory acidosis	CO_2 retention due to: hypoventilation (CNS depression or pain); excess of pulmonary secretions; pneumonia; atelectasis; chronic obstructive pulmonary disease (COPD)	Improve ventilation and gas exchange: coughing and deep breathing exercises; incentive spirometry; postural drainage; respiratory stimulants; narcotic antagonists; steam inhalations; nebulised saline; bronchodilators; pain control; intermittent positive pressure breathing (IPPB); artificial ventilation
Respiratory alkalosis	Excessive CO_2 loss due to hyperventilation in anxiety/hysteria or poorly adjusted mechanical ventilation	Sedation; relaxation exercises; adjust mechanical ventilation
Metabolic acidosis	Accumulation of ketones in diabetes (ketoacidosis); excessive loss of base in diarrhoea; failure to secrete H^+ in renal disease	Treat underlying cause
Metabolic alkalosis	Prolonged vomiting; nasogastric aspiration; excess intake of antacids	Anti-emetics; reduce antacids

Patient 7

$pH = 7.37$, $PaCO_2 = 7.7\,kPa$, $[HCO_3^-] = 36\,mmol/l$
1 pH is on the low side of normal – acidosis
2 $PaCO_2$ is high
3 $[HCO_3^-]$ is high
Conclusion: compensated respiratory acidosis

Finally, it is not within the scope of this book to describe the management of acid–base imbalance. However, Table 4.6 summarises the interventions which might be made by the multidisciplinary team. Note that not all the interventions listed are appropriate in each case. Doctors, nurses and physiotherapists are all involved in the management of acid–base imbalance, but some interventions are appropriate to one professional group alone – for example, the prescription of drugs by doctors.

Summary

1 Acids are substances which donate hydrogen ions during a chemical reaction.
2 Bases are substances which accept hydrogen ions during a chemical reaction.
3 Acids and bases are described as weak or strong depending upon the extent of their dissociation.
4 Acids and bases react together to produce a salt and water.
5 The concentration of hydrogen ions is described in terms of pH.
6 A buffer is a solution which resists a change in pH.
7 The body's acid–base balance is maintained by buffer systems, respiratory regulation and renal regulation.
8 Acid–base imbalances are either acidosis or alkalosis and may be further described as respiratory or metabolic.

Self-test questions

4.1 Which one of the following is produced by the stomach?
  a Citric acid
  b Amino acid
  c Hydrochloric acid
  d Folic acid
4.2 Which one of the following is *not* an important component of the diet?
  a Folic acid
  b Amino acid
  c Ascorbic acid
  d Acetylsalicylic acid
4.3 Which one of the following may a hydrogen ion also be referred to as?
  a An electron
  b A proton

c A neutron

d An atom

4.4 Match the acid on the left with the appropriate description on the right.

a Carboxylic acid	**i** formed from the reaction between carbon dioxide and water
b Carbonic acid	**ii** a component of fats
c Ascorbic acid	**iii** carries genetic information
d Deoxyribonucleic acid	**iv** also known as vitamin C

4.5 Which acid has the chemical formula CH_3COOH?

a Ethanoic acid

b Citric acid

c Deoxyribonucleic acid

d Hydrochloric acid

4.6 Neutrality on the pH scale is taken to be which of the following?

a 4

b 14

c 7

d 17

4.7 A fall in blood pH as a consequence of the excessive production of ketones in diabetes mellitus is described as which of the following?

a Respiratory acidosis

b Respiratory alkalosis

c Metabolic acidosis

d Metabolic alkalosis

4.8 Which of the following represents a normal $PaCO_2$?

a 4.0 kPa

b 5.0 kPa

c 7.0 kPa

d 8.0 kPa

4.9 Which of the following represents a normal $[HCO_3^-]$?

a 10 mmol/l

b 15 mmol/l

c 25 mmol/l

d 35 mmol/l

4.10 Which one of the following group of figures represents compensated respiratory acidosis?

	pH	$PaCO_2$/kPa	$[HCO_3^-]$/mmol/l
a	7.33	5.5	18
b	7.3	7.5	23
c	7.31	7.5	33
d	7.33	4.2	17

Answers to self-test questions

4.1 c	**4.6** c
4.2 d	**4.7** c
4.3 b	**4.8** b
4.4 a ii	**4.9** c
b i	
c iv	
d iii	
4.5 a	**4.10** c

Further study/exercises

4.1 The next time that you visit a clinical area where patient blood results are available, look over the values and determine if there is an acid–base imbalance. Discuss what you decide with a clinical supervisor.

4.2 Make notes on the following nursing interventions:
a Coughing and deep breathing exercises
b Airway aspiration (suctioning)
c The use of inhalers and nebulisers
d Postural drainage.
Alexander M. F., Fawcett J. N. and Runciman P. J. eds. (1994). *Nursing Practice, Hospital and Home: The Adult*. Edinburgh: Churchill Livingstone.

5 Pressure, fluids, gases and breathing

Learning outcomes

After reading the following chapter and undertaking personal study you should be able to:

1 Define the term *pressure*.
2 List some of the units used to measure pressure, especially those used in nursing practice.
3 Describe how to measure arterial blood pressure using a sphygmomanometer and stethoscope.
4 Apply an understanding of Pascal's principle to static fluids within the body.
5 Describe the factors which influence blood pressure.
6 Apply an understanding of Boyle's law to breathing.
7 Apply an understanding of the laws of Dalton and Henry to gas exchange in the lungs.
8 Outline the factors which influence blood flow through a vessel.
9 Describe oxygen therapy and distinguish between fixed and variable performance oxygen masks.
10 Describe the operation of a nebuliser.

Introduction

In this chapter we are going to consider the concept of **pressure** and apply an understanding of it to fluids in the body, such as blood, and to gases and breathing. Most people will have some idea that pressure has something to do with **force**. Even in everyday conversation we might say that someone who compels another to act in a certain way 'places pressure on them' or 'forces them'. So even if you have not studied science before, perhaps the concept of pressure is not so alien after all! However, when the word *pressure* is used in a scientific context something quite specific is meant. Pressure is defined as force per unit area:

$$\text{pressure} = \frac{\text{force}}{\text{area}}$$

Therefore pressure is not the same as force, but depends upon both force and the surface area over which the force is applied. For example, it is not usual for an

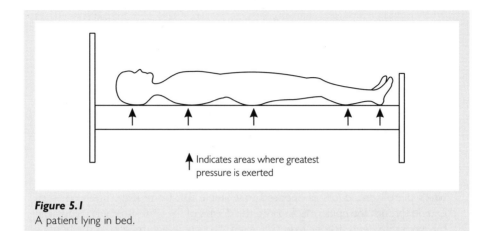

Indicates areas where greatest
pressure is exerted

Figure 5.1
A patient lying in bed.

individual of average weight to damage a tiled floor simply by walking on it. This is because the body weight (the force) is spread over the surface of the soles of the shoes. However, if stiletto heels are worn the same force is applied over a much smaller area and the effect resembles small hammer blows to the surface of the tiles. In a similar way the point of a hypodermic needle readily penetrates the skin, but the same force produces much less pressure when applied over the larger surface area of the fingertip.

Even such a simple understanding of pressure and its relationship to force and surface area has immediate practical applications for health practitioners. For example, suppose we were to take a solid block of wood some 70 kg in weight and place it on a table. If we now had some means of measuring the pressure beneath the block we would find that, provided the surfaces of both block and table were smooth, then the pressure would be the same at any point we might care to measure. Now let us compare this with a 70 kg patient lying supine (flat on the bed). The surface of the patient's body is certainly not flat, and some parts, like the hollow of the back, are not in contact with the bed at all. This means that a much greater pressure than might otherwise be expected is exerted upon those parts of the body that do have contact. This is especially true of the bony prominences, as illustrated in Figure 5.1.

Practice point 5.1: pressure sores

Throughout both wakefulness and sleep some part of the body is subjected to pressure from its own weight as a consequence of lying, sitting or leaning. Movement ensures that the pressure exerted upon the soft tissues does not remain over one area for very long. Indeed, even during sleep the weight of the body against the skin is continually redistributed because of movement.

What happens if mobility is restricted?

Experience shows that one consequence of sustained low pressure against the body is damage to the integument and the development of a *pressure sore*.

Can you think of some of the reasons for reduced mobility?

You may have realised that possible causes include physical illness which necessitates spending longer than usual in bed or restricting therapies such as intravenous infusions or a plaster cast. Some people may be too frail to move or find changing position too painful, while others may be sedated or unconscious.

What do pressure sores look like?

Like any other kind of wound, pressure sores can differ in severity. An alternative name for them is *decubitus ulcers*, and the word *ulcer* implies an erosion or wound in which there is tissue loss as opposed to a simple cut. Commonly, pressure sores extend through the epidermis and into the dermis of the skin, although it is possible for them to extend deeper than this, even through fat and muscle to the bone.

But how can the weight of our own bodies result in such severe wounds?

To answer this question we need to go back and consider once again the individual lying in bed. Remember that bodies are not flat, like a wooden block, and that the pressure of the body against the bed is greatest over bony prominences, such as the sacrum, heels, elbows, shoulders and occiput (back of the head). In fact, if we were to measure the pressure applied to the surface of the body by its own weight at such points we would find it to be in excess of the pressure of blood in capillaries (mean capillary blood pressure is only about 25 mmHg), thus compressing them and depriving the soft tissues of a blood supply. The consequence is that soft tissue deprived of a blood supply dies and a pressure sore results. Thus pressure sores do not result from the direct effects of pressure, but a low sustained pressure results in cell death through the mechanism of a compromised blood supply. A second process involved in pressure sore formation is damage to capillaries as a consequence of friction and shearing forces, which occur when incorrect patient moving techniques are employed and the skin and underlying tissues are dragged over a resistant surface such as the bedclothes. There is a great deal more to learn about pressure sores, but for now this brief introduction has served to illustrate how important a consideration of pressure is to nursing practice.

Thus far we have only considered pressure and solid objects, but when we consider pressure in the body we are often thinking about gases and breathing or liquids such as blood. In addition, we need to consider the units of measurement of pressure. Let us do this by first of all thinking about atmospheric pressure.

Atmospheric pressure

Most people know something about atmospheric pressure, even if it is only through watching the weather forecast or from taking a trip in an aeroplane. We do not think too much about the pressure exerted by the atmosphere against our bodies because, for most of the time, we cannot feel it. Nonetheless, the mixture of gases which surrounds us and which forms the atmosphere does exert quite a pressure on our bodies. Perhaps the time when we do become aware of it is when the atmospheric pressure changes as is the case when ascending or descending a steep hill in a car or during take-off or landing in an aeroplane. At these times we feel uncomfortable because of the distortion of the ear drum (tympanic membrane) caused by a difference in pressure between the middle and outer ears. You can find a diagram of the ear in Chapter 12 (Sound and hearing). The middle ear is vented by a tube (eustachian tube) and swallowing helps to open this tube and equalise the pressure in the middle ear, thus preventing the distortion of the drum, which can be painful.

> *Can you think of a common cause of inability to vent the middle ear which then results in considerable pain when flying?*

You may have worked out that the eustachian tube may become blocked during a cold. For this reason flying is not then recommended. The purpose of noting this is simply to point out that we all have some experience of atmospheric pressure on our bodies.

Practice point 5.2: relieving the pain of flying

If swallowing helps to vent the middle ear through the eustachian tube what could be done to relieve the discomfort of take-offs and landings?

You probably already know that sucking a sweet may help.

But what about babies?

Have you noticed that babies often wake up and cry during take-offs and landings? This is because they are experiencing discomfort too, but they cannot reach for a bag of sweets!

What could you advise about this?

It is a good idea to have a bottle of water prepared, and if the baby can be encouraged to suckle the middle ear will be vented and the discomfort reduced.

Let us now return to our consideration of atmospheric pressure.

> *If we wanted to measure this pressure how could it be done?*

This problem also faced Evangelista Torricelli in the 17th century. His solution was to take a glass tube filled with mercury and invert it into a container also containing mercury. This is illustrated in Figure 5.2.

This experiment could easily be repeated today, but since mercury is poisonous its use is best confined to closed systems such as thermometers. Examine Figure 5.2 and consider what happened. When inverted, the heavy column of mercury sank, creating a near vacuum (v) in the closed end of the tube. The height of the column of mercury above that in the container is indicated by the letter h.

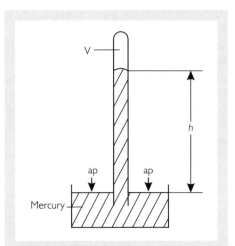

Why did the mercury column not fall further?

Figure 5.2
Torricelli's barometer.

You may have worked out that it was prevented from doing so by the atmospheric pressure (ap) pressing on the mercury in the container and pushing the column up.

Can you work out what would happen if we were to take Torricelli's device up a mountain?

As atmospheric pressure decreases with altitude the mercury column would fall as we ascend. Consequently, we could measure atmospheric pressure in terms of the height of the column. Torricelli used millimetres (mm) and since the chemical symbol for mercury is Hg, the unit mmHg (millimetres of mercury) was born. This is significant, since this unit of measurement of pressure is still used by health professionals today. However, it is worth pointing out that pressure is not a length! The height of the column depends upon both the weight of the mercury being supported in the tube and the cross-sectional area of the tube. The height of the column is a measurement of the ratio of force to area.

Finally, suction (negative pressure) can be measured in terms of negative units. For example -50 mmHg suction may be applied to a chest drain in order to facilitate the evacuation of the pleural space. This is described in a little more detail in a practice point later in the chapter.

Other units of measurement of pressure

The measurement of pressure is complicated by the use of a number of different units. For example, the mmHg is sometimes renamed the torr after Torricelli (1 mmHg = 1 torr). When atmospheric pressure is measured at sea level it is found to be 760 mmHg, and this value is sometimes referred to as a standard atmosphere (atm). Of course, water could be used in a measuring device instead of mercury, but since it is much less dense than mercury a very tall tube would be needed to measure atmospheric pressure. However, the units cmH_2O are sometimes used by health professionals in the measurement of central venous pressure, which is much lower than arterial blood pressure. Other units of measurement of pressure include pounds per square inch (psi) – important to know this when inflating the tyres of a car (!) – and the SI unit of pressure the pascal (Pa). Since this unit is commonly used it is worthwhile describing it in a little more detail.

The SI unit of area is the square metre (m^2) and that of force is the newton (N). At this point it is not essential to understand how the SI unit of force came into being, but we measure pressure in terms of newtons per square metre (N/m^2 or Nm^{-2}) – a unit subsequently renamed the pascal in honour of the French scientist Blaise Pascal. It is, however, a very small unit of pressure, so in practice units of a thousand pascals (kilopascal or kPa) are used.

It is not without purpose that some time has been spent in looking at the units of measurement of pressure, since you will encounter them in clinical practice. Indeed it is sometimes necessary to convert between units. With this in mind the following data is given:

$1\,kPa = 7.5\,mmHg = 10.2\,cmH_2O$

$1\,mmHg = 0.133\,kPa = 1.36\,cmH_2O$

$1\,cmH_2O = 0.098\,kPa = 0.735\,mmHg$

Pressure-measuring devices

The term *manometer* refers to a device in which pressure is measured by means of a vertical column of fluid. Therefore Torricelli's device was a manometer, and manometers are commonly used by health professionals too. For example, central venous pressure is measured using a saline manometer and arterial blood pressure using a mercury-filled **sphygmomanometer** (*sphygmo* is Greek for pulse). Devices such as the sphygmomanometer actually measure pressure in reference to atmospheric pressure. Consequently, when we measure systolic blood pressure as 120 mmHg what we really mean is that it is 120 mmHg above atmospheric pressure.

The term *manometer* should not be confused with the term *barometer*, which is a device for measuring atmospheric pressure. Of course, a barometer may use a column of fluid, as is the case in Torricelli's device, but this is not necessarily so. For example, anaeroid barometers measure atmospheric pressure based upon the extent of the distortion of an evacuated box. Anaeroid devices may also be used to measure arterial blood pressure. They are easily identified by the presence of a dial and needle instead of a graduated mercury column. Strictly speaking, such devices are not manometers, but they are commonly referred to as such since they perform the same role in the measurement of blood pressure.

Practice point 5.3: measuring blood pressure

Owing to the beating action of the heart, arterial blood pressure has two values – recorded during systole (contraction) and diastole (relaxation). Two pieces of equipment are needed to measure arterial blood pressure non-invasively (without inserting a needle into an artery): a sphygmomanometer and a *stethoscope*. The sphygmomanometer consists of a mercury (or anaeroid) manometer, a cloth cuff enclosing an inflatable bladder and an inflating bulb with release valve, as shown in Figure 5.3.

Pressure created by the inflation of the bladder pushes the column of mercury upward against gravity and the pressure is measured in terms of the rise in height of the column as described previously (in mmHg). Accurate readings can be obtained by looking at the meniscus (level of mercury) at eye level. Looking at the meniscus from above or below will give inaccurate readings. It is a good idea to check the manometer before use. Before inflation of the cuff the meniscus should be at zero and the scale should not be obscured by oxidised mercury. The cuff, bladder, tubing, connections, inflating bulb and valve should all be in good condition. Cuffs

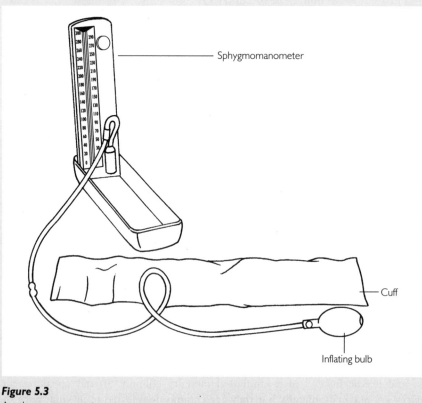

Sphygmomanometer

Cuff

Inflating bulb

Figure 5.3
A sphygmomanometer.

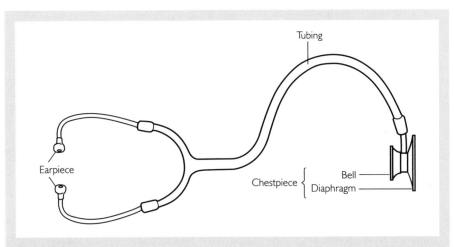

Figure 5.4
A stethoscope.

are available in different sizes for patients of different ages or build. The width of the inflatable bladder within the cloth cuff should be 40% of the circumference of the midpoint of the arm, while its length should be at least 80% of the circumference. In children over 5 years old the bladder should be over 12 cm long, while in normal and lean adult arms a bladder length of 35 cm is recommended as an alternative to the 23 cm long bladders which are often used. A bladder 42 cm long is reserved for muscular and obese arms.

The stethoscope is a simple instrument for *auscultation* – that is, listening to sound waves originating from within the body. It is illustrated in Figure 5.4.

The stethoscope is basically a closed 'cylinder' used to amplify the sound waves so that they can be heard more readily. The earpieces should be placed in the ears and angled forward so that they follow the angle of the auditory canal. The chest piece is then placed against the body. This part of the stethoscope is so called because the instrument is commonly used to listen for respiratory sounds. However, in blood pressure measurement it is placed above the artery in which the pressure is to be measured. The diaphragm is the circular, flat part of the chest piece, which is covered by a plastic disc. This is good for listening to high-pitched sounds such as bowel and lung sounds. The bell transmits low pitched sounds such as heart and vascular sounds and it is best used here. The operator should rotate the chest piece so that the appropriate part is operational. However, some simple stethoscopes have no bell but a diaphragm only.

In most cases it makes little difference whether the patient is sitting or lying during a blood pressure measurement, but the patient should be at rest, not anxious or stressed by recent activity. However, pregnant women sometimes experience a fall in blood pressure when lying supine (on their back) because raised abdominal pressure impairs venous return. The blood pressure of such women should be taken sitting or lying on the side. Furthermore, some patients, such as those taking medicines to treat high blood pressure (antihypertensives), experience a fall in blood pressure when standing, so measurement of blood pressure should be made in both supine and standing positions.

Whichever position is used, the arm should be supported in a comfortable horizontal position at heart level and the sphygmomanometer should be at the same level. If the arm is lower than the heart the result will be artificially high and if above the measurement will be artificially low. The patient should be warm and restrictive clothing removed. The point of maximal pulsation of the brachial artery just above the antecubital fossa is identified and the appropriate size of cuff fitted to the upper arm so that the centre of the bladder lies over the line of the artery and the edge of the cuff 2–3 cm above the point where the pulse was felt. This is shown in Figure 5.5.

Before using the stethoscope, however, an estimation of systolic pressure is made. This is achieved by palpating the brachial pulse and then inflating the cuff until the pulse disappears. The point at which it does so represents the approximate value of the systolic blood pressure.

Can you work out why this is so?

You may have realised that when the pressure in the cuff exceeds systolic blood pressure, blood flow in the artery ceases. The cuff is now deflated.

Following this an auscultatory measurement of systolic and diastolic blood pressure is made. The stethoscope is held gently over the brachial artery at the point of maximal pulsation. The cuff is now inflated to 30 mmHg above the estimated systolic blood pressure, and following this the valve is opened a little so that the cuff deflates slowly at 2–3 mmHg per second. The operator should now listen carefully. The point at which a tapping sound can be heard is an accurate measurement of systolic blood pressure.

Can you work out why?

As soon as the cuff pressure is reduced to systolic blood pressure, blood is forced through the compressed artery during systole. The tapping noise (first Korotkoff sound, or phase one) is the result of this pulsating turbulence. A mental note of the systolic blood pressure is made as the cuff continues to deflate. The point at which the tapping sound disappears (fifth Korotkoff sound, or phase five) is also noted, and this is an accurate measurement of diastolic blood pressure.

Can you work out why this is so?

When the pressure in the cuff reaches diastolic blood pressure blood flows freely in the artery; therefore the tapping sound disappears. However, in some groups, such as children, pregnant women, anaemic or elderly patients, the tapping sound never completely disappears. In these cases the point at which the tapping sound becomes muffled (fourth Korotkoff sound, or phase four) is recorded. The point of muffling is usually higher than the true diastolic pressure, and if this is recorded it should be

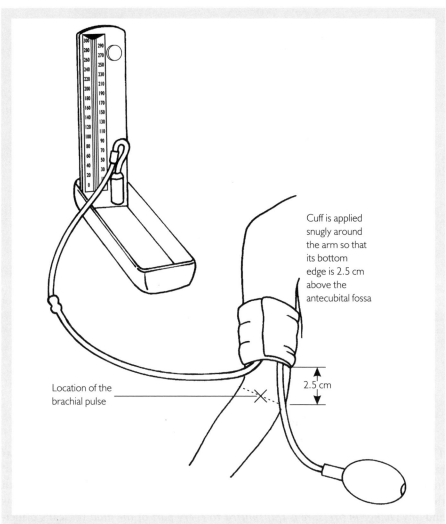

Cuff is applied
snugly around
the arm so that
its bottom
edge is 2.5 cm
above the
antecubital fossa

2.5 cm

Location of the
brachial pulse

Figure 5.5
The positioning of the cuff in blood pressure measurement.

made clear that the second reading is phase four and not phase five. Phase four is,
however, used routinely in pregnant women.

The values of blood pressure are usually recorded as though they are a fraction
– for example 120/80 mmHg. The first figure corresponds to the systolic measurement and the second to the diastolic. If the first, fourth and fifth sounds are
all recorded then the result might be 120/100/80 mmHg.

You may be interested to note that the Korotkoff sounds are so called because
a Russian surgeon of that name first noted them in 1905. Not all the sounds which
he described have been noted here; only the first, fourth and fifth, which are important in blood pressure measurement.

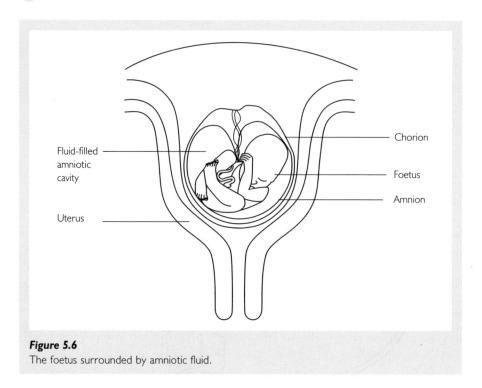

Figure 5.6
The foetus surrounded by amniotic fluid.

Pressure in static liquids

The term *static liquids* refers to fluids in enclosed containers. It does not mean that individual molecules are static, but rather that there is no net movement of the fluid. Individual molecules are indeed moving, and in colliding with each other and with the sides of the container they exert a pressure. This pressure acts in all directions and *any change in pressure is transmitted to every part of the fluid* (Pascal's principle). This behaviour of fluids has biological significance, since a number of body cavities contain fluid. An obvious example is amniotic fluid in the amniotic cavity, which surrounds the foetus (Figure 5.6). Clearly a blow to the foetus could be damaging, but since it is surrounded by amniotic fluid any rise in pressure is transmitted uniformly throughout the amniotic fluid (Pascal's principle) and this serves to protect the foetus from localised damage.

Fluids in containers

Of course the nurse is not only interested in fluid in compartments within the body – we regularly have to deal with fluids in containers and sometimes pressure is an important consideration here. The pressure at a point below the surface of a fluid

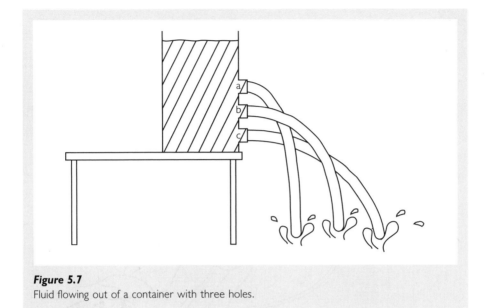

Figure 5.7
Fluid flowing out of a container with three holes.

in a container is dependent upon two factors. The first is the height of the column of liquid. Consider the diagram of a container with three holes (Figure 5.7).

Through which hole is water flowing at the fastest rate?

You should have been able to work out that water is flowing at the fastest rate through hole c – that is, the lowest hole.

Why is this?

This is because the pressure is greatest at this point, as this is the lowest hole and the height of the column of water is greatest above it. The other factor which determines the pressure of fluid in a container is the density of the fluid. Consequently, the pressure beneath a column of mercury is greater than that beneath a column of water of the same height.

Practice point 5.4: regulating intravenous infusions

One obvious application to this understanding of the pressure of fluids in containers is to intravenous infusions. A patient with an intravenous infusion is shown in Figure 5.8.

What would be the effect upon the infusion flow rate if the height of the intravenous bag were to be raised?

By raising the height of the intravenous bag the pressure of fluid would rise at the point where the cannula enters the patient and the flow rate would increase.

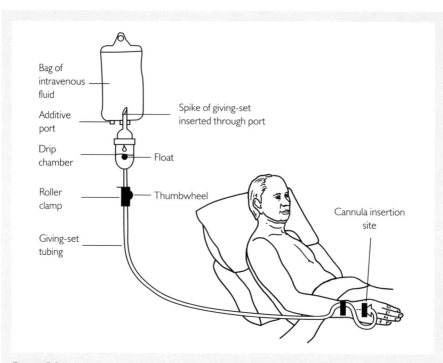

Figure 5.8
A patient with an intravenous infusion.

Perhaps this is a little obvious, but who said science had to be difficult? The important thing is to note that the height of the intravenous bag above the cannula is one factor which influences the intravenous flow rate. It is also worthwhile noting another practical point. Suppose you have regulated the flow rate of an infusion so that the prescribed volume will be infused in the time desired but the patient now gets up from the supine position (lying flat on the back) and begins to walk around.

How might this influence the flow rate?

In the standing position the height of the intravenous bag above the cannula is decreased and so the flow rate will be reduced. For this reason you will have to check the flow rate at intervals so as to ensure that the infusion is completed in the prescribed period of time.

Of course to use variations in the height of the intravenous bag above the cannula as the means of controlling infusion rates would be a little inconvenient. Once the bag is hung from the drip stand the flow rate is instead adjusted by a roller clamp, which compresses the intravenous tubing.

So how is the flow rate actually determined?

To work this out you need to know how much fluid is to be given (for example, 500 ml in 4 hours) and how many drops of the drop counter of the giving-set there are to each millilitre (ml). This information is given on the giving-set packaging. In an adult giving-set it is usually 20 drops per ml for crystalloid fluids (clear fluids such as dextrose or saline) and 15 drops per ml for blood. Paediatric giving-sets usually give 60 drops per ml. So if we were to be giving 500 ml of normal saline in 4 hours through an adult giving-set the drop rate calculation would be as follows:

drop rate (in drops per minute)

$$= \frac{\text{volume to be given} \times \text{number of drops per ml}}{\text{number of hours} \times 60}$$

$$= \frac{500 \times 20}{4 \times 60}$$

$$= 42 \text{ drops per minute}$$

One other point to note is that we have moved from talking about static fluids to considering fluid flowing in an intravenous infusion. While it is true that in the static state, when the fluid is not flowing, the pressure at a given point is proportional to the height of the column of fluid and the density of the fluid, there are other considerations when fluid flowing in a tube is considered. In the case of the intravenous infusion, there is a pressure drop between the beginning of the tubing and the point of attachment of the tubing to the cannula.

Why is this?

This drop in pressure is due to the friction between the fluid and the vessel walls.

What factors influence the extent of this friction?

You may have been able to work out that among the factors which influence the extent of friction between a fluid and the walls of a tube in which it is flowing are the viscosity of the fluid and the diameter of the tube. There are other factors too, as you will see when we move on to consider blood and blood vessels.

Blood flow

The circulatory system is the important transport system of the body, and like all transport systems it has three components. Firstly, there is a means of carrying the various substances which need to be transported. This role is performed by the blood. Then there is a means of propulsion – the heart undertakes this role. Finally, there is a system of structures through which the transport medium passes. These are the blood vessels.

Why does blood flow?

This may sound like an obvious question – you might say that blood flows because it is pumped by the heart – but let us look at blood flow in a little more detail. Fluids like blood flow for one reason – there is a *pressure gradient*. Consider the arterial blood pressure for a moment.

Remember that the normal is about 120/80 mmHg. What does this actually mean?

Recall that the lower figure refers to the pressure in arteries during diastole – that is, when the ventricles of the heart are not contracting. The higher figure is the systolic pressure generated when the heart contracts. So blood flows from the heart during systole because the contraction of the ventricles generates a pressure gradient down which the blood flows. In fact, if we look at the various vessels throughout the circulatory system we can see that the pressure drops as we travel from the arterial to the venous side and this accounts for the flow of blood. Note that it is not the blood pressure at a particular point which itself determines blood flow but the difference in pressure between two points.

Resistance to blood flow

We have already noted that friction between a fluid and the vessel in which it is flowing opposes the flow of blood, and we have noted some of the factors which account for this friction. Let us look at these in a little more detail.

Blood viscosity

The more viscous a fluid the greater the friction between it and the vessel wall and the greater the resistance to blood flow.

What factor influences blood viscosity the most?

Blood is not a simple solution of dissolved substances. It is in fact a tissue, and contains cells. It is the proportion of cells which has the greatest impact on blood viscosity, and since red blood cells (erythrocytes) are the most abundant type of blood cell it is these which are the most significant here.

Can you think of situations in which the number of red blood cells might vary?

If you have not studied biological science before you might have found this question difficult, but anaemia involves a reduced number of red blood cells, and as a consequence a reduced blood viscosity, which in turn results in rapid blood flow. In contrast, people with an excessive number of red blood cells (polycythaemia) have an elevated blood viscosity and sluggish blood flow.

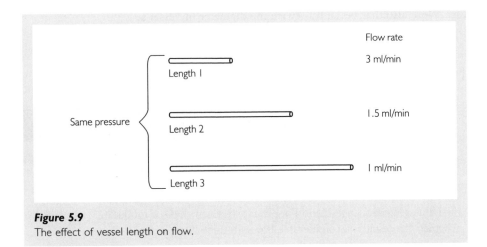

Figure 5.9
The effect of vessel length on flow.

Vessel length

The longer the vessel the greater the inner surface and thus the greater the friction too. For this reason resistance to blood flow is directly proportional to the length of the vessel. This is illustrated in Figure 5.9. Here there are three vessels of different lengths, and despite the fact that the pressure of fluid entering each vessel is the same the flow rates are different. In fact, the vessel which is three times the length of the shortest has only a third of its flow.

Vessel diameter

The friction between a fluid and the vessel wall in which it is flowing is clearly an important consideration and leads to slow flow rates along the inner surface of the vessel. However, fluid flowing down the middle of the vessel lumen flows much faster. This effect means that flow in large-diameter vessels is very much greater than in narrow vessels, due to this rapid flow through the central portion of the vessel. In fact, flow rate is directly proportional to the fourth power of the diameter provided that all other factors remain constant. This is illustrated in Figure 5.10. Note that the pressure of fluid entering the vessels is the same in each

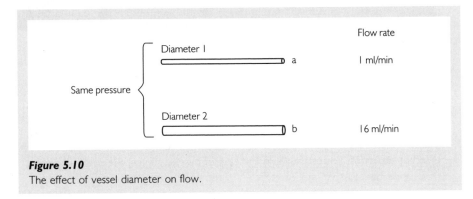

Figure 5.10
The effect of vessel diameter on flow.

case. Note also that vessel b is twice the diameter of vessel a.

> *How much greater is the flow rate in vessel b compared with that in vessel a?*

The flow rate in vessel b is 16 times that of the flow rate in vessel a.

Practice point 5.5: atherosclerosis

We will look at this disease process again in the next chapter, but for now it is enough to know that it involves the focal accumulation of fats and other molecules in the walls of arteries and that it is associated with poor diet and smoking. These *plaques* of fatty substances causes a reduction in the internal diameter of the vessel, and as a consequence blood flow is also reduced. Even a moderately narrowed lumen can result in a considerable flow rate reduction.

Can you recall why this is so?

It is of course because the blood flow is proportional to the fourth power of the diameter. Therefore if the diameter is reduced by half the flow rate is reduced by one sixteenth.

Atherosclerosis can affect arteries in all sites of the body, but its effects on the heart are probably best known.

Do you know what the effects of atherosclerosis on the heart might be?

You may have already encountered the patient with angina. This is a crushing central chest pain which is associated with exercise or emotional stress. It results when the blood flow through atherosclerotic arteries cannot meet the heart's increased demand for oxygen when its workload is increased, as it is during activity.

Practice point 5.6: the relationship between pressure, flow and resistance

We might summarise what we have said so far by saying that pressure tends to increase flow and resistance tends to decrease it. We can also state this as a formula:

$$\text{Blood flow} = \frac{\text{Pressure}}{\text{Resistance}}$$

We can see the body's attempts to regulate these three parameters in a number of different situations. For example, in the case of pyrexia (fever) the body attempts to increase heat loss.

What change could be made to the resistance offered by vessels in the skin in order to increase heat loss?

You have probably been able to work out that the resistance offered by the vessels in the skin is reduced by vasodilation (the muscle in their walls relaxes and they become wider). This increases the blood flow through the skin (as seen by flushing) and as a consequence heat loss is increased too.

There are other applications too. Suppose an individual suffers considerable blood loss and arterial blood pressure starts to fall. Looking at our equation we can see that if blood pressure starts to fall the blood flow will also fall and the vital organs will be deprived of a blood supply. Clearly the body needs to do something to maintain the blood pressure.

But just what could the body do to maintain blood pressure?

To answer this question it might be helpful to rewrite our equation, this time making blood pressure the subject:

Blood pressure = Blood flow × Resistance

In our current scenario blood flow is reduced because of haemorrhage, so in order to maintain blood pressure resistance will have to rise; this is achieved through *vasoconstriction* (the muscle in the vessel walls constricts and the vessel becomes narrower). This sounds fine so far, but if the blood vessels to important structures like the brain, heart and lungs were to constrict, thus maintaining the blood pressure, the blood flow to these organs would be reduced and the body's efforts would be thwarted.

So just what could be done?

If you think about this for some time you might realise that the response of the body to blood loss is to cause the blood vessels in certain areas such as the skin to constrict, thus helping to maintain the blood pressure, but the vessels to other structures remain unaffected so that blood flow to the brain, heart and lungs is ensured. This example of the importance of a consideration of pressure, flow and resistance is a little harder to understand than the first, but no doubt it will begin to make a great deal of sense when you actually look after a patient who has suffered blood loss. Incidentally, not all blood loss is visible: it is possible to haemorrhage internally, and for this reason taking the blood pressure even when there is no frank bleeding might be important.

Are there other observations to be made too?

Yes, there are a number. One is to look at the patient's complexion. We have seen that one method which the body employs to maintain blood pressure is to cause a cutaneous vasoconstriction, so pallor (pale complexion) may be a significant observation.

Gases and breathing

The behaviour of gases is explained in a number of laws which are named after the scientists who first described them. Upon initial examination you may not first appreciate the relevance of these laws to nursing practice, and this section seeks to relate them to the function of the lungs and to interventions such as oxygen therapy.

For example, the relationship between the volume of a gas and pressure is described in Boyle's law, which states that, *provided the temperature of a gas does not change, the pressure of a fixed amount of gas will increase as the volume decreases (and vice versa)*. That is, pressure varies inversely with volume, and this relationship may be given in the form of a formula (where the symbol $\propto$ means proportional to):

Pressure $\propto$ 1/Volume

Suppose we were to take a volume of gas in a sealed container at constant temperature and we had some means of changing the volume of the container – a piston perhaps. This arrangement is illustrated in Figure 5.11.

When the volume of the sealed container is increased by pulling on the piston what happens to the pressure?

According to Boyle's law it varies inversely, that is it falls.

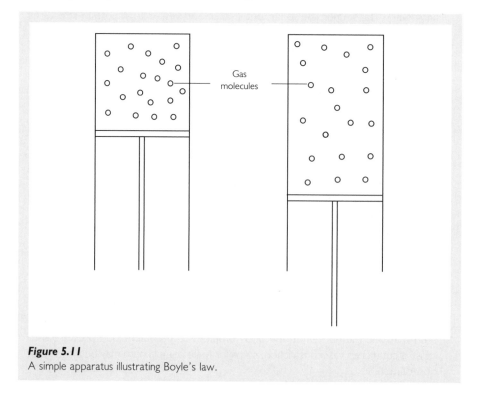

Figure 5.11
A simple apparatus illustrating Boyle's law.

Practice point 5.7: the mechanism of breathing

We could relate our consideration of Boyle's law to breathing. The thoracic cavity (chest) is rather like our sealed container, and it does have the capacity to change size, although not by the movement of a piston! Inspiration takes place when the diaphragm (the muscle which separates the thoracic cavity from the abdominal cavity) contracts and becomes flattened, descending towards the abdomen. This results in an increase in the vertical dimension of the chest. At the same time the external intercostal muscles (between the ribs) also contract, and these lift the ribs upwards rotating them outwards. This results in an increase in the chest in the lateral and anterior–posterior planes. So then, the size of our 'box' (the chest) is increased.

According to Boyle's law what now happens to the pressure of gas inside the chest?

Just like in the box previously described, the pressure tends to fall. The expression 'tends to fall' is important here because the thoracic cavity is not actually a sealed box. There is one way in and out – the airway! Therefore, as the pressure begins to fall, air is drawn into the lungs via the mouth or nose until the increased space within the lungs is filled with air at atmospheric pressure. This sequence of events is illustrated in Figure 5.12.

When inspiration ceases and expiration begins the thoracic cavity returns to its previous dimension by passive recoil. That is, it is not an active process – muscle

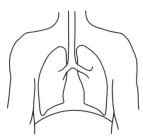 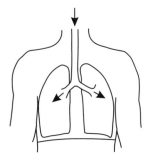

Before inspiration.

During inspiration the diaphragm becomes flattened and descends towards the abdomen, the ribs are pulled upwards and outwards and the chest expands. The pressure within the lungs falls and air is drawn in.

Figure 5.12
Inspiration.

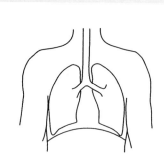

The chest returns to its previous
dimensions by passive recoil.

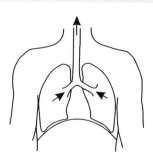

Air is forced out of the lungs.

Figure 5.13
Expiration.

contraction is not involved. As the volume of the cavity falls you might have worked
out from Boyle's law that the pressure in the lungs tends to rise and air is now
forced out of the airway. Expiration is illustrated in Figure 5.13.

Practice point 5.8: Boyle's law and intercostal drains

Intercostal drains are also referred to as chest drains, pleural drains or underwater
seal drains, although this latter term refers to the type of drainage system rather
than to the site of the drain. Intercostal drains present particular problems for
nurses, who sometimes become confused about the management of such systems
when they are uncertain about the science behind them. First of all, it is important
to note that the lung is covered by a membrane called the pleura. This word may
sound familiar if you have ever encountered someone with inflammation of the
pleura (pleurisy or pleuritis) – perhaps because of an infection. The pleura is in
fact a double membrane, as illustrated in Figure 5.14.

The innermost or visceral pleura is adherent to the lung surface, but at the
point where the airway enters the lung (hilum) it is reflected back upon itself to
form the outer pleura (parietal pleura). The parietal pleura is adherent to the
chest wall, but the two layers of the pleura are not normally adherent to each
other. In diagrams the two membranes are often drawn as though there is a
space between them, but this is not actually the case either. In fact, there is a potential space between the two layers – that is, they touch but they are not adherent,
and they slide over each other but can be pulled apart. The pleura is described as a
serous membrane – it secretes a fluid which resembles serum and which helps the
two membranes to slide over each other. As an illustration of how small the pleural
space is you might be interested to know that in the adult the volume of pleural fluid
forming the film between the pleurae of one lung is only approximately 2 ml.

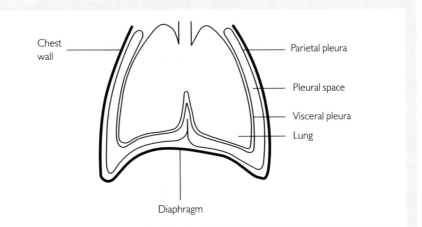

Figure 5.14
The pleura.

Next it should be pointed out that the lung is filled with elastic tissue which produces a tendency to collapse, and since the visceral pleura is adherent to the lung it is pulled away slightly from the parietal pleura.

So what actually prevents the lung from collapsing?

Let us think about this. Since the two pleurae are continuous, as the visceral is pulled away from the parietal, the space between them is enlarged, and according to Boyle's law the intrapleural pressure drops. It is this slight negative pressure in the pleural space which just balances the tendency of the lung to collapse. We might say that the lungs are held up by suction!

Of course, since the pleural space is on the inside of the chest it is subject to a change in pressure as we breathe in and out. However, unlike the intra-alveolar pressure (pressure inside the alveoli – microscopic air sacs of the lungs) the intra-pleural pressure does not oscillate between negative and positive but between more negative and less negative as shown below:

	Intra-alveolar pressure (mmHg)	Intrapleural pressure (mmHg)
Inspiration	$^-3$ mmHg	$^-8$ mmHg
Expiration	$^+3$ mmHg	$^-2$ mmHg

Now let us consider what would happen if either of the pleurae were to be breached. The visceral pleura might rupture as a consequence of a structural weakness, while trauma, such as a knife attack, may result in penetration of the parietal pleura and probably the visceral pleura and lung as well. The consequence would be that atmospheric air would enter the pleural space and the negative pressure would be lost.

What would now happen to the affected lung?

Since there would no longer be a negative pressure in the pleural space of the affected lung the natural elasticity of the lung would no longer be balanced and it would collapse, with consequent respiratory complications. This condition is referred to as pneumothorax, and you might like to read more about it in a nursing text.

It is not too difficult to work out that the treatment of pneumothorax must involve releasing air from the pleural space and the obvious way to do this is to intubate it – that is, place a tube within the space. However, there is a problem.

Can you work out what it is?

A simple tube passing from the outside and into the pleural space would allow air in as well as out! What is needed is some kind of valve which would allow air to leave the pleural space but not to enter it. This is most easily achieved by connecting the drain to a tube, the end of which is dipped under water as illustrated in Figure 5.15.

In this system air is pushed out of the pleural space as it is compressed during inspiration. The air has only one route through which to pass and it bubbles out in the water and escapes from the bottle via the vent. It cannot of course re-enter the chest because of the presence of the water; hence the term 'underwater seal' is used. In some cases, evacuation of the pleural space may be enhanced by attaching the vent to a vacuum and a pressure of up to $^-13$ kPa may be used.

Once again you may like to look at a nursing text to find out more about intercostal drains. For now it is sufficient to point out that a simple understanding of Boyle's law can help nurses comprehend normal physiology, altered physiology (such as pneumothorax) and interventions, such as intercostal drainage.

Dalton's law of partial pressures

This law helps us to understand mixtures of gases such as the atmosphere and it states that *in a mixture of gases, the total pressure is a sum of the pressures exerted by each of the gases alone.*

That sounds like a bit of a mouthful, but let us see if we can make some sense of it. We have already thought about atmospheric

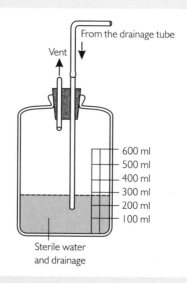

Figure 5.15

An underwater seal drainage system. (Redrawn from Foss M. A. (1989). *Thoracic Surgery*. London: Austen Cornish.)

Table 5.1 The partial pressures of gases in atmospheric air.

Gas	Proportion of atmospheric air (%)	Partial pressure	
		mmHg	kPa
N_2	78.62	597	79.6
O_2	20.84	159	21.2
H_2O	0.5	3.85	0.51
CO_2	0.04	0.15	0.02

pressure, and a value for it of 760 mmHg has been given. The air around us is of course a mixture of gases, mainly nitrogen, oxygen, carbon dioxide and water vapour.

> How much does each of these gases contribute to the net atmospheric pressure?

Table 5.1 shows typical figures for atmospheric air. The first column identifies the percentage of the atmosphere which the gas forms, while the second and third columns show the partial pressures of the individual gases in mmHg and kPa, respectively.

Let us now examine what these figures mean. For example, nitrogen is the most abundant gas in the atmosphere. In fact 78.6 % of the atmosphere is nitrogen, and it exerts 78.6% of the atmospheric pressure; that is 79.1 kPa or 597 mmHg. This value is termed the **partial pressure** of nitrogen in atmospheric air. Therefore a partial pressure is the fraction of the pressure of a mixture of gases which one of the gases in the mixture exerts.

Henry's law

Gases like oxygen and carbon dioxide are of course very important in the body. Oxygen is required for the energy-producing reactions of cellular respiration, while carbon dioxide is a waste product of these processes. Therefore the body must be able to transport these compounds. You will of course realise that the principal transport system of the body is the bloodstream and that oxygen and carbon dioxide are not transported as free gases, but instead are dissolved in solution. Oxygen is subsequently attached to a special carrier molecule – haemoglobin – while carbon dioxide undergoes a chemical reaction in the blood. Nonetheless, first of all they are both dissolved in the plasma.

> What influences how much gas will dissolve in liquids such as water or plasma?

The answer to this question is given in Henry's law, which states that *the mass of a gas that will dissolve in water at a given temperature is proportional to the partial pressure of the gas and to its solubility coefficient.* It may be given in the form of an equation:

$$C_x = k_x \times pp_x$$

where C_x is the concentration of the dissolved gas k_x is its solubility coefficient and pp_x is the partial pressure of the gas x in the atmosphere above the liquid. The solubility coefficient is a constant which is a property of the individual gas – that is, some gases are more soluble than others.

Practice point 5.9: oxygen transport

We have just noted that some gases are more soluble than others. Let us look at a practical example in the body. Carbon dioxide is highly soluble in water, whereas oxygen is much less soluble. In fact, our blood cannot transport enough dissolved oxygen to meet the needs of our cells.

So how is oxygen transported by the blood?

From what has been noted before you will no doubt have realised that our bodies do not rely upon dissolved oxygen. Instead, once oxygen has been dissolved it is in effect removed from solution by combination with the haemoglobin molecule. This then 'makes way' for more oxygen to be dissolved, which is in turn bound to the haemoglobin molecule and so on until all, or nearly all, the haemoglobin molecules have oxygen bound to them – that is they have become oxyhaemoglobin. The proportion of the haemoglobin molecules which have oxygen bound to them is termed the *saturation*, which in arterial blood is normally 98–100%.

When might you need to know the value of a patient's saturation?

The saturation may fall when a patient has a respiratory disease which affects the ability of the lungs to take up oxygen from the atmosphere.

Where in the blood is this important molecule haemoglobin found?

You probably know already that it is found within red blood cells (erythrocytes). So then the amount of haemoglobin and the number of red blood cells are also important measurements, and you might like to find out what the normal values are.

But what happens if someone is deficient in haemoglobin or red blood cells?

This condition is called *anaemia* and it results in a reduced ability to transport oxygen, one manifestation of which is tiredness.

Having made these practical points we do not need to consider the solubility coefficient in any more detail. It is sufficient to note that, at a given temperature, the variable which influences how much of a gas dissolves is the partial pressure of that gas in the atmosphere. As the partial pressure increases, so does the amount of gas dissolved.

Practice point 5.10: the effect of altitude

We have previously noted that Torricelli found a fall in atmospheric pressure with increasing altitude: this is an effect of gravity – the atmosphere is most dense at sea level. The overall fall in net atmospheric pressure reflects a fall in the partial pressures of each of the atmospheric gases including oxygen. If we now apply Henry's law we are able to work out that the effect of this falling partial pressure is a reduction in the amount of oxygen which will dissolve in blood. It is this reduction which produces the effects of altitude sickness. This is clearly a potential problem for mountain climbers and those who travel to cities at high altitudes. You will no doubt realise that the body does adapt to high altitudes given time, and you might like to think about this and undertake further study using a physiology text.

Gases in solution and their movement

In order to have a fuller understanding of gases in solution and how they move between alveolar air and blood and between blood and the cells of the body we have to think a little more about the physical principles of gases in solution and diffusion. Figure 5.16 shows a container half-filled with water with atmospheric air above it. Molecules of the gases which form the atmosphere are continually bouncing against the surface of the water and some of them become dissolved

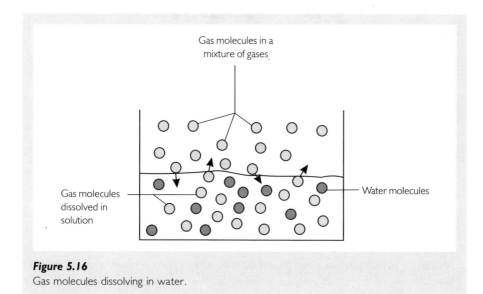

Figure 5.16
Gas molecules dissolving in water.

in it. These dissolved molecules continue to bounce among the water molecules and so exert a pressure against them. Some of these molecules reach the surface of the water once again and bounce back into the air. After some time the number of gas molecules dissolving in the water becomes equal to the number moving in the opposite direction, and we say that a state of equilibrium has been reached.

An important point to note here is that molecules of gas dissolved in solution exert a pressure against water molecules just as they did against other gas molecules before they dissolved. This means that we can speak of the partial pressure of gases dissolved in solution as well as the partial pressure of gases in a gaseous mixture. The practical relevance of this becomes obvious when we consider the exchange of gases in the lungs. Figure 5.17 is a diagrammatic representation of a single alveolus and alveolar capillary on which is noted the partial pressures of oxygen and carbon dioxide in alveolar air, blood entering the capillary (venous blood) and blood leaving the capillary (arterial blood). The partial pressure of a gas in alveolar air is given as P_A, in venous blood entering the capillary as P_v and in arterial blood leaving the capillary as P_a.

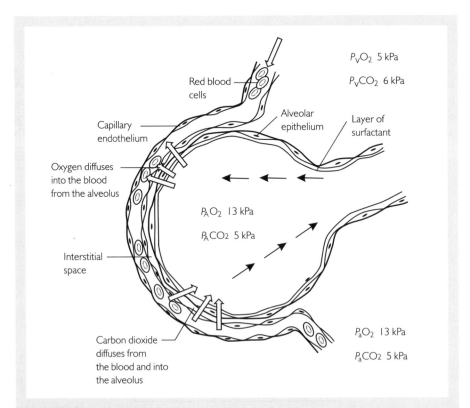

Figure 5.17
An alveolus and capillary. (Redrawn from Foss M. A. (1989). *Thoracic Surgery*. London: Austen Cornish.)

It should be noted that the composition of alveolar air is not the same as that of atmospheric air for a number of reasons. Firstly, the air which we inhale is warmed and humidified on its passage through the upper airways, and this changes the gas composition. Secondly, carbon dioxide is eliminated by the lungs, so the partial pressure of this gas is greater in alveolar air than it is in atmospheric air.

From this diagram we can see that the partial pressure of oxygen in alveolar air (13 kPa) is greater than that in venous blood (5 kPa). Consequently, oxygen diffuses from alveolar air and into the blood. In contrast, the partial pressure of carbon dioxide in alveolar air (5 kPa) is less than in venous blood (6 kPa), so carbon dioxide diffuses from venous blood and into alveolar air. The net result of this diffusion of gases is that blood leaving the lungs has a partial pressure of oxygen of 13 kPa and of carbon dioxide of 5 kPa.

Oxygen therapy

Oxygen therapy involves the administration of oxygen in concentrations higher than those in the normal atmosphere for the treatment of various conditions which result in **hypoxaemia**. This is a state of reduced partial pressure of oxygen in arterial blood and it may be caused by a number of respiratory illnesses such as a chest infection, asthma or chronic bronchitis. Oxygen may be administered in a number of different ways, but the most obvious is the face mask, one example of which is illustrated in Figure 5.18.

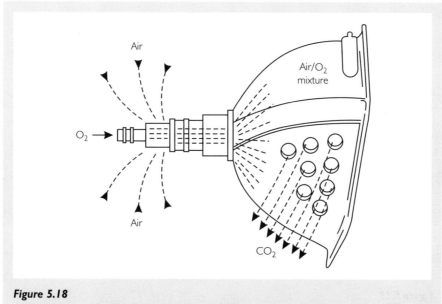

Figure 5.18
An oxygen mask. (Redrawn from Foss M. A. (1989). *Thoracic Surgery*. London: Austen Cornish.)

How is oxygen therapy effective?

When an oxygen mask is applied over someone's face the mixture of gases which the individual inhales is changed. The proportion of oxygen given increases, and as a consequence so does the partial pressure of oxygen in the inhaled gas mixture. This is subsequently reflected in an increase in P_AO_2. In fact, if 100% oxygen were to be inspired the P_AO_2 would rise from 13 kPa to 80 kPa.

> Remember that, according to Henry's law, the partial pressure of a gas is one factor which determines the mass of gas which will dissolve.

As a consequence the P_aO_2 also rises; that is, hypoxaemia is alleviated.

The Bernoulli effect

It is worthwhile considering the design of different oxygen masks, since their effectiveness varies. In order to do this we need to understand the *Bernoulli effect*. In the 18th century Daniel Bernoulli made the experimental observation that the pressure in flowing liquids is lowest where the speed of flow is fastest. This observation subsequently became known as the Bernoulli effect, and it is illustrated in Figure 5.19.

Note that in Figure 5.19(a) fluid is flowing at a constant rate in a horizontal tube. A fall in pressure can be seen between points A to D as shown in the diminishing heights of the columns. Now consider Figure 5.19(b). Here there is a constriction

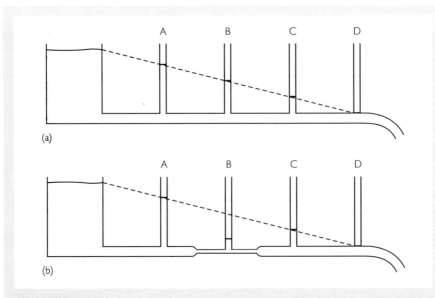

Figure 5.19
An experiment to demonstrate the Bernoulli effect.

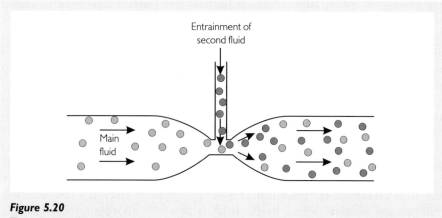

Figure 5.20
The principle of entrainment.

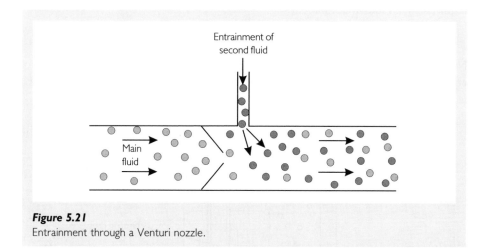

Figure 5.21
Entrainment through a Venturi nozzle.

which causes a drop in pressure at point B. This drop in pressure can be used to draw in (entrain) a second fluid, as illustrated in Figure 5.20.

Later experiments by Giovanni Venturi led to the development of a nozzle for the purpose of entrainment, and Figure 5.21 shows how such a nozzle works.

The importance of this is that the Bernoulli effect can be used not only to entrain one fluid into another but also to entrain one gas into another or a fluid into a jet of gas. These latter two examples are illustrated in the case of oxygen masks and nebulisers.

Oxygen masks

Oxygen masks fall in to one of two broad types, *variable performance devices* and *fixed performance devices*. Variable performance masks are typically of small volume and low gas flow rate. When using such a mask, the gas inspired consists of a mixture of oxygen, air from outside the mask and (carbon dioxide-rich) exhaled breath retained in the body of the mask. Furthermore, this mixture depends upon the flow rate created when the patient inspires, and this is not constant – it varies with each breath. Since this is so, the gas mixture also varies and this accounts for the use of the term *variable performance*. Put simply, the nurse cannot rely upon the mask to give a constant percentage of oxygen. Such masks have limited use.

In contrast, the mask illustrated in Figure 5.18 is of the fixed performance type, and such masks are distinguished by being of large volume and having a venturi barrel.

In a venturi barrel a relatively low flow rate of oxygen is forced through a narrow opening which results in a very fast jet of gas entering the mask. The barrel has side-holes and the fast-moving jet of oxygen causes a drop in pressure and air from outside the mask is drawn in through these holes at a fast rate. The resultant gas mixture created by the mask is at a rate above inspiration, so the gas mixture is constant. Slight differences in jet design will produce different gas mixtures, so it is possible for the doctor to specify the specific oxygen concentration required by an individual patient.

Nebulisers

Nebulisers are devices in which a pressurised gas, such as oxygen or air, is used to created droplets of a fluid, perhaps containing a drug, which are then inhaled. The

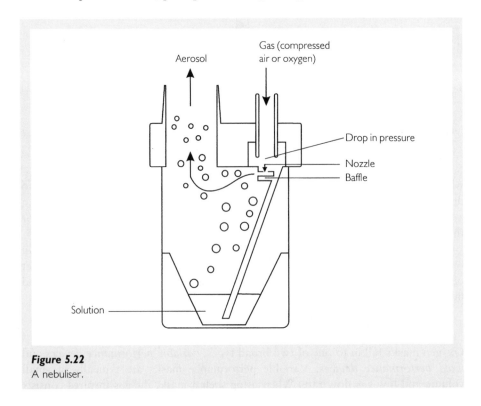

Figure 5.22
A nebuliser.

use of nebulisers in order to administer drugs is common in patients with respiratory diseases. Figure 5.22 will be used to illustrate how nebulisers work.

The pressurised gas is forced into the nebuliser through a narrow opening, through which it accelerates creating a drop in pressure inside the nebuliser – the Bernoulli effect once again. As a consequence of this droplets of fluid are drawn into this stream of gas, which then leaves the nebuliser and is inhaled by the patient. The fluid used in the nebuliser might be normal saline, which may be of use in loosening tenacious secretions, or a drug for the treatment of a respiratory illness. You might like to find out more about the kinds of drugs which are used in nebulisers when you meet a patient who is using such a device. For now it is sufficient to note that an understanding of the Bernoulli effect will help you to understand how two common items of medical equipment function.

Summary

1 Pressure is defined as force per unit area.
2 Commonly used units of pressure include pascals (Pa), millimetres of mercury (mmHg) and centimetres of water (cmH_2O).
3 Normal arterial blood pressure is considered to be 120/80 mmHg.
4 The protection afforded to various body structures by being surrounded by fluid is explained on the basis of Pascal's principle.
5 Fluid flows between two points when there is a pressure difference between them.
6 Resistance to the flow of a fluid is directly proportional to the length of the vessel in which the fluid is flowing.
7 Flow rate is directly proportional to the fourth power of the diameter of a vessel provided that other factors remain the same.
8 Pressure = Flow × Resistance.
9 Boyle's law, which states that, provided the temperature of a gas does not change, the pressure and volume of a fixed amount of gas are inversely related, is used to explain inspiration and expiration.
10 Henry's law identifies the importance of the partial pressure of a gas in determining the mass of gas that will dissolve in water at a given temperature.
11 Gas exchange in the lungs can be explained in terms of the partial pressure of oxygen and carbon dioxide in alveolar air and venous blood.
12 The Bernoulli effect, in which the pressure in a liquid is lowest when the speed of flow is fastest, is used to explain the operation of a nebuliser.
13 The Bernoulli effect is used to produce entrainment in the venturi barrel of an oxygen mask.

Self-test questions

5.1 Suppose that 20% of a gas mixture is oxygen and the pressure of the mixture of gases is 50 kPa. What is the partial pressure of oxygen?
   a 1 kPa

b 10 kPa
c 20 kPa
d 50 kPa

5.2 Which one of the following would be considered to be a normal value for systolic blood pressure in mmHg?

a 60
b 80
c 120
d 240

5.3 When using a sphygmomanometer and stethoscope to measure blood pressure the diastolic value is normally taken to be the point at which

a There is a continuous rushing sound.
b A tapping sound can first be heard.
c Muffling of the sound.
d Disappearance of the sound.

5.4 According to Boyle's law, pressure is proportional to which one of the following? (T = temperature and V = volume).

a $1/T$
b $1/V$
c V
d VT

5.5 Which one of the following represents a normal P_aO_2 in kPa?

a 30
b 3
c 1.3
d 13

5.6 In the case of a container of liquid which of the following statements about pressure at a point in the liquid are true?

a It depends upon the depth of the point below the surface.
b It acts in all directions.
c It depends upon whether the point is in the middle of the container or near the sides.
d It depends upon the volume of liquid above the point.

5.7 A patient with an intravenous infusion is prescribed 1 litre of fluid to be administered over eight hours. Assuming that there are 20 drops to a millilitre, what rate, in drops per minute, should the infusion be regulated to?

a 12
b 24
c 32
d 42

5.8 Suppose that the diameter of a blood vessel is reduced to half; provided the blood pressure remains the same, the flow rate through the vessel would be reduced to:

a 1/4
b 1/2
c 1/8
d 1/16

5.9 Concerning gaseous exchange in the lungs, which of the following are true?
    a P_AO_2 is greater than P_vO_2
    b P_vO_2 is greater than P_aCO_2
    c P_vCO_2 is greater than P_aCO_2
    d P_aO_2 is greater than P_AO_2

5.10 Concerning an oxygen mask with a venturi barrel, which of the following statements are true?
    a Atmospheric air is entrained through the venturi barrel.
    b The mask may be described as a variable performance device.
    c Relatively low oxygen flow rates produce a high flow rate through the mask.
    d The gas mixture delivered to the patient is constant in composition.

Answers to self-test questions

5.1 b	**5.6** a and b
5.2 c	**5.7** d
5.3 d	**5.8** d
5.4 b	**5.9** a and c
5.5 d	**5.10** a, c and d

Further study/exercises

5.1 Suppose you wish to identify which of your patients are at risk of developing pressure sores. How could this be done? (*Hint*: find out about pressure sore risk calculation scores such as those of Norton and Waterlow). In addition, by what means are pressure sores prevented?
    Collier M. and Cole A. (1997). Pressure area care: knowledge for practice. *Nursing Times*, **93**(4), 1–4
    Collier M. and Cole A. (1997). Pressure sore care: the role of the nurse. *Nursing Times*, **93**(5), 5–8
    Daley C. (1997). Pressure ulcer prevention – the UK perspective. *Dermatology Nursing*, **9**(2), 108–113
    O'Dea K. (1997). Wound care. Equipped to care . . . pressure sore prevention and equipment provision. *Nursing Times*, **93**(16), 81
    Phillips J. (1997). Pressure sore risk assessment tools. *Professional Nurse*, **12**(5), 321

5.2 Arterial blood pressure can be measured non-invasively using a sphygmomanometer and stethoscope, as we have seen. However, central venous pressure (pressure in the great veins and right atrium of the heart) has a much lower value and has to be measured invasively. What is the procedure for measuring central venous pressure with a simple saline manometer?
    Cornock M. (1996). Making sense of central venous pressure. *Nursing Times*, **92**(40), 38–39

Gourlay D. A. (1996). Surgical nurse. Central venous cannulation. *British Journal of Nursing*, **5**(1), 8–15

Haynes S. (1991). Central venous pressure monitoring. *Professional Nurse*, **6**(12), 727–729

McDermott M. (1995). Central venous pressure. *Nursing Standard*, **9**(35),.54

Woodrow P. (1992). Monitoring central venous pressure. *Nursing Standard*, **6**(33), 25–29

Further reading

Jolley A. (1991). Taking blood pressure. *Nursing Times*, **87**(15), 40–43

Mumford S. (1986). Using jet nebulisers. *The Professional Nurse*, **1**(4), 95–97

Nolan J and Nolan M. (1993). Can nurses take accurate blood pressure? *British Journal of Nursing*, **2**(14), 724–729

Thompson D. R. (1981). Recording patient's blood pressure. *Journal of Advanced Nursing*, **6**, 283–290

Wells D. (1990). A case for accuracy: monitoring blood pressure. *Professional Nurse*, **6**(1), 30, 32

6 Biological molecules and food

Learning outcomes

After reading the following chapter and undertaking personal study, you should be able to:

1 Identify the seven components of the diet.
2 Outline the chemical structure of carbohydrates, fats and proteins.
3 Describe how carbohydrates, fats and proteins are digested and outline their use within the body.
4 Outline the roles of vitamins and minerals within the body.
5 Describe a healthy diet.
6 Outline the relationship between the different nutritional groups, health and illness.

Introduction

Aspects of basic chemistry were considered in Chapter 2, and a distinction was made between *organic* chemistry and *inorganic* chemistry.

> *How is organic chemistry defined?*

It is indeed the study of compounds that contain carbon. Some of these were discussed in Chapter 2 but we now concentrate on foods.

Diet and food

The components of the diet may be divided into seven groups:

carbohydrates
proteins
fats
vitamins
minerals
water
fibre roughage

Although all the above may be identified as important substances in food, there are important differences between these components of the diet. For example, not all the above are organic compounds. Water is not, and neither are the minerals. Some, such as the carbohydrates, proteins and fats, have to be digested (broken down) before they can be absorbed, while vitamins, minerals and water are absorbed unchanged. Finally, roughage is neither digested nor absorbed but provides bulk to food and is important in the maintenance of a formed stool which is readily passed through the intestines.

Carbohydrates

The word *carbohydrate* indicates that this type of substance contains carbon, hydrogen and oxygen. There are three classes of carbohydrate:

   monosaccharides   (simple sugars)
   disaccharides   (double sugars)
   polysaccharides   (complex sugars)

Monosaccharides

The word **monosaccharide** means 'one sugar', and it is used of those compounds which cannot be broken down into more simple sugars. The ratio of carbon, hydrogen and oxygen in monosaccharides is $1:2:1$, as is the case in glucose – $C_6H_{12}O_6$. Glucose is found naturally in honey along with another monosaccharide, fructose, which is the sugar found in fruit. Indeed, the word *fructose* means 'fruit sugar'. Galactose is a monosaccharide found in milk, and like fructose and glucose its molecular formula is $C_6H_{12}O_6$. If you have already read Chapter 2 (The physical world and basic chemistry) you will realise that these three sugars are in fact **isomers** – compounds with the same molecular formula but different structural formulae. The structural formulae of all the monosaccharides are not given, but that of glucose can be seen in Figure 6.1.

When ingested, glucose, fructose and galactose are absorbed into the body through the ileum of the gut (see Figure 6.2) without the need to be broken down into simpler substances. However, fructose and galactose are subsequently converted into glucose, since only glucose can be used as a source of energy.

The cells of the body are dependent upon receiving a constant supply of glucose, and sometimes a pinprick sample of blood is obtained from a patient and tested for glucose concentration. The normal is

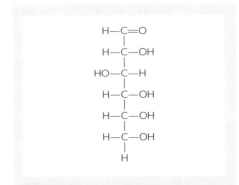

Figure 6.1
The structural formula of glucose.

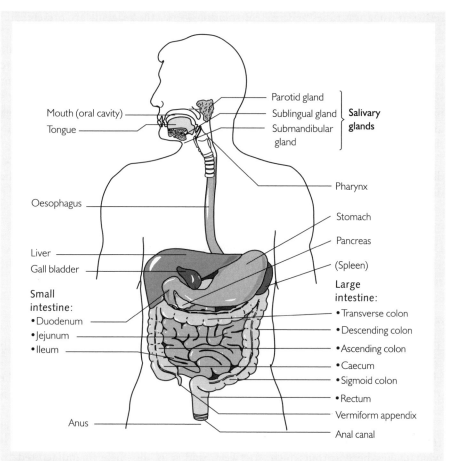

Figure 6.2
The gastrointestinal tract. (Redrawn from Coleman G. J. and Dewar D. (1997) *The Addison-Wesley Science Handbook for Students, Writers and Science Buffs*. Don Mills, Ontario: Addison Wesley.)

3.9–5.6 mmol/l. In the condition diabetes mellitus blood sugar is elevated (**hypergly-caemia**) and the body begins to excrete glucose in the urine (**glycosuria**). Glucose is not normally present in the urine, but if diabetes is suspected a patient's urine may be tested for it. Since one form of diabetes can go undetected for years, testing the urine of apparently healthy individuals is a useful health screening measure.

The traditional test for glucose involves observing the formation of a red copper (I) oxide precipitate after heating, in a boiling water bath, a solution of glucose to which pale blue Benedict's solution has been added. The precipitate results from the reduction of copper (II) sulphate in the Benedict's solution by the glucose. Consequently, glucose is referred to as a *reducing* sugar. Colour changes which occur as a consequence of a chemical reaction are still used in the detection of the presence of glucose in blood or urine. However, in the clinical setting the nurse does not have to resort to using test tubes and reagent solutions, since the reagents are presented in pads on a plastic strip.

In the analysis of blood glucose a drop of blood is placed on the pad, while urinalysis reagent strips are dipped in a fresh sample of the patient's urine before the colour change is read. (Urinalysis reagent strips contain a number of different reagent pads.) Whichever test is being performed, following the manufacturer's instructions and observing the correct time lapse before reading the colour change are most important. Perhaps you should take the time to familiarise yourself with these tests. A member of clinical staff may help you.

There are sugars which are not capable of performing the reduction reaction described above, and these are referred to as *non-reducing* sugars. Most of the disaccharides which you will encounter are non-reducing sugars.

Disaccharides

Disaccharides are formed from a **condensation reaction** between two mono-saccharides. A condensation reaction is one in which water is eliminated from the reaction thus:

$$C_6H_{12}O_6 + C_6H_{12}O_6 \longrightarrow C_{12}H_{22}O_{11} + H_2O$$

Three disaccharides are important. *Sucrose* is what we buy in packets at the shops. It is supplied as crystals of different sizes (table sugar, icing sugar and so on), but it is all sucrose. Sucrose was first extracted from sugar cane and came to Britain in the 18th century following the colonisation of the West Indies. However, today it is extracted from sugar beet.

Maltose is the sugar which results when a plant seed converts its long-term energy store of starch for the purpose of germination and growth. Maltose is what makes bean sprouts taste so sweet, and it is the sugar manufactured by grains such as barley when they are given warmth and moisture in the malting house. You will have seen this process if you have ever visited a brewery or distillery. Finally, *lactose* is the disaccharide found in milk.

Of the disaccharides mentioned only maltose is a reducing sugar, and therefore it is the only one which will give a positive result with the Benedict's test described previously. Sucrose and lactose will not, but it is possible to break these disaccharide molecules to form two monosaccharides and then re-test with the Benedict's reagent and obtain a positive result. The breaking of the disaccharide molecule and incorporation of water to form two monosaccharides is referred to as **hydrolysis** – meaning to break with water. The reaction is given below:

$$C_{12}H_{22}O_{11} + H_2O \longrightarrow C_6H_{12}O_6 + C_6H_{12}O_6$$

In the laboratory this hydrolysis may be performed by adding dilute hydrochloric acid (HCl_{aq}) to the disaccharide solution and heating it in a boiling water bath for at least five minutes. After cooling the solution is then neutralised with sodium hydrogen carbonate ($NaHCO_3$) before the Benedict's test is repeated.

The hydrolysis of disaccharides also takes place in the body, although here it does not involve boiling with acid! Instead the body uses **enzymes** to perform the hydrolysis (enzymatic hydrolysis). Remember that enzymes are protein **catalysts** which speed up chemical reactions by lowering their **activation energy**. The enzymes involved in disaccharide digestion are found in the duodenum – part of the small intestine (see Figure 6.2). Disaccharides have to be broken down into monosaccharides because they are too large to be absorbed unchanged.

In the naming of enzymes the prefix identifies the **substrate** (substance acted upon) and the suffix 'ase' indicates that an enzyme is being referred to. Consequently, maltase is the enzyme which acts upon maltose. The reactions for the enzymatic hydrolysis of maltose, sucrose and lactose are given below:

$$\text{maltose} \xrightarrow{\text{maltase}} \text{glucose} + \text{glucose}$$

$$\text{sucrose} \xrightarrow{\text{sucrase}} \text{glucose} + \text{fructose}$$

$$\text{lactose} \xrightarrow{\text{lactase}} \text{glucose} + \text{galactose}$$

Note that in the digestion of one disaccharide molecule two monosaccharide molecules result, and one of these is always glucose. In addition, maltose yields a second glucose molecule, sucrose a molecule of fructose and lactose a molecule of galactose (both lactose and galactose mean milk sugar).

Practice point 6.1: lactose intolerance

If you are involved in caring for infants you may encounter babies who fail to produce lactase and who cannot therefore digest lactose. The condition is known as *lactose intolerance* and it is characterised by diarrhoea and failure to thrive. Lactose intolerance is a serious illness, since lactose is present in all animal milk and this is of course the only source of food in early life.

Can you think what the affected infant might be given instead?

Soya milk is used.

Polysaccharides

Polysaccharides consist of chains of monosaccharides – *poly* means 'many'. The polysaccharide chain may be elongated and the molecule described as *fibrous*. Such chains are often employed for the purpose of giving strength and support. An example is the plant polysaccharide cellulose, which is important in plant cell walls. Alternatively, the polysaccharide chain may be wound up like a ball and the expression *globular* is used. Examples of globular polysaccharides are energy storage molecules such as starch in plants and **glycogen** in animals. The naturally occurring anticoagulant (impairs blood clotting) heparin is also a globular polysaccharide.

After ingestion polysaccharides are broken down in the gastrointestinal tract. The digestion of cooked starch, such as bread and potatoes, begins in the mouth. The enzyme salivary amylase (from the salivary glands) begins the breakdown of cooked starch, but since food spends relatively little time in the mouth the process is not completed there. The effect of salivary amylase is to convert some of the starch in a meal to substances called dextrins, which are intermediate molecules between starch and maltose. You may have performed a simple test using iodine to show this. Iodine turns cooked starch blue/black. Try it on some bread. Now take another piece of bread and chew it for a while before repeating the test. You will now find a red colour, with iodine indicating the presence of dextrins. Carbohydrate digestion does not continue in the stomach because of the low pH there. However, the environment in the duodenum is alkaline and the conversion of starch to maltose is completed by further amylase – this time from the pancreas (see Figure 6.1). Finally, the maltose which results is digested into two molecules of glucose, as previously described.

In summary, disaccharides and polysaccharides are broken down to the three monosaccharides which are absorbed via the cells of the ileum. Here fructose and galactose are converted into glucose before absorption into the blood. Therefore it is glucose which is the end product of carbohydrate digestion.

Practice point 6.2: how does the body keep a constant blood glucose level?

After glucose has been absorbed into the blood it travels in the hepatic portal vein to the liver. Some of this glucose is converted by liver cells (hepatocytes) into the storage product **glycogen**. This process is known as **glycogenesis** (literally meaning the making of glycogen). The hormone insulin and a number of enzymes are important in this process. If blood sugar subsequently becomes lowered (**hypoglycaemia**) then glycogen in the liver can be broken down into glucose once more. Have you worked out that the word **glycogenolysis** refers to this? Adrenaline (epinephrine) and glucagon are the important hormones in this process.

Glycogen is also stored in skeletal muscle, but here it cannot be converted back to glucose for the purpose of raising blood sugar. Glycogen in muscle is, however, available to the muscle as a source of energy.

In addition, the body is able to manufacture glucose from non-carbohydrate sources such as amino acids. This process is described as **gluconeogenesis**, meaning the manufacture of 'new' glucose.

Energy production

Of course the ultimate fate of glucose in the body is to be broken down so that the energy 'locked' in its molecules is released. This is described in Chapter 7 (Energy in the body).

Practice point 6.3: diabetes mellitus

This is a disorder of carbohydrate metabolism, one form of which (type I) is characterised by an inadequate production of the hormone insulin by the beta cells of the Islets of Langerhans in the pancreas. Insulin is required for the uptake of glucose by most cells of the body except brain cells. Therefore, when insulin production is inadequate the cells are improperly nourished no matter how much carbohydrate is eaten. One of the features of diabetes mellitus is hyperglycaemia.

Can you work out the reason for this?

You will probably have been able to figure out that in the absence of insulin glucose absorbed into the blood from the intestines has 'nowhere to go', and consequently the blood sugar level rises. This high blood sugar level has a number of effects, including increasing the susceptibility to infection. In the long term there may also be damage to capillaries such as those of the retina at the back of the eye, and after many years blindness could result. However, while the blood sugar level is high the cells do not have access to this source of energy due to the lack of insulin. Not surprisingly, then, the patient often complains of weakness. The blood glucose level may rise so high that glucose begins to appear in the urine (**glycosuria**).

In glomerular filtrate glucose exerts an osmotic effect, and this results in increased urine output. You may need to think about this point a little more, but for now it is enough to know that polyuria (increased urine output) is another feature of diabetes mellitus.

Can you work out what the effect of an elevated urine output might be on the thirst?

Perhaps this question is a little more obvious. The patient complains of thirst and drinks a lot (polydipsia). There is of course much more to learn about diabetes mellitus, but in summary it is characterised by a lack of insulin, and this leads to hyperglycaemia, weakness, susceptibility to infection, polyuria, polydipsia and in the long term to capillary damage and possibly blindness.

Proteins

Proteins consist of chains of molecules called **amino acids**. Like carbohydrates they contain carbon, hydrogen and oxygen, but amino acids also contain nitrogen and sometimes sulphur too. The general structure of amino acids is given in Figure 6.4 in which R represents a hydrogen atom or chain of carbon atoms with hydrogen atoms attached.

In proteins, amino acids are joined together by peptide bonds formed following a condensation reaction, as illustrated in Figure 6.3.

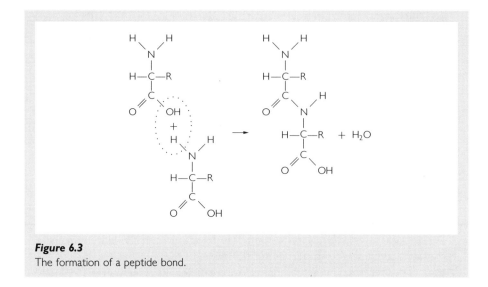

Figure 6.3
The formation of a peptide bond.

The molecule formed from the combination of between two and ten amino acids is termed a *peptide*, while longer chains of up to a hundred amino acids are referred to as *polypeptides*. Only chains of more than a hundred amino acids are called **proteins**. Twenty different amino acids are important in health. They can be divided into two groups based upon dietary requirements. *Non-essential* amino acids can be synthesised by the body (provided other

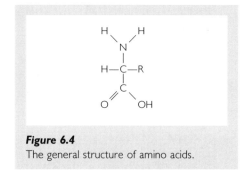

Figure 6.4
The general structure of amino acids.

amino acids are present), and as a consequence they do not need to be present in the diet. However, eight amino acids cannot be synthesised and so must be present in food. These are termed *essential* amino acids.

Based upon the relative amounts of amino acids that they contain, protein in the diet can be classified as *complete* (high quality) or *incomplete* (low quality). Complete proteins provide all the essential amino acids in adequate amounts and are only obtained from animals (meat, fish, eggs and dairy products). In contrast, plant proteins are incomplete in as much as they are deficient in one or more amino acids. However, because different plant proteins lack different essential amino acids they can be combined in the diet so as to complement each other. That is, even when all sources of animal protein are avoided (pure vegetarian/vegan diet) the consumption of a combination of grains, nuts and seeds of leguminous plants (pulses such as beans, peas and lentils) along with other vegetables ensures that sufficient quantities of all the essential amino acids are consumed.

Protein structure

The order of amino acids in a protein chain is termed the *primary structure* (Figure 6.5(a)). We might draw an analogy between this and the order of coloured beads in a child's necklace. However, few proteins exist as simple chains. The *secondary structure* of a protein results from hydrogen bond formation between different areas of the chain. The most common form of secondary structure is the α-helix (Figure 6.5(b)), which results from a tight coiling of the amino acid chain. In contrast, the β-pleated sheet results when hydrogen bonds link chains that lie side-by-side (Figure 6.5(c)). A higher level of complexity, referred to as the *tertiary structure*, occurs when α-helices or β-pleated sheets fold upon one another to produce a ball-like (globular) molecule (Figure 6.5(d)). Finally, when several protein units,

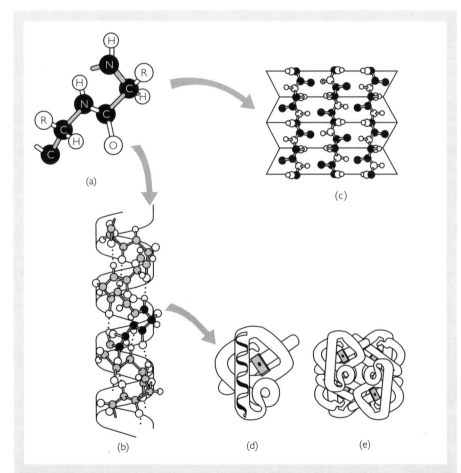

Figure 6.5

The (a) primary, (b) and (c) secondary, (d) tertiary and (e) quaternary structure of proteins. (Adapted from Sackheim G. I. (1996). *An Introduction to Chemistry for Biology Students*, 5th edn. Belmont, CA: Benjamin/Cummings.)

each with its own primary, secondary and tertiary structure, combine to form a more complex molecule, the term *quaternary structure* is applied (Figure 6.5(e)). For example, the protein haemoglobin has a quaternary structure of four sub-units.

In addition, proteins can also be classified as *globular* or *fibrous* in a similar way to carbohydrates previously described. The chains of globular proteins are wound up like a ball. They have functional uses rather than structural ones – that is, they perform functions other than forming part of the structure of the body. Enzymes are globular proteins, as are the plasma proteins **albumin** and fibrinogen and the red blood cell pigment haemoglobin. In contrast, the chains of fibrous proteins are extended like fibres in a rope, and these have important structural functions. For example, the protein keratin confers the property of impermeability on the epidermis of the skin. Compare the appearance of the inside of your mouth with the skin covering the body. The skin covering the body is keratinised and it is much tougher and less permeable than the mucus membrane of the mouth. Keratin is also an important component of hair and nails. Other examples of fibrous proteins include the helical molecules collagen and elastin. These are also important in the skin, but this time in the connective tissue dermis. Collagen confers toughness, while elastin ensures that the dermis has a degree of elasticity. To check this out pinch a fold of skin and note how quickly it springs back when you let go. If you were to repeat this simple test on the skin of an elderly person the fold will remain for some time after pinching. This is because there is less elastin in the skin of the elderly compared with the young, and as a consequence the skin of the elderly is more susceptible to trauma. You will need to understand this when caring for elderly people.

Protein digestion

Protein digestion begins in the lumen of the stomach, where the enzyme precursor pepsinogen is converted in to the active enzyme pepsin by hydrochloric acid. Pits in the lining of the stomach (gastric pits) contain a number of secretory cells. Some produce mucus to protect the tissues, while zymogenic (chief) cells secrete pepsinogen and parietal (oxyntic) cells secrete hydrochloric acid.

> *Can you work out why pepsin itself is not secreted but is instead formed from the action of hydrochloric acid on pepsinogen?*

The answer is that, since pepsin is a protein-digesting enzyme, secretion of it in its active form may result in the digestion of the stomach cells themselves! When performing laboratory experiments with pepsin take care not to splash it in to the eyes. *Wear safety glasses or shields whenever you perform an experiment involving chemicals.*

The action of pepsin upon protein chains is to break them into a number of smaller (peptide) chains. The peptides which result from the digestion of a protein are sometimes referred to as peptones. This process may be summarised as follows:

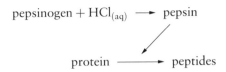

Protein digestion continues in the duodenum with two enzymes from the pancreas. Trypsin breaks peptide chains into smaller peptide chains. Like pepsin, trypsin is not secreted in an active form but results from the action of the enzyme enterokinase, found in intestinal secretions, upon the enzyme precursor trypsinogen:

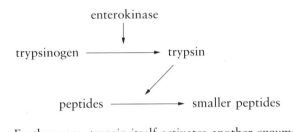

Furthermore, trypsin itself activates another enzyme precursor present in pancreatic secretions:

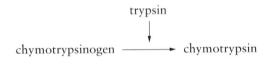

Chymotrypsin has a similar action to trypsin.

Finally, a group of enzymes called peptidases, some of which are present in pancreatic secretions and some in intestinal secretions, break off amino acids from the small peptide chains. These amino acids then diffuse into the bloodstream via the intestinal villi and pass to the liver in the hepatic portal vein.

Amino acids and the liver

The liver uses its supply of amino acids to synthesise a number of proteins, including the serum proteins albumin, fibrinogen and prothrombin. The latter two are important in blood clotting. Other amino acids are circulated and used by cells elsewhere in the body, while the liver also plays a role in breaking down unwanted amino acids. Damaged components of the body are constantly being replaced, and this includes protein.

What happens to unwanted amino acids?

In the liver the nitrogen-containing part of amino acids (amino group) is removed – a process known as **deamination**. This leads to the formation of **ammonia**, which is extremely toxic, but the liver converts this into the less toxic substance urea in the ornithine citrulline cycle. Urea is of course eventually excreted in the urine, as is creatinine, which is the waste substance produced from the breakdown of muscle.

> **Practice point 6.4: disturbances of protein metabolism**
>
> A number of conditions affect our ability to digest or use protein. For example, cystic fibrosis is a genetic (autosomal recessive) disease in which the pancreas and other exocrine glands become blocked with abnormally viscous mucus. Consequently, pancreatic enzymes cannot reach the duodenum and protein digestion is impaired. Fortunately, these enzymes can be added to food just before eating, but this does not prevent the effects of the disease elsewhere in the body.
>
> Secondly, phenylketonuria is a condition caused by an enzyme deficiency which results in failure to break down the amino acid phenylalanine. The accumulation of phenylalanine in the infant leads to mental retardation, so a diet low in this amino acid must be consumed.

Nitrogen balance

The term *nitrogen balance* refers to the state in which the intake of nitrogen (protein) equals its loss. A negative nitrogen balance is the state in which protein intake is less than expenditure, as is the case in poor diet, starvation and anorexia nervosa. In contrast, in some conditions the intake of protein is actually restricted.

> *Can you work out when this might be the case?*

Protein intake is often restricted in conditions affecting the liver or kidneys.

Lipids

The terms **lipids** and *fats* are used synonymously for a group of substances which contain the elements carbon, hydrogen and oxygen but not in a fixed ratio, as is the case in monosaccharides. Furthermore, lipids often contain phosphorous and nitrogen too. In fact, quite a number of different molecules are classified as lipids because they share the common physical property of being insoluble in water but soluble in organic solvents such as acetone and chloroform. Four classes of lipid are described:
1 glycerides
2 sterols
3 phospholipids
4 prostaglandins
Each will be examined here, but we shall concentrate on the most abundant group – the glycerides.

Glycerides

Glycerides are also known as **neutral fats** and are the substances which first come to mind when the word *fat* is used. Glycerides are stored in the adipose tissue of the body, where they form an important long-term energy store and act as a heat

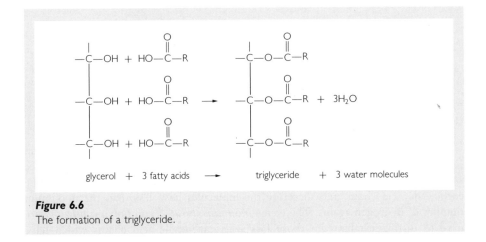

Figure 6.6
The formation of a triglyceride.

insulator. Adipose tissue also gives shape to the body and provides protection for vital structures such as the kidneys.

Glycerides are composed of two types of molecule – **carboxylic acids** (fatty acids) and **glycerol**. You have already encountered carboxylic acids in Chapters 2 and 5. Remember that they conform to the general formula RCOOH and undergo a reaction with alcohols to form esters. Glycerides are esters of the alcohol glycerol and different carboxylic acids. Remember that glycerol possesses three hydroxyl groups with which to react with carboxylic acids. If only one hydroxyl group reacts with a carboxylic acid the resultant ester is referred to as a *monoglyceride*. If two react a *diglyceride* results and if all three react the result is a *triglyceride*. Most of the fat within the body occurs as triglycerides. The formation of triglycerides is illustrated in Figure 6.6.

Saturated and unsaturated fat

If you are to understand health promotion advice relating to saturated and unsaturated fat then you must first of all understand the meaning of these terms and something of the chemistry of fats too. Saturated fats are those in which all the carbon atoms of the component carboxylic acids have the maximum number of hydrogen atoms attached. That is, there are no double bonds in the molecule. An example is palmitic acid ($C_{15}H_{31}COOH$), the structure of which is shown in Figure 6.7.

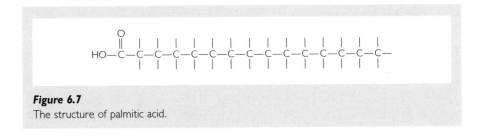

Figure 6.7
The structure of palmitic acid.

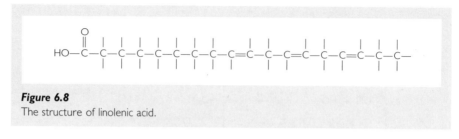

Figure 6.8
The structure of linolenic acid.

Saturated fats are solid at room temperature and are mainly of animal origin (think of lard and butter). However, some plants, such as coconuts, also produce saturated fat. Unsaturated fats are those in which double bonds occur in the carboxylic acid chain, and therefore the molecules have less than the maximum number of hydrogen atoms. The terms *monounsaturated* and *polyunsaturated* are used to distinguish between those fats with carboxylic acid chains which have one double bond (monounsaturated) and those with more than one double bond (polyunsaturated). An example of a polyunsaturated carboxylic acid is linolenic acid ($C_{18}H_{29}COOH$), the structure of which is shown in Figure 6.8.

Unsaturated fats are liquids or soft solids at room temperature and are of plant extraction (think of sunflower oil or vegetable oil). Sometimes a more solid presentation of vegetable fat is required, as is the case in margarine. In order to achieve this, hydrogen atoms are added to the carboxylic acid molecules which are then described as hydrogenated. Indeed, if you look at the information on margarine cartons you will often find this referred to.

Practice point 6.5: how do we decide who is too fat or too thin?

Most people believe that there is a weight which is 'about right for them' and that being either excessively underweight or overweight is unhealthy. Nonetheless, they may be unsure as to what their 'ideal' weight is.

How do we determine who is too fat or too thin?

Clearly, we cannot rely upon what we intuitively think is an acceptable weight since this would be influenced by social expectation and have more to do with what we perceive as an individual to be attractive and little to do with health. Furthermore, we would have to compare an individual's weight with other factors, such as height, if we want a realistic idea of the appropriateness of that person's weight.

This is in fact what is done in the calculation of the *body mass index*, the formula for which is given below:

$$\text{body mass index} = \frac{\text{weight in kg}}{(\text{height in m})^2}$$

You might like to pause here and calculate your own body mass index. Of course, you will need to know your height and weight in metric units. What figure did you arrive at?

There are no units for body mass index (an index is by definition a figure without units), but the value of the figure is helpful when compared with the values given below:

Body mass index	Description
20–25	Ideal
30	Obese
40	Gross obesity

Why do we consider a body mass index of 20–25 to be ideal?

The reason is that a body mass index of 20–25 is associated with the lowest mortality risk.

What is the effect on mortality risk of a body mass index of less than 20 or greater than 25?

You will no doubt have realised that outside the range 20–25 mortality risk is elevated.

You might use body mass index in a number of different situations. It might play a part in your role as health promoter – perhaps you are working with patients who are obese or who have diseases such as type II diabetes mellitus or hypertension (high blood pressure) which are sometimes associated with obesity. On the other hand, you might be working with clients who are severely undernourished – they might have a serious acute illness such as cancer of the oesophagus or perhaps they have mental health problems such as anorexia nervosa or bulimia nervosa. Finally, pressure sore risk assessments may involve determining whether the patient is an appropriate weight for their height. Body mass index is clearly useful in this situation too.

Digestion and absorption of glycerides

The small intestine is the only site of fat digestion since the pancreas is the only source of the fat-digesting enzyme lipase. However, since glycerides and their break-down products are all insoluble in water, special treatment of fat is necessary before digestion can take place. In water fats aggregate in large globules with a relatively small proportion of the lipid molecules on the surface and exposed to the water-soluble lipase molecule. However, upon entry to the duodenum fatty globules are acted upon by bile salts. Bile salts are molecules with a **hydrophobic** (repels water) end and a **hydrophilic** (attracts water) end. The hydrophobic end adheres to glyceride molecules, while the hydrophilic end repels other bile salt molecules. The result is that small droplets of fat are pulled off the larger globules, and as a consequence many more glyceride molecules are exposed to the enzyme lipase. Such a solution of small suspended droplets is called an **emulsion**. In the biological science component of your course you may perform an experiment to show how ineffective lipase is unless lipids are first of all emulsified. You should then be able to relate this to clinical practice. For example, suppose the flow of bile were to be obstructed as might be the case in the presence of gallstones. How would this affect the patient? Think about this and you may be able to work out that fat present in the diet could not be digested, and so would pass out of the body in bulky, fatty faeces which float. Furthermore, the discomfort of gallstones increases when fat is present in the meal, so the patient may be advised to take a low fat diet.

The action of lipase upon triglycerides is to break off two of the carboxylic acid chains, thus yielding free carboxylic acids and monoglycerides, both of which are insoluble in water. Free carboxylic acids and monoglycerides quickly become associated with bile salts and lecithin (a phospholipid in bile) to form micelles. These are collections of insoluble substances surrounded by bile salts with the hydrophilic ends oriented towards the outside. The hydrophobic core consists of monoglycerides, carboxylic acids, cholesterol and fat soluble vitamins. Micelles are similar to but much smaller than emulsion droplets and can diffuse between the microvilli of the cells of the ileum. When in close contact with the cells, the lipids, their breakdown products and fat soluble vitamins diffuse through the lipid-containing plasma membrane and enter the cells of the intestinal epithelium. Here the free carboxylic acids and monoglycerides are resynthesised into triglycerides. The triglycerides are then combined with small amounts of phospholipid, cholesterol and free carboxylic acids and are coated with protein to form water-soluble **lipoprotein** droplets called **chylomicrons**.

Chylomicrons are too large to enter blood capillaries and so diffuse into the more permeable lymph capillaries (lacteals). Thus the route for the absorption of the end products of lipid digestion is quite different to that for glucose and amino acids, which are small enough to enter blood capillaries. However, chylomicrons eventually appear in the circulation as they drain in turn through lymph vessels and the thoracic duct, which empties its fatty, milky contents (chyle) into the left subclavian vein in the base of the neck.

Practice point 6.6: so why do some people become fat while others remain lean?

We have previously looked at how we decide whether a person is overweight by using body mass index. Now we turn to look briefly at obesity itself.

Is obesity a disease?

If obesity is a disease it should be possible to formulate a theory which would explain why it develops in some people and not in others. In addition, for our theory to be accepted it must be reliable – that is it must explain all, or at least the majority of, the cases of obesity. No such unifying theory for obesity exists and therefore we cannot consider it to be a disease. Nonetheless there are some endocrine conditions, such as hypothyroidism and hypersecretion of the adrenal cortex (Cushing's disease) which do involve weight gain, the important thing to note here is that when such conditions do exist, weight gain is not the only symptom.

When you have a moment you might like to find out a little about these two diseases and some of the symptoms associated with them.

For the moment let us continue by considering the equation below:

energy intake = energy expenditure ± changes in body energy stores

Since obesity represents an increase in the body's energy stores (that is, fatness) then surely it must result from either an excessive energy intake or a reduced energy expenditure.

So does obesity result from overeating?

Surprisingly, most large scale surveys have failed to demonstrate overeating by the obese, although we might question the reliability of the dietary record as a method of investigation. It is possible that those who take part in such surveys simply under-report what they eat, or perhaps they actually eat less for the period of investigation.

What about energy expenditure then? Are the obese simply underactive?

Once again, research has failed to demonstrate conclusively that this is the case. Perhaps you are beginning to realise that obesity is more complicated than it first appears! The fact is that different people are obese for different reasons and there are more explanations than those outlined here. Nonetheless, overeating and too little exercise are possible causes.

In fact, whatever the cause of obesity, and provided disease is absent, reducing the amount of food consumed and increasing the amount of exercise taken must result in some weight loss.

Does the type of food taken rather than the amount influence the development of obesity?

Clearly the answer to this question is yes. For example, empty calorie foods such as sugar provide a good deal of calories in a small volume. In addition, sugar stimulates the appetite. Perhaps that's why desserts appear so 'more-ish' even when we feel full after a main course! In contrast, sources of complex carbohydrates, such as wholemeal bread and potatoes, provide calories in a much more filling form. In addition, they also contain other nutritional substances, such as vitamins, minerals and roughage. Finally, studies on congenitally obese mice suggest that they become fat faster when more of the energy is provided as fat than as carbohydrate. Think about this a little more. Two groups of mice are given isoenergetic diets (the same amount of energy) but one group gets 80% of the energy as fat while the other gets 80% of the energy as carbohydrate. The first group gets fat faster.

Can you think of how to relate this to health promotion in humans?

It makes sense to recommend that, if the amount of energy consumed by the individual is about right, 80% of this energy is provided in the form of carbohydrate and not as fat.

What kind of carbohydrates would you recommend to replace some of the fat?

You will probably have worked out that complex carbohydrates rather than empty calorie foods should be chosen.

Sterols

The structure of this group of lipids is based around the circular sterol molecule. Included in this group are cholesterol, bile acids and steroid hormones such as oestrogen, progesterone, testosterone, cortisone and aldosterone. (Note that the names of steroid hormones usually end in 'one'.)

Phospholipids

Phospholipids are similar to glycerides, but one of the carboxylic acid molecules is replaced by a molecule containing phosphate. You will encounter phospholipids as important components of cell membranes.

Prostaglandins

Prostaglandins are lipids produced by most cells of the body. They may be released as part of a response to injury (inflammation) and they play an important role in pain perception and temperature regulation. Aspirin (acetylsalicylic acid) inhibits prostaglandin synthesis and it is therefore important as an anti-inflammatory drug, as an analgesic (reduces pain) and as an antipyretic (reduces fever).

Energy production and lipids

Carboxylic acids and glycerol can both be oxidised in the body to yield energy. In fact, 1 gram of fat yields more energy than 1 gram of carbohydrate or protein. This is more fully explained in Chapter 7 (Energy in the body).

Practice point 6.7: fats and health

The transportation of fats as lipoproteins has already been described. Lipoproteins may be classified according to density. Low-density lipoproteins (LDL) and very low-density lipoproteins (VLDL) have a greater proportion of lipid than do high-density lipoproteins (HDL). LDL and VLDL are implicated in the development of the arterial disease atherosclerosis. The term *atherosclerosis* means a gruel-like hardening and refers to the focal accumulation of lipids and other large molecules in the intima and media of artery walls. The consequence is a narrowed vessel lumen (opening) and compromised blood supply. If the coronary arteries are affected then angina and myocardial infarction (heart attack) may result, while atherosclerosis of cerebral arteries may lead to cerebrovascular accident (CVA/stroke).

A number of factors contribute to the development of atherosclerosis, including high circulating levels of LDL and VLDL.

Do you know what kind of diet is responsible for high levels of these types of lipoprotein?

Perhaps you have realised that it is a diet high in saturated fat. Consequently, current health promotion advice is to reduce the amount of animal fat in the diet and consume polyunsaturated fat instead. The health benefits of consuming mono-unsaturated fat such as olive oil and peanut oil are less certain.

In contrast, HDL appears to be protective against atherosclerosis – removing fat from arterial walls and transporting it to the liver for metabolism and excretion in the bile. The discovery that fish oil is associated with increased HDL levels has added a new impetus to recommending its consumption instead of animal meat.

Vitamins

The word *vitamin* is derived from the Latin for 'life' and serves to indicate the importance to health of this group of organic molecules. Vitamins must be obtained from food since few are synthesised in the body, or at least not in sufficient quantities. Having noted this, the amount of each vitamin required daily is very small. Vitamins are often destroyed by heat, as is the case when food is overcooked, and some are sensitive to light. However, they are not broken down in the digestive tract but are absorbed unchanged. Vitamins are used by the body in their original or slightly modified forms. They function as co-factors for enzymes and are therefore important in regulating metabolic reactions. Despite the fact that what we now know as vitamin C was a key factor in the therapeutic trial into the treatment of scurvy conducted by Lind on HMS *Salisbury* in 1747, no vitamin was isolated until this century. Indeed, vitamin C was not isolated in its pure form until 1932. The system of identifying vitamins devised by Hopkins involves identifying each by a letter, vitamin A being the first to be discovered. Subsequently, this simple system has become a little more complex with some vitamins being identified by a letter unrelated to the order of discovery and some being more commonly referred to by name. In addition, some vitamins are actually a group of closely related substances which may be collectively referred to as a single vitamin, while in other cases the individual substances are distinguished by a letter and a number. However, the first step in classifying vitamins is to distinguish two groups on the basis of solubility in fat or water.

Fat-soluble vitamins

The fat-soluble vitamins are A, D, E, and K. They are not absorbed directly into the gut, but first pass into the lymphatic system. Since they are not soluble in water they can only be transported as components of lipoproteins. Fat-soluble vitamins are readily stored by the body and so do not need to be consumed every day. The principal sites of storage are the liver and adipose tissue. If consumption is excessive, fat-soluble vitamins are not readily eliminated and toxicity may be a problem when dietary supplements are used. This is especially true of vitamins A and D.

Vitamin A (retinol)

Vitamin A was identified in 1913. The alternative name *retinol* is obviously derived from the name of the structure in which it plays a major role – the retina of the eye. Retinol is a pigment important in night (black and white) vision and in a preformed state it is only found in animal foods, especially liver. However, the body can synthesise vitamin A from a number of pigments (carotenes) found in plants and some animal fats. Carotenes are yellow pigments, their name indicating the vegetable in which they are clearly visible. However, carotenes are present in a number of vegetables, but their colour is often masked by the darker green pigment chlorophyll. Sources of carotenes include leafy vegetables, broccoli, carrots, peaches and apricots. In addition, vitamin A is added to margarine.

The presence of vitamin A in a wide range of foods means that nutritional deficiency in developed countries is unlikely. Even if vitamin A were eliminated from the diet, extensive stores mean that manifestations of deficiency would not become obvious for a number of years. When vitamin A is lacking, night blindness and xerophthalmia may result. In addition to its role as a visual pigment, vitamin A is important for the health of secreting epithelia (lining and covering tissue). Xerophthalmia is the name given to the effects of vitamin A deficiency on the conjunctiva and cornea of the eye, which occurs mostly in east Asia and Africa. Manifestations include ulceration and necrosis (keratomalacia). Subsequently, the cornea becomes scarred and opaque, and blindness ensues. About half a million new cases of xerophthalmia occur each year, and about half of those affected become blind. This is so despite the fact that the association between eye lesions and nutritional deficiency in African natives was first made by David Livingstone as long ago as 1857, and that xerophthalmia is relatively easy to prevent with high-dose vitamin A capsules given twice a year.

In contrast, the problem of vitamin A toxicity is more prevalent in developed countries and is related to the consumption of supplements rather than to diet. Manifestations range from dry skin, brittle nails and hair loss to liver damage and death. Finally, the consumption of liver by pregnant women is best avoided, since the large amounts of vitamin A concentrated there by herbivores fed animal products pose a threat to the normal development of the foetus.

Vitamin D

Dietary sources of vitamin D are few, but do include liver, fish liver oils and eggs. Vitamin D is also added to margarine. In contrast to other vitamins, most vitamin D is synthesised in the body – a process which involves the skin, the liver and the kidneys. Ultraviolet light converts 7-dehydrocholesterol present in the skin to cholecalciferol (D_3). In the liver an enzyme converts D_3 to 25-hydroxycholecalciferol, which in turn undergoes enzymatic conversion to 1,25 dihydroxycalciferol (calcitrol) in the kidneys. Ergocalciferol (D_2) is the artificially produced form of vitamin D used as a vitamin supplement.

Vitamin D is essential for the absorption of calcium from the gastrointestinal tract and for the effective use of calcium in the formation of bone. Both D_2 and D_3 are active, but calcitrol is the most potent form of the vitamin. Because calcitrol is synthesised in one organ but has functional activity elsewhere it may be regarded as a hormone. Vitamin D deficiency results in defective utilisation of calcium in bone leading to rickets in children and osteomalacia in adults. Since the amount of vitamin D synthesised depends upon the area of the skin exposed, the length of exposure and the characteristics of the skin (light-coloured skin synthesises vitamin D more readily than dark-coloured skin), it is not difficult to work out that dark-skinned people who live in cities of the northern hemisphere are particularly susceptible. Rickets and osteomalacia may of course be prevented by exposure to sunlight and by dietary supplements.

However, excessive consumption of vitamin D also presents a danger. Overdose leads to hypercalcaemia (elevated blood calcium), the manifestations of which include thirst, polyuria (increased urine output), calcification of soft tissue, renal calculi (stones) and renal damage.

Vitamin E

Vitamin E is really a group of related substances called tocopherols. They are relatively widespread, being found in vegetable oils and products such as margarine which are made from vegetable oil. In contrast, animal fats are poor sources of vitamin E. The amount of vitamin E required by the body is proportional to the intake of polyunsaturated carboxylic acids, since it protects them from oxidation. Because cell membranes, especially those of erythrocytes, are rich in polyunsaturated carboxylic acids, vitamin E is important in maintaining cell membrane integrity. However, dietary deficiency of vitamin E is unlikely except where fat digestion and absorption are impaired (for example, in cystic fibrosis). Somewhat more likely is deficiency in premature babies, since vitamin E does not easily cross the placenta. Haemolysis (red blood cell destruction), is the major manifestation in infancy.

Compared with vitamins A and D, vitamin E appears relatively non-toxic, which is perhaps fortunate since unsupported claims for its efficacy have led to it being taken as a remedy for such diverse conditions as heart disease, cancer, infertility and skin disorders.

Vitamin K

The designation of the letter K to this vitamin is based upon its role in blood coagulation (German *koagulation*). Two forms of vitamin K exist. K_1 (phylloquinone/phytomenadione) is found in dark green vegetables, while the K_2 vitamins (menaquinones) are synthesised by intestinal bacteria. Animal foods do of course provide both the above forms of vitamin K. The regular consumption of vitamin K in the diet is important since intestinal synthesis alone is inadequate to meet the body's needs and stores of vitamin K in the liver are not extensive. Nonetheless, deficiency is uncommon except where malabsorption exists (such as in obstructive jaundice). Since vitamin K is important in the synthesis of the coagulation protein prothrombin, deficiency is manifest by hypoprothrombinaemia (low blood pro-thrombin level) and bleeding. Consequently, patients who are to have surgery for biliary obstruction should receive a parenteral supplement of phytomeniadone (Konakion), and in some countries it has become customary for the prevention of haemolytic disease of the newborn to administer a dose of vitamin K to the infant at birth. Haemolytic disease of the newborn is caused by vitamin K deficiency as a consequence of a sterile gut and low levels of vitamin K in cord blood and in breast milk.

The anticoagulant warfarin owes its therapeutic action to its antagonism of vitamin K. Therefore Konakion is the antidote to overdose and should be kept in stock whenever warfarin is administered.

Water-soluble vitamins

The water-soluble vitamins are the B group and vitamin C. Water-soluble vitamins are absorbed directly into the bloodstream through the gut wall without the need for specialised carrier molecules. They are readily filtered by the kidney and lost in the urine when consumed in excess. However, the body is unable to store large quantities of water-soluble vitamins and therefore they should be supplied daily in the diet.

Vitamin B$_1$ (thiamine)

Thiamine is abundant in unrefined grains and cereals, pulses, seeds, nuts and pork, but it is present in other vegetables and in meats too. The thiamine content of food is reduced by cooking and alkalis – important if sodium hydrogen carbonate (sodium bicarbonate) is added to vegetables to maintain colour during cooking. The body's stores of thiamine are not extensive and deficiency becomes manifest about a month after the exclusion of thiamine from the diet.

Thiamine is an important component of an enzyme involved in the conversion of glucose to energy. It is also needed for protein and fat metabolism and the normal function of the nervous system. Severe thiamine deficiency, known as beri beri, is rare even in the Orient, where it was once common. Here it is related to reliance upon milled rice as a principal component of the diet. Rice is not deficient in thia-mine, but refining reduces the thiamine content.

Beri beri is characterised by fatigue, muscle weakness (including heart failure), paralysis, irritability, emotional instability, depression and confusion. It is unlikely that many who read this book will witness beri beri at first hand, but thiamine deficiency related to alcohol abuse is more common. Chronic excessive use of alcohol affects the intake, absorption and metabolism of thiamine and results in Wernicke–Korsakoff syndrome. In 1880 Wernicke described alcoholic encephal-opathy characterised by apathy, stupor and ataxia, while in 1887 Korsakoff described memory loss, inability to retain new memories and confabulation. The combination of observations made by Wernicke and Korsakoff provides a picture of the typical chronic excessive alcohol abuser.

Vitamin B₂ (riboflavin)

There are no body stores of this vitamin, so it must be supplied in the diet on a daily basis. Riboflavin is widely distributed in foods but only in small amounts. Much of the riboflavin in the British diet is supplied in milk and cheese, but it is also found in meat, fish and fortified grains. Riboflavin has a number of important functions within the body. It is a component of a number of enzymes involved in the energy-producing reactions of the cells and it is required for the manufacture of erythrocytes and corticosteroids. The biochemical functions of riboflavin do not easily explain the manifestations of deficiency, which include dermatitis (inflammation of the skin), normocytic anaemia (low red blood cell number, but the cells are normal), cheilosis (cracks in the lips), angular stomatitis (cracks in the skin in the corners of the mouth) and glossitis (inflammation of the tongue). The recognition of deficiency is complicated by the fact that many of the manifestations also occur in other deficiency states, but fortunately riboflavin is rarely absent from the diet. Those at risk of inadequate intake are those whose calorie intake is poor, including adolescent females, the elderly and alcoholics. In addition, some groups have increased riboflavin requirements, including pregnant women, people with thyrotoxicosis and people taking the drugs chlorpromazine (antipsychotic/ neuroleptic), imipramine and amitriptyline (antidepressants). These nutrient–drug interactions are obviously important considerations for professionals working in a mental health context. Riboflavin is used pharmacologically only to treat deficiency states. It has no known toxic effects.

Vitamin B₆ (pyridoxine)

A number of closely related substances are included under the name of this vitamin and they are found widely in meat and fish as well as in whole grain cereals, peanuts and eggs. Consequently, deficiency is rare, although in 1954 an outbreak of convulsions in babies was attributed to insufficient B₆ in a modified cow's milk formula. Vitamin B₆ forms part of a co-enzyme involved in amino acid metabolism, so the body's requirement for it is directly related to protein intake.

Vitamin B₁₂ (cobalamin)

This vitamin is found only in animal products, and this fact is of special importance to vegans. Food sources rich in vitamin B₁₂ are offal, meat, eggs, fish and dairy products. B₁₂ plays a role in the synthesis of RNA (**ribonucleic acid**) and DNA (deoxyribonucleic acid), and it is required for the growth and maturation of red blood cells. Primary dietary deficiency is rare except in vegans who fail to take oral vitamin B₁₂ supplements. Even then the signs of deficiency may take up to 10 years to develop due to the presence of extensive B₁₂ stores in the body. Vitamin B₁₂ is absorbed in the terminal ileum, but this requires the presence of *intrinsic factor* – a glycoprotein secreted by the stomach. Consequently, the absence of intrinsic factor leads indirectly to B_{12} deficiency, which is characterised by hyperchromic, megaloblastic anaemia (pernicious anaemia – the red blood cells are large have an abundance of the pigment haemoglobin). Signs of this include pallor, dyspnoea (the subjective experience of a difficulty breathing), weakness and fatigue as well as gastrointestinal symptoms such as glossitis (inflammation of the tongue), anorexia (loss of appetite) and indigestion. Pernicious anaemia is also manifest in resectional gastrointestinal surgery which bypasses the ileum and of course in gastrectomy (surgical removal of the stomach). Vitamin B_{12} deficiency caused by an inadequate intake can be corrected by oral supplements, but deficiency due to impaired absorption requires intramuscular B_{12} injections.

Niacin (nicotinic acid)

Although there are few body stores of this vitamin, it is found in a wide variety of foods, including meat, fish, milk, yeast and wholemeal grains. In addition, niacin can be synthesised by the body from the amino acid tryptophan. Consequently, niacin deficiency is rare except in parts of Africa and Asia, where maize is the staple food. Niacin forms parts of a number of co-enzymes required by all living cells for the release of energy from carbohydrates, proteins and fats, including nicotinamide adenine dinucleotide (NAD – see Chapter 7). Early manifestations of deficiency include anorexia, apathy, weakness and indigestion. Severe deficiency is called pellagra (literally 'rough skin') and classic symptoms are the four 'd's'; dermatitis, diarrhoea, dementia and, if untreated, death. Niacin deficiency is readily treated with supplements, although toxicity can occur and includes flushing, itching, nausea, vomiting, diarrhoea, hypotension, tachycardia, fainting, hypoglycaemia and liver damage.

Folic acid (folate)

The name of this vitamin comes from the latin for leaf, but it is found in a wide range of foods which come from both animal and plant sources, including liver, kidney, spinach, broccoli, peanuts, oranges and wholemeal bread. Body stores are not extensive, so if the diet is deficient in folic acid symptoms can develop quickly. However, deficiency because of malabsorption or drug antagonism are more common. Alcohol is the most common antagonist. Deficiency results in a macrocytic (large cell) anaemia and consequent fatigue, weakness and pallor. Other symptoms include weight loss diarrhoea and glossitis. Treatment is undertaken with folic acid supplements.

Vitamin C

This is almost certainly the most familiar vitamin to members of the public and most people will be able to identify blackcurrants and citrus fruits as good sources. However, a large percentage of the vitamin C intake in Britain is provided by potatoes – not because potatoes are especially rich in this vitamin but because large quantities of potatoes are eaten. A number of vegetables, such as cauliflower and broccoli, are also sources of vitamin C, as are liver and milk. The association of vitamin C and citrus fruits goes back to an experiment on the treatment of scurvy conducted by James Lind on board HMS *Salisbury* in 1747, although this vitamin was not actually isolated until 1932. Vitamin C is required for collagen synthesis and symptoms of deficiency (scurvy) reflect defects in the manufacture of this important body component, including swollen, inflamed gums which bleed easily, pinpoint haemorrhages in the skin, malformations of bone and fractures, delayed wound healing and secondary infection. Vitamin C deficiency can be treated with oral supplements; in addition, taking these supplements at considerably greater than nutritional levels as a means of preventing certain types of cancer and the common cold are advocated by some.

Minerals

Important minerals include the following:

sodium	potassium	calcium
iron	phosphorous	iodine
magnesium	zinc	

Minerals are inorganic substances which are necessary for the normal function of the body. Some are present in the body as ions dissolved in solution and in this state they are referred to as **electrolytes**. The functions of minerals within the body may be summarised as follows:

1 Some combine with organic molecules to form important substances within the body. For example, phosphorous combines with lipids to form phospholipids, which are important components of cell membranes.
2 Some, such as monohydrogen phosphate ions and dihydrogen phosphate ions, function as **buffers** (see Chapter 4 (Acids, bases and pH balance)).
3 Some act as **co-enzymes**.
4 Sodium is an important osmotic regulator.
5 Sodium, potassium and calcium are important in the generation of nervous impulses and in muscle contraction.
6 Calcium and phosphorous are important structural components of bones and teeth.

Sodium (Na/Na$^+$)

The element sodium is a highly reactive soft grey metal found in Group I of the periodic table. The stable and unreactive sodium ion is found widely in fruits and vegetables as well as in meats and fish. As the compound sodium chloride (common salt/NaCl) it is used as a preservative in tinned foods and cured meats and it is often added to food at the table. Consequently, the average daily intake in British diets of 10–15 g is considerably in excess of the required 1 g per day. The normal blood plasma concentration of sodium is 135–145 mmol/l, and in the blood sodium ions function as an important osmotic regulator. In addition, sodium ions also play an important role in the generation of electrical impulses in nerves (**action potentials**) and also in muscle contraction.

A low serum sodium is referred to as hyponatraemia and an elevated serum sodium as hypernatraemia. However, sodium and water are usually gained and lost together. For example, both are lost in excessive sweating, vomiting, diarrhoea and during increased diuresis (urine output). If sodium and water depletion is great there is a reduction in the volume of extracellular (outside cells) fluid, of cardiac output and of blood pressure. Fatigue and faintness are the result. In contrast, sodium and water retention leads to an expansion of the blood volume and to **oedema** (an excess of interstitial fluid). It is common in cardiac, renal and liver disease. The swollen ankles of the elderly (due to oedema) are commonly a manifestation of sodium and water retention due to cardiac failure.

Potassium (K/K$^+$)

Potassium is also a highly reactive metal found in Group I of the periodic table, but its ion is stable and unreactive. Potassium is the principal intracellular (inside the cell) ion, while plasma levels are much lower than those of sodium at only 3.0–5.5 mmol/l. Potassium plays an important role in nerve conduction and muscle contraction. A low plasma potassium level (hypokalaemia) may result from diarrhoea and the use of certain diuretics (drugs which cause an increase in urine output) such as frusemide. It is made manifest by weakness and possibly mental symptoms such as apathy and confusion.

High serum potassium (hyperkalaemia) may occur in renal failure, when there has been an excessive consumption of potassium supplements or when an intravenous infusion containing potassium is allowed to run too quickly. An elevated serum potassium may result in ventricular **fibrillation** or asystole – these are both forms of cardiac arrest.

Calcium (Ca/Ca^{2+})

Calcium is a metal found in Group II of the periodic table and calcium ions are an important component of the diet. They may be found in cereals, vegetables and fish, while dairy products are very good sources. The absorption and utilisation of calcium requires vitamin D, while the hormones calcitonin and parathormone, from the thyroid and parathyroid glands respectively, are responsible for regulating the serum calcium level, which is normally 2.0–2.5 mmol/l. Calcium is an important component of bones and teeth and it plays an important role in blood coagulation, the conduction of nervous impulses and muscle contraction.

A low serum calcium level is referred to as hypocalcaemia and it may result when there is a deficiency in parathormone. One example of such a deficiency occurs when the parathyroid glands, which are embedded in the thyroid glands, are removed as part of a thyroidectomy (surgical removal of the thyroid gland). Hypocalcaemia may result in tetany – that is, sustained muscular contraction. An example of this is carpo-pedal spasm, which is sustained involuntary contraction of the wrist and ankle. In the case of a longstanding deficiency of calcium, the body attempts to maintain serum calcium levels at the expense of calcium stores in the bones. This results in a negative calcium balance and loss of bone mass, referred to as osteoporosis. Osteoporosis is also associated with a lack of physical activity and with a reduction of oestrogens in post-menopausal women. You might wish to learn more about this by undertaking further study in physiology, pathology and nursing texts.

A high serum calcium level is referred to as hypercalcaemia and it may be recognised by nausea, vomiting, weakness, constipation, polydipsia (increased thirst) and polyuria (increased urine output). If present over a prolonged period, hypercalcaemia may result in renal calculi (kidney stones). Possible causes of hypercalcaemia include tumours of the parathyroid glands with excessive production of parathormone or excessive ingestion of vitamin D.

Iron (Fe/Fe^{2+} and Fe^{3+})

This element, which is identified in the periodic table as a transition metal, is familiar to everyone and has been extracted from ferrous rocks since the so-called Iron Age. It is also an important component of our bodies and therefore an important nutritional element too. Iron in the diet is present in two forms. That which forms part of the haem molecule is referred to as haem iron, and meat is a good source. Nonhaem iron is obtained from plant sources, such as grains, vegetables and nuts. Only 10% of the iron in the diet is absorbed, and sometimes nurses have to consider how a patient may be helped to improve the availability of this important mineral. Haem iron is better absorbed than nonhaem iron, but the latter accounts for a larger percentage of the total iron intake. In addition, the rate of haem iron absorption (about 15–35%) varies only with the body's need, while the rate of absorption of nonhaem iron (about 3–8%) is influenced by a number of other factors. For example, vitamin C enhances the absorption of nonhaem iron. Consequently, when patients have an iron deficiency and are taking iron supplements, nurses have traditionally encouraged the consumption of a diet rich in vitamin C as well as in iron. Similarly, when iron supplements are prescribed, compounds containing iron which are readily absorbed, such as ferrous sulphate, are chosen.

The body contains about 3–5 g of iron, but only a very small amount is found in the plasma, where it is bound to the protein transferrin, which, as its name suggests, is responsible for the transport of this important mineral. Most of the iron in the body (70%) is in the haem part of the haemoglobin molecule, which is the pigment of erythrocytes (red blood cells) responsible for oxygen transport. A further 5% is in a similar molecule found in muscle called myoglobin. Body iron stores in the liver, bone marrow and spleen in the form of a molecule called ferritin account for another 20% of the body's iron, while the remaining 5% appears in cellular enzyme systems.

Practice point 6.8: iron deficiency anaemia

One important measurement which you should be familiar with is the amount of haemoglobin in the blood. It is approximately 11–13 g/dl (a decilitre or dl is 100 ml) in women and 13–16 g/dl in men. Iron deficiency leads to one form of anaemia (the condition of a reduction in the amount of haemoglobin or the number of erythrocytes). In iron deficiency anaemia the problem relates to a lack of haemoglobin because of the deficiency of this important component of the haemoglobin molecule. Therefore the erythrocytes are deficient in pigment (hypochromic) and as a consequence they are small too (microcytic). A low haemoglobin level combined with hypochromic and microcytic erythrocytes is diagnostic of iron deficiency anaemia. In addition, the patient might complain of pallor, fatigue and breathlessness.

Can you think why?

Glossitis (inflammation of the tongue), gingivitis (inflammation of the gums) and angular stomatitis (inflammation of the corners of the mouth) may also occur. Have you seen these features in the elderly, where they are not at all uncommon? You may wish to be alert for these since the diet of the elderly may be deficient in good sources of iron. Dietary deficiency of iron may also occur in the prolonged breast-feeding of infants because milk is a poor source of iron.

Other causes of iron deficiency anaemia include chronic blood loss in peptic ulceration (stomach and duodenal ulcers), haemorrhoids and when the menses are heavy. In addition, pregnancy may result in an increased requirement for iron and iron supplements may be prescribed. In the case of severe anaemia a blood transfusion may be given. You might like to find out about this procedure and the role of the nurse by studying a nursing text.

Iron toxicity is uncommon because the body regulates iron uptake according to need. However, the genetic condition haemochromatosis involves a defect in iron metabolism and results in the accumulation of iron in the tissues and eventually in multiple organ failure. Toxicity due to excessive consumption of iron supplements can occur, but it is children who are usually affected. Important advice to parents who are prescribed iron supplements is to store them where children cannot reach them. Manifestations of iron poisoning include gastrointestinal pain, nausea, vomiting, convulsions and coma.

Phosphorous (P)

This is the second most abundant mineral in the body – along with calcium, phosphorous forms the mineral matrix of bone. It is not surprising therefore that 85% of the body's phosphorous is found in bone. Nonetheless, phosphorous is important elsewhere too. For example, phosphorous combines with lipids to form phospholipids. These are important in a number of places in the body, but perhaps most importantly in the cell membrane. Here a double layer of phospholipids (phospholipid bilayer) is an essential component of the membrane and this structure is described in the fluid mosaic model of the cell membrane. You might like to read more about this in a physiology text. Phosphorous also forms an important part of our genetic makeup, where it forms part of the sugar–phosphate 'backbone' of DNA (deoxyribonucleic acid). The phosphate ion is formed from phosphorous and oxygen (PO_4^-) and phosphate ions are also important in the body's chemical form of energy **adenosine triphosphate** (ATP). More about ATP can be found in Chapter 7 (Energy in the body). Phosphorous is also an important component of the monohydrogen phosphate ion (HPO_4^{2-}) and the dihydrogen phosphate ion ($H_2PO_4^-$), the two of which form an important buffering system for the regulation of acid–base balance in intracellular fluid and in renal tubules. You will find more about this in Chapter 4 (Acids, bases and pH balance).

Phosphorous is widespread in the diet and dietary deficiency is uncommon. Vitamin D is involved in the absorption of phosphorous and parathormone in the regulation of plasma levels. Hypophosphataemia (low blood phosphorous level) may occur with the excessive use of the antacids aluminium hydroxide or magnesium hydroxide, both of which decrease phosphorous absorption. However, this is rare, as is hyperphosphataemia (high blood phosphorous level).

Iodine (I)

Iodine belongs in Group VII of the periodic table – the halogens. Iodine is found in a number of different body tissues, but it is perhaps most familiar as an essential component of the thyroid hormones thyroxine (T_4) and triiodothyronine (T_3). Seafood is a good source of iodine, as are plants grown in iodine-rich soils. A deficiency of iodine leads to underproduction of the thyroid hormones and you may wish to think what the manifestations of this might be. Over a prolonged period of time the thyroid gland swells in response to this iodine deficiency – a condition known as a goitre.

> Can you work out why this swelling occurs? (Hint: – you will need to think about the action of thyroid-stimulating hormone (TSH or thyrotropin)).

TSH stimulates the growth of the thyroid gland in an attempt to increase thyroxine production. However, thyroxine levels remain low due to iodine deficiency. In the past, such a goitre in Britain was sometimes referred to as Derbyshire neck because the low levels of iodine in the water of this hilly county meant that goitre was common there.

> Can you work out why this is no longer the case?

You may have worked out that in industrial societies people are no longer confined to eating local produce.

Apart from its role in the diet you may encounter iodine in a number of other contexts. When dissolved in water iodine forms a useful antiseptic solution or hand-washing preparation. When dissolved in alcohol it may be used as a skin preparation solution prior to surgical incision – the alcohol ensures that it dries quickly. You should, however, be aware that some people are allergic to iodine and this should be ascertained prior to its use. This is especially important when radiopaque (opaque to X-rays) iodine solutions are injected into the blood as part of a radiological investigation (investigation using X-rays). Finally, radioactive iodine preparations may be used in the investigation and treatment of thyroid disorders.

Magnesium (Mg/Mg^{2+})

This metal is found in group two of the periodic table and has a number of uses in the body, including forming part of the mineral matrix of bone. Because magnesium is an important component of the plant pigment chlorophyll, it is abundant in the diet when vegetables are regularly consumed. A dietary deficiency of magnesium has not been reported in people consuming a mixed diet.

Zinc (Zn/Zn^{2+})

This metal is an important component of nucleic acids, insulin and a number of enzymes. It is believed to be important in wound healing. Zinc is found in meat and seafood, milk, egg yolks, legumes and wholegrain cereals, although the zinc from animal sources is better absorbed. There is little evidence to suggest that zinc deficiency is common in the general population.

Fibre (roughage/non-starch polysaccharide)

The words *roughage* and *fibre* are often used without precise definition. They are commonly used synonymously for plant non-starch polysaccharide that cannot be digested by human enzymes. This material includes cellulose, an important component of plant call walls, lignin, a component of the woody parts of plants, pectins and mucilages, which give many fruits their ability to hold large volumes of water, as well as some other molecules. Of these molecules, only cellulose has a fibrous structure, so the use of the term *fibre*, although increasingly common, is perhaps not strictly accurate.

Fibre can be divided into two types on the basis of its water solubility. *Soluble fibre* includes mucilages and pectins, while cellulose and lignin are examples of *insoluble fibre*. The effects of these two groups of substances are summarised below:

Soluble fibre	Insoluble fibre
Decreases intestinal transit time	Decreases intestinal transit time
Slows gastric emptying	Absorbs water
Lowers serum cholesterol	Increases faecal bulk
Delays glucose absorption	Reduces pressure in colon

The two different types of fibre are found in different proportions in different types of plant material, but a diet which contains a variety of vegetables, including pulses, as well as fruit and whole grain cereals, will ensure that both soluble and insoluble fibres are consumed. The consumption of both types of fibre is important for health. For example, both decrease intestinal transit time (the length of time that ingested material remains in the gut), but insoluble fibre has the additional effect of absorbing water and increasing faecal bulk. The larger, softer stools are more readily moved through the gut by peristalsis (the wave-like contraction of the intestine) and this reduces the incidence of constipation and probably of a number of other intestinal conditions such as diverticular disease and colonic cancer. In contrast, soluble fibre has effects on serum cholesterol and also on glucose absorption. The former is an important consideration in dietary advice for the reduction of athero-sclerosis, while the delayed absorption of glucose may help to improve glucose tolerance in diabetics.

Summary

1 There are seven important components of the diet.
2 Carbohydrates are divided into three classes – monosaccharides, disaccharides and polysaccharides.
3 Diabetes mellitus is one example of a disorder of carbohydrate metabolism.
4 Proteins consist of chains of amino-acids joined by peptide bonds.
5 Proteins have primary, secondary, tertiary and quaternary structures.
6 Important classes of lipid include glycerides, sterols, phospholipids and prostaglandins.
7 Glycerides may be described as monoglycerides, diglycerides or triglycerides.
8 Lipids may be described as saturated, monounsaturated or polyunsaturated.
9 The process of digestion of carbohydrates, proteins and fats involves enzyme hydrolysis.
10 Vitamins are organic compounds which function as co-factors for enzymes.
11 Vitamins are either fat-soluble or water-soluble.
12 Minerals are inorganic substances which are necessary for the normal function of the body.
13 The terms *roughage* and *fibre* refer to plant material which cannot be digested by human enzymes.

Self-test questions

6.1 Match the substances on the left with the appropriate descriptions on the right.

a glucose	**i**	a monosaccharide
b lactose	**ii**	table sugar ,
c glycogen	**iii**	a disaccharide found in milk
d sucrose	**iv**	a polysaccharide

6.2 Match the minerals on the left with the appropriate descriptions on the right.

a sodium	**i**	involved in oxygen transport
b calcium	**ii**	the most abundant extracellular ion
c iron	**iii**	important in the formation of thyroxine
d iodine	**iv**	important component of bone

6.3 Match the vitamins on the left with the appropriate descriptions on the right.

a thiamine (B_1)	**i**	not present in any plant source
b D	**ii**	deficiency common in alcohol abuse
c K	**iii**	required for the formation of prothrombin
d B_{12}	**iv**	made in a process involving sunlight

6.4 Match the enzymes on the left with the appropriate descriptions on the right.

a pepsin   **i** responsible for starch digestion

b amylase   **ii** secreted by stomach

c lipase   **iii** proteolytic enzyme of the pancreas

d trypsin   **iv** responsible for fat digestion

6.5 Match the substances on the left with the appropriate descriptions on the right.

a amino acid   **i** vitamin C

b carboxylic acid   **ii** building block of protein

c ascorbic acid   **iii** B group vitamin

d folic acid   **iv** building block of triglycerides

6.6 Match the substances on the left with the appropriate descriptions on the right.

a mucilages   **i** consist of long chain of amino acids

b cellulose   **ii** soluble fibre

c tocopherols   **iii** vitamin E

d polypeptides   **iv** insoluble fibre

6.7 Match the substances on the left with the appropriate descriptions on the right.

a minerals   **i** always contain nitrogen

b carbohydrates   **ii** elements or inorganic substances

c vitamins   **iii** organic substances absorbed undigested

d proteins   **iv** contain carbon, hydrogen and oxygen only

6.8 Match the structures on the left with the appropriate descriptions on the right.

a stomach   **i** B_{12} is absorbed here

b terminal ileum   **ii** pepsinogen secreted here

c colon   **iii** contains enzymes secreted by parotid glands

d mouth   **iv** the principal site of water absorption

6.9 Which *one* of the following *best* describes the effect of consuming a diet rich in marine fish oil?

a Reduces serum level of low-density lipoprotein

b Increases serum level of low-density lipoprotein

c Increases serum level of high-density lipoprotein

d Reduces serum level of high-density lipoprotein

6.10 Which *one* of the following *best* describes the effect of consuming a diet high in polyunsaturated fat and low in saturated fat?

a Reduces serum level of low-density lipoprotein

b Increases the serum level of low-density lipoprotein

c Reduces the serum level of high-density lipoprotein

d Increases the serum level of high-density lipoprotein

Answers to self-test questions

6.1 a i
     b iii
     c iv
     d ii

6.2 a ii
     b iv
     c i
     d iii

6.3 a ii
     b iv
     c iii
     d i

6.4 a ii
    b i
    c iv
    d iii

6.5 a ii
    b iv
    c i
    d iii

6.6 a ii
    b iv
    c iii
    d i

6.7 a ii
    b iv
    c iii
    d i

6.8 a ii
    b i
    c iv
    d iii

6.9 c

6.10 a

Further study/exercises

6.1 When you next sit down for a main meal, consider the foods which you are about to eat. Identify the sources of carbohydrates, fats, proteins, vitamins, minerals and roughage. Later reminder yourself of the processes involved in the digestion of carbohydrates, fats and proteins.

6.2 What methods could be used to determine whether an individual is an appropriate weight? How would you decide if someone were overweight or obese? (*Hint:* we have already considered body mass index, but now explore other techniques such as anthropometric measurement.)

Goodison S. M. (1987). Assessment of nutritional status. *Professional Nurse*, **2**(11), 367–369

Goodison S. M. (1987). Anthropometric assessment of nutritional status. *Professional Nurse*, **2**(2), 388–393

Goodison S. M. (1987). Biochemical assessment of nutritional status. *Professional Nurse*, **3**(1), 8–9, 11–12

Goodison S. M. (1987). Assessing nutritional status: subjective methods. *Professional Nurse*, **3**(2), 48–51

7 Energy in the body

Learning outcomes

After reading the following chapter and undertaking personal study you should be able to:

1 Define the term *energy*.
2 List various forms of energy and distinguish between kinetic and potential energy.
3 Describe the principle of the conservation of energy and identify examples of energy conversions.
4 Identify the units used to measure energy and use kilojoules (kJ) to compare the energy value of different foods and the energy expended in different activities.
5 Describe the ATP (adenosine triphosphate) molecule and identify its importance in the body.
6 Outline the energy producing reactions of the body (cellular/internal respiration) including the following:
   a glycolysis
   b the conversion of pyruvic acid to acetyl Co A
   c the Krebs cycle
   d the electron transport chain
7 Outline fat and protein catabolism and briefly describe some of the potential consequences of using fat as a principal source of energy.

Introduction

This chapter begins with a general consideration of **energy** before moving on to look at energy in the body and especially at the energy-liberating reactions of cells. Energy is an important concept in many of the subsequent chapters of the book and consequently its introduction here is significant.

The concept of energy

In general, when introducing new ideas, it is usual to begin with a definition of the topic. In the case of energy it is probably the definition which is the trickiest part of the study! Nonetheless, we use the word 'energy' all the time. For example, if we are trying to control our weight we might ask about the energy value of a particular food, or when tired we might describe ourselves as having no energy.

> *So if the concept of energy is so familiar why is it difficult to define? Have you ever wondered what energy is?*

See what we mean about tricky? The concept is familiar and yet intangible. We seem to know what people mean when they use the word, but a convincing definition is elusive.

> *Why is this?*

Perhaps one of the reasons is that it is difficult to conjure a definition of energy which takes into account all its different forms. We have to apply a definition to such diverse forms of energy as light, heat, sound, a moving object and the energy 'locked up' in a chemical such as glucose. Quite a challenge!

> *So just how do scientists define energy?*

Defining energy

Scientists define energy as *the ability to do work*. After what we have said previously about the difficulties of definition, the above seems extraordinarily simple. Perhaps it even appears unsatisfactory. After all, such a definition does not tell us anything about the nature of any specific form of energy such as light or heat. But this is to be expected when dealing with such a broad concept. Perhaps an example will illustrate how our definition holds true. Take the case of someone pulling a sledge uphill. Energy is being expended in muscular contraction and work is being done – the sledge is heavy and it feels like hard work! This is using everyday language of course. To the scientist the word work has a specific meaning – it is the product of **force** and distance moved. So we can apply our definitions of energy and work to the case above.

> *But what about heat and light? How does our definition apply to them? How do they do work?*

The validity of the application of our definition of energy to heat and light becomes clear when we think of these forms of energy being converted into other forms. The steam engine is a good example of a device which uses heat to produce movement through the expansion of steam. And what about the solar panel? Here light is converted into electrical energy which may be used to perform all kinds of work – power a water pump perhaps. So then, the scientist's definition of energy appears disappointing at first, but it can be applied to energy in all its forms.

Forms of energy

We have already noted the fact that energy exists in a number of forms.

Table 7.1 **Different forms of energy.**

1 Mechanical	5 Chemical
2 Heat	6 Electrical
3 Electromagnetic (for example, light)	7 Nuclear
4 Sound	

Which forms have we noted so far? Can you add others to your list?

Table 7.1 is a list of the different forms of energy.

Some forms, such as light, heat and sound are very familiar to us because we experience them in everyday life. There are separate chapters on each of these later in this book and indeed on other forms, such as electricity, too. In fact, since most chapters of the book refer to energy to a lesser or greater degree you could say that much of this book is actually about energy! Some forms of energy are less tangible than others and perhaps we don't think about them very often. After all, when did you last have a conversation about **infrared** radiation? Still other forms are only given attention when our perspective is changed by some event. For example, when out shopping we are certainly concerned about the freshness and price of food. However, if we have to lose weight energy value becomes important too. Similarly, the prospect of a holiday in a hot country may prompt us to think about **ultraviolet** radiation, sunburn and skin cancer! Thus energy is definitely important in our bodies.

Try making a note about the importance to the body of each of the forms which we have listed.

Now compare your notes with Table 7.2.

Kinetic and potential energy

It is sometimes helpful to describe energy as either **kinetic energy** or **potential energy**. *Kinetic* means 'movement', so kinetic energy refers to the energy of movement. Do you remember that in Chapter 3 (Water electrolytes and body fluids) we referred to randomly moving smoke particles as having kinetic energy?

Table 7.2 **Different forms of energy and the body.**

1 Mechanical: muscle contraction when walking; the pumping of the heart
2 Heat: temperature homeostasis
3 Electromagnetic: light and vision; ultraviolet and tanning, sunburn and skin cancer
4 Sound: hearing; ultrasound scans
5 Chemical: the energy value of food
6 Electrical: nervous impulses
7 Nuclear: nuclear scans

> *Now think about what happens to the pressure of a gas in a sealed container as it is heated.*

Have you worked out that as the container is heated the pressure rises?

> *Why is this?*

It is because the kinetic energy of the gas molecules has increased through heating, and as a consequence they experience more frequent and forceful collisions. If you are not sure about this you might need to review Chapter 5 (Pressure, fluids, gases and breathing). For now it is sufficient to realise that in heating the container energy is transferred to the gas molecules, and this results in an increase in their kinetic energy.

Other forms of energy are described as potential. Remember the example of the sledge being pulled uphill? We could describe the stationary sledge at the top of the hill as being in a higher energy state – that is, it has greater potential energy than it did at the bottom of the hill. In order to demonstrate this simply jump on, lift your feet and be taken down the slope with no effort at all. You could say that you have recouped some of the energy expended in dragging the sledge to the top in the first place. It is a good example of a conversion from potential to kinetic energy.

The energy 'locked in' chemicals is another example of the potential kind. For example, when plants make **carbohydrates** (such as starch in potatoes) from carbon dioxide and water, they use energy from the Sun in the form of light. If we eat these plant products our bodies are able to break the bonds between atoms of the carbohydrate molecules in order to release the energy for our own use. In other processes there is a very long time between the manufacturing process by the plant and human use. For example, when we burn fossil fuels, such as coal or peat, we are in reality making use of the remains of plants which died thousands of years ago!

The conservation of energy

We are not referring to a 'green' issue here but to an important principle which states that, in ordinary chemical reactions, *energy cannot be created or destroyed*, although it may be changed from one form to another. Consequently, if we talk about 'using' energy or the energy-'producing' reactions of the body we are not strictly correct. It is rather that one form of energy is being converted into another. The expression 'ordinary chemical reactions' is used here in order to exclude certain nuclear reactions in which energy and matter are inter-converted.

Energy is converted from one form to another in our bodies and in devices made by us for this very purpose.

> *Can you think of some examples?*

Now check your list with Table 7.3.

Note that, in some cases, a number of energy conversions actually take place. For example, in a gas-burning power station the chemical potential energy of the gas molecules is converted to heat, which is used to heat water, and the steam produced is used to drive a steam turbine which produces electricity. In a similar way. the energy produced from the breakdown of food in our bodies is actually used to regenerate a molecule called ATP (**adenosine triphosphate**), which is the source of energy for activities such as muscle contraction. We should also note

Table 7.3 **Some examples of energy conversions.**

In the body
Chemical → heat: all cells especially liver cells
Light → electrical: the eyes
Sound → electrical: the ears
Chemical → mechanical: muscles

Artificial devices
Mechanical → electrical: generator
Electrical → mechanical: electric motor
Electrical → heat: electric heater
Chemical → electrical: battery
Electrical → light: light bulb

that energy-converting reactions are generally inefficient. For example, the light bulb produces a good deal of heat as well as light, while electric heaters commonly produce light too. The heat produced by the light bulb and the light produced by the heater might be regarded as a kind of 'waste'. Similar observations can also be made of the body. If you are not sure about this, simply perform a few minutes of vigorous exercise. What do you feel like? Hot and sweaty and possibly exhausted! Intensive muscular activity does generate an excess of heat, and we have to lose this to the environment if we are to avoid 'overheating' – more about this in the following chapter. For now it is enough to note that we too produce 'waste' energy.

Now let us further apply the principle of the conservation of energy to our bodies. In view of what we have learned so far, energy taken into our bodies in terms of the food we consume cannot disappear. It must be expended in some form of work or otherwise stored (mostly as fat). Consequently, the following equation must be true:

energy intake = energy expenditure ± changes in body energy stores

From the above you will see that if energy intake exceeds energy expenditure there will be an increase in energy stores and that the reverse must also be true. We might even begin to talk about the body's energy balance in a similar way to that of water balance, and this is clearly important when we consider weight control. Furthermore, if we are to consider the body's energy balance we must have some way of measuring energy. This is what we consider next.

Units of energy

When we are concerned about the energy value of particular foods and ask 'how many Calories does it contain?' we clearly expect a numerical answer. We obviously assume that energy can be measured.

What is a Calorie, and are there other units of energy?

The Calorie (spelt with a capital letter and abbreviated Cal) is mentioned here because, even if you have not studied science before, you will probably have heard of it. The calorie (without a capital letter and abbreviated cal) is the amount of energy required to raise the temperature of 1 g of water by one degree Celsius (1 °C). It is a rather small unit of energy, so the unit of the kilocalorie (kcal − a thousand calories) is more convenient. The energy value of food is sometimes given in nutritional Calories, which are actually the same as kilocalories. Apart from having the potential to confuse, calories are a rather old-fashioned unit and we shall use the SI (Système Internationale) unit of energy, the joule (J).

Do you remember how we defined energy?

We said that it is the ability to do work, and we defined work as the product of force and distance. We could express this in the form of a formula:

work = force × distance

Distance is no problem − we know that the SI unit of distance is the metre.

What about force − how is it measured?

The SI unit of force is the newton − it is named after Sir Isaac Newton. We shall consider it further in Chapter 9 (Force, mechanics and biomechanics). For now it is enough to know that force has units of newtons and distance has units of metres. If we now look back at our equation we should be able to work out that the SI unit of work is therefore the newton metre. It sounds a little cumbersome, so you will be glad to know that is has been renamed the joule, after James Joule. In order to have some idea as to the magnitude of the joule it is perhaps worth knowing that it takes 4183 J to raise the temperature of 1 kg of water by 1 °C. Once again, one joule is therefore a rather small unit of energy and it is more convenient to use the larger unit of one thousand joules − the kilojoule (kJ).

How many kilojoules would it take to raise the temperature of 1 kg of water by 1 °C?

It would take 4.183 kJ. If you need to use both calories and joules you should find it accurate enough to consider that 1 cal = 4.2 J.

But do we use joules in clinical practice?

As you might expect, the answer is that we do − that is the reason for explaining this unit here. Have a look at Table 7.4. You will see that the energy value of some foods is given in kilojoules per hundred grams (kJ/100 g).

Table 7.4 The energy value of selected foods. (Taken from Bender D. A. (1997). *Introduction to Nutrition and Metabolism, 2nd edn. London: Taylor & Francis, pp. 314–323.)*

Food	Energy value (kJ/100 g)	Food	Energy value (kJ/100 g)
Cornflakes	1523	Boiled potatoes	356
Porridge	197	Chips	1100
Chocolate biscuits	2239	Rice	544
Digestive biscuits	2026	Spaghetti	511
Brown bread	984	Butter	3097
White bread	1030	Margarine	3051
Chapattis	1461	Cheddar cheese	1716
Flaky pastry	2394	Cottage cheese	276
Milk chocolate	2252	Stewing beef	712
Apples	150	Beef sausages	1272
Bananas	360	Lamb chops	1569
Grapefruit	92	Liver	632
Oranges	167	Chicken	607
Brussels sprouts	75	Turkey	444
Cabbage	33	Cod	310
Carrots	92	Trout	569
Cauliflower	59	Herring	1000

Practice point 7.1: the energy value of food

Have a look at Table 7.4 again.

Which foods have the lowest energy value?

Fruit and vegetables have the lowest energy values, but it does depend upon how they are cooked. Compare boiled potatoes and chips, for example.

Which foods have the highest energy value?

High energy value foods include chocolate and pastry. You probably knew this already.

Are there any energy value surprises?

Look at butter and margarine. Their energy values are similar. The important thing to note here is that the value to health of foods cannot simply be equated with their energy value. Margarine is said to be a healthy substitute for butter because it is made from unsaturated fats rather than saturated fats. Fats have already been discussed in Chapter 6 (Biological molecules and food).

What about meat and fish?

Table 7.5 Energy expended in kJ/min for various exercises. (Taken from Barker H. M. (1991). *Beck's Nutrition and Dietetics for Nurses*, 8th edn. Edinburgh: Churchill Livingstone.)

Activity level	Example	Energy expended (kJ/min)
Very light	Typing, driving	<10
Light	Bed making	10–20
Moderate	Cycling, gardening	20–30
Heavy	Digging, football	30–40
Very heavy	Boxing, using an axe	40–50

Looking at Table 7.4 once again, you can see that fish, chicken and turkey have the lowest energy values. Nonetheless, there are a few surprises here too. For example, stewing beef has a lower energy value than herring! Once again, do not confuse energy value with value to health. Fish oils have particular health benefits, while the consumption of high levels of animal fat pose a challenge to health.

Now have a look at Table 7.5, which shows energy consumption in kilojoules per minute (kJ/min) for various activities.

Are the figures much as you expected? Work out the length of time it takes in moderate exercise, say 20 kJ/min, to expend the energy contained in 100 g of flaky pastry compared to 100 g of grapefruit.

You should get approximate figures of two hours for the pastry and four and a half minutes for the grapefruit! Clearly exercise may play an important role in the control of weight, but it may take a good deal of exercise to 'burn off' the energy contained in higher energy foods.

It is also worthwhile pointing out that we have considered the energy value of foods with energy reduction in mind. This is because excess energy consumption in relation to activity is a problem in developed countries and is part of the cause of obesity. On the other hand, if you work in a developing country the opposite problem often exists, and you may need to consider ways of increasing energy intake.

The ATP molecule

ATP (adenosine triphosphate) is often described as the energy currency of the cell.

What is this term trying to express?

Clearly the body gains all its energy from the food we eat, but breaking down nutrient molecules to release energy is quite a complex biochemical process and

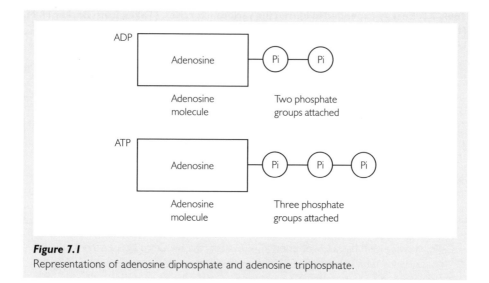

Figure 7.1
Representations of adenosine diphosphate and adenosine triphosphate.

the energy is not immediately available. What is required is a molecule which is able to temporarily store energy but which is then able to release that energy quickly in a one-step reaction. This is the role of ATP. ATP is an organic compound consisting of adenine, an organic base also found in DNA (deoxyriboucleic acid) and RNA (**ribonucleic acid**), and a pentose (five-carbon atom) sugar. ATP also contains three inorganic phosphate groups (Pi). ATP is synthesised from ADP (**adenosine diphosphate**) in all cells of the body. ATP and ADP are represented in Figure 7.1. Note that these are for the purpose of illustration – they do not represent the molecular structure of these compounds!

> *How many phosphate groups does ADP possess?*

It has two. The conversion of ADP to ATP therefore requires the addition of one phosphate group – a process called **phosphorylation**. This process requires energy (for example, from the breakdown of glucose), and it can be summarised in the following way:

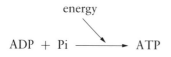

Incidentally, because the energy for the above reaction is derived from the **oxidation** of food the term *oxidative phosphorylation* is applied.

The third phosphate group is readily removed from ATP to recycle ADP and Pi. In this process energy is released again as shown below:

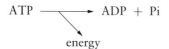

Of course, in neither of these processes is energy actually being created or destroyed. Remember that important principle? Rather, the energy released by the breakdown of glucose is used to synthesise ATP, and when ATP is itself broken down the energy released is used to perform some form of work, such as muscular contraction.

> *What are the processes by which glucose is broken down to provide the energy for the synthesis of ATP?*

This is what we consider next.

The energy-liberating reactions of cells

Before proceeding further it might be worthwhile clarifying some terms. Clearly there are a great many reactions taking place within the body, and the word **metabolism** is used in reference to them all. Some of these reactions are of the synthesis type and are described as *anabolic*, while others involve breaking down substances and are called *catabolic*. The catabolic cellular reactions which result in the liberation of energy are also referred to as internal or cellular respiration. In carbohydrate catabolism the starting point is glucose, which may be derived from carbohydrates in the diet or from the carbohydrate storage molecule **glycogen**. In the breakdown of glucose, each glucose molecule is completely oxidised to carbon dioxide and water, and in the process a maximum of 38 molecules of ATP are generated. We can summarise the reaction as shown in the equation below:

$$C_6H_{12}O_6 + 6O_2 \longrightarrow 6CO_2 + 6H_2O$$

$$38\ ADP \qquad 38\ ATP$$

We should note that the chemical potential energy of the products of this reaction is considerably less than that of the original glucose molecule. Clearly some energy has been lost along the way!

> *In what form has this energy been lost?*

From what has been said previously you have realised that this energy has been lost as heat. Keep this in mind when you study the next chapter.

In order to proceed further we need to remind ourselves of some of the points covered in Chapter 2 (The physical world and basic chemistry).

> *What is oxidation?*

No doubt you have noted that oxidation may involve a gain of oxygen, but did you also remember that the loss of hydrogen or an electron are also included?

> *Now what about reduction – how is it defined?*

Reduction may involve a loss of oxygen, the gain of hydrogen or of an electron. Remember that oxidation and reduction always occur together. In a reaction which involves the oxidation of one substance another substance will be reduced.

In the reactions which are described a little later, two molecules (**co-enzymes**) appear a number of times. They are nicotinamide adenine dinucleotide (NAD), a derivative of the vitamin niacin, and flavin adenine dinucleotide (FAD), a derivative of vitamin B_2. From now on we shall refer to them only by their abbreviations NAD and FAD – the full names are too long to fit easily into diagrams! Let us make life even simpler and not look at their chemical formulae – we do not need to for our purpose here. What we do need to know is that both these molecules exist in two forms.

Do you know what these forms are?

You may have guessed that the two forms are the oxidised form and the reduced form. The oxidised form of NAD is NAD^+ and it reacts with two hydrogen atoms to form NADH and H^+ (hydrogen ion/proton). This reaction is shown below:

$$NAD^+ \quad + \quad 2H \quad \longrightarrow \quad NADH \quad + \quad H^+$$

| oxidised form | 2 hydrogen atoms | reduced form | hydrogen ion |

Why do we say that NAD^+ has been reduced to NADH?

It is because the molecule has gained an hydrogen atom and an electron. Remember that both of these are forms of reduction. Because of the above we sometimes describe the role of NAD as carrying or transferring hydrogen atoms. FAD behaves in a similar way, as shown below:

$$FAD \quad + \quad 2H \quad \longrightarrow \quad FADH_2$$

| oxidised form | 2 hydrogen atoms | reduced form |

If both NAD^+ and FAD are reduced as part of the energy-liberating reactions of the cell, are other substances being oxidised?

Yes, and you will see what these substances are in a moment.

Overview of carbohydrate catabolism

There are four stages involved:
1 glycolysis
2 the conversion of pyruvic acid to acetyl Co A (acetyl co-enzyme A)
3 the Krebs cycle
4 the electron transport chain.

These four stages are represented in Figure 7.2.

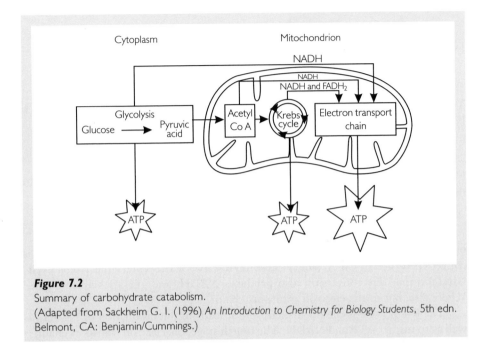

Figure 7.2
Summary of carbohydrate catabolism.
(Adapted from Sackheim G. I. (1996) *An Introduction to Chemistry for Biology Students*, 5th edn.
Belmont, CA: Benjamin/Cummings.)

Study Figure 7.2 for a moment. Note that glycolysis occurs in the **cytoplasm** of the cell while the Krebs cycle and the electron transport chain both take place in the **mitochondria**. The conversion of pyruvic acid to acetyl Co A is an intermediate stage which also takes place in the mitochondria but which does not involve the liberation of energy. Glycolysis is described as an **anaerobic** process.

> *What does this mean?*

It means that it does not require oxygen. In contrast, the electron transport chain does require oxygen.

> *What word is used to describe this?*

It is described as an **aerobic** process.

> *What about the Krebs cycle?*

The Krebs cycle itself does not require oxygen, but since it is closely linked to the electron transport chain the two are often dealt with collectively and described as aerobic cellular respiration. You need to bear this in mind when looking at different textbooks as it is a point at which students can become confused.

Let's have a brief look at all the stages in energy production before considering each one in more detail. You will need to refer to Figure 7.2 in order to work out what is happening. In glycolysis the starting point is glucose.

What is glucose converted to?

In glycolysis, glucose is converted to pyruvic acid.

Is there anything else produced by glycolysis?

Yes – ATP (remember that this is the body's chemical form of energy) and NADH. NADH transfers hydrogen elsewhere in the cell.

Where does NADH transfer hydrogen to?

Follow the arrow in Figure 7.2 and you will see that NADH transfers hydrogen to the electron transport chain of the mitochondria.

What happens to pyruvic acid?

In the second stage, pyruvic acid diffuses into the mitochondria, and here it is converted to acetyl Co A (**acetyl co-enzyme A**), which enters the Krebs cycle. Note too that this conversion also produces NADH, and guess where this goes? Where else, but to the electron transport chain?

The Krebs cycle is the third stage, and it leads to the production of $FADH_2$ as well as to further ATP and NADH. The fourth stage, the electron transport chain, is a series of reactions which lead to the production of many more molecules of ATP than does glycolysis or the Krebs cycle. Consequently, Figure 7.2 shows a larger 'explosion' symbol at this point!

Glycolysis

Figures 7.3 and 7.4 give more detail of glycolysis. In these diagrams the length of the carbon chain of each molecule is represented as a chain of black dots and a phosphate group as P enclosed in a circle. Note that this is not an attempt to draw the chemical structure of the compounds – it is simply for the purpose of illustration. Looking at Figures 7.3 and 7.4 you will see that glycolysis involves 10 steps. Each step is catalysed by a different enzyme, but for the sake of simplicity the enzymes are not named here. Note once again that the starting point is glucose.

How many carbon atoms does glucose have?

It has six. In step 1, a phosphate group is added to glucose (phosphorylation) to produce glucose 6-phosphate.

Where does this phosphate group come from?

It comes from ATP. Thus, energy is actually used in this initial step! In step 2, glucose 6-phosphate is changed to fructose 6-phosphate.

Do you remember that glucose and fructose are isomers?

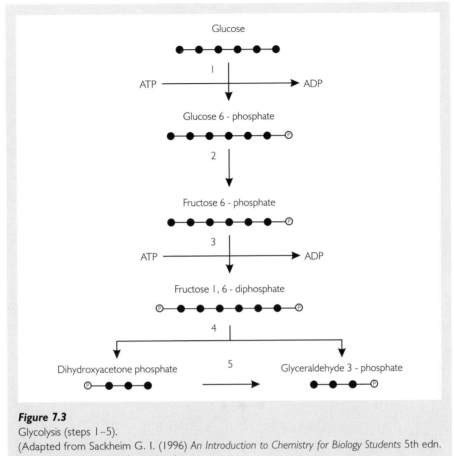

Figure 7.3
Glycolysis (steps 1–5).
(Adapted from Sackheim G. I. (1996) *An Introduction to Chemistry for Biology Students* 5th edn. Belmon, CA: Benjamin/Cummings.)

In step 3, a second phosphate molecule (again provided by ATP) is added to produce fructose 1,6-diphosphate. Note that the length of the carbon atom chain is still six. In step 4, fructose 1,6-diphosphate is broken-down to two smaller molecules – dihydroxyacetone phosphate and glyceraldehyde 3-phosphate (G3P).

> *How many carbon atoms does each of these molecules possess?*

Each has three. In step 5, the dihydroxyacetone molecule is converted to a second G3P molecule. This means that steps 6–10 are repeated, and this is important when we come to add up the products of glycolysis. In step 6, a second phosphate molecule is added and a molecule of NAD^+ is reduced to NADH.

> *So what has happened to G3P?*

It has been oxidised to 1,3-diphosphoglyceric acid. In step 7, a phosphate molecule is lost in the production of 3-phosphoglyceric acid.

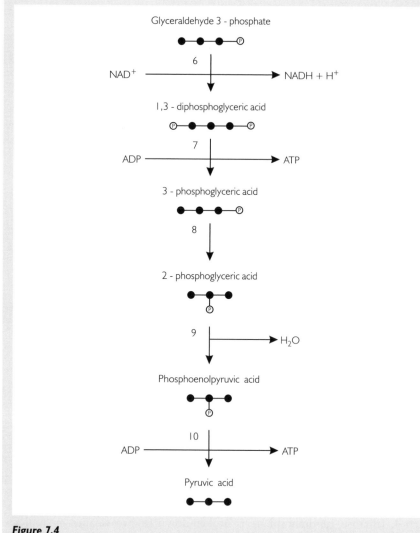

Figure 7.4
Glycolysis (steps 6–10).
(Adapted from Sackheim G. I. (1996) *An Introduction to Chemistry for Biology Students* 5th edn. Belmon, CA: Benjamin/Cummings.)

What is this phosphate molecule used for?

It is used to regenerate a molecule of ATP. Step 8 involves the transfer of a phosphate group from the third carbon atom of 3-phosphoglyceric acid to the second carbon atom. This results in 2-phosphoglyceric acid, which in step 9 loses a water molecule to produce phosphoenolpyruvic acid. Finally, in step 10, phosphoenolpyruvic acid loses its remaining phosphate group to recycle another ATP and produce pyruvic acid.

> *How many molecules of pyruvic acid are produced per glucose molecule?*

If you think two you are right – remember that steps 6–10 are repeated with the second G3P molecule.

> *What about ATP molecules – how many of these are produced in total?*

You should have noted that for each glucose molecule four ATP molecules are produced – one in step 7 and one in step 10, but since steps 6–10 are repeated that makes four molecules in all.

> *What is the net ATP gain?*

It is not four but two, since a molecule of ATP is required in both steps 1 and 3.

> *What about molecules of NADH – how many of these are produced?*

One is indeed produced at step 6, which of course occurs twice, so that makes two in all.

In summary then, for every molecule of glucose entering glycolysis, two molecules each of pyruvic acid, ATP and NADH are produced.

The conversion of pyruvic acid to acetyl Co A

Pyruvic acid produced by glycolysis diffuses into the mitochondria and, in the presence of oxygen, is converted in to acetyl Co A, as shown in Figure 7.5.

> *Remind yourself how many carbon atoms there are in pyruvic acid.*

There are three.

> *What about an acetyl group – how many carbon atoms does it possess?*

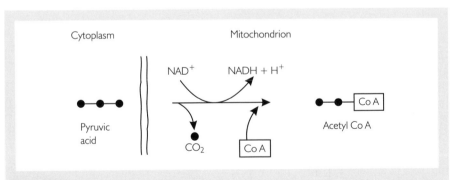

Figure 7.5
The conversion of pyruvic acid to acetyl Co A.
(Adapted from Sackheim G. I. (1996) An *Introduction to Chemistry for Biology Students* 5th edn. Belmon, CA: Benjamin/Cummings.)

It has two; therefore one carbon atom has been lost.

Where has the lost carbon atom gone?

Have a look again at Figure 7.5 and you will see that it can be accounted for by the production of one molecule of carbon dioxide (CO_2). The acetyl group from pyruvic acid is joined by a Co A molecule (derived from the B group vitamin pantothenic acid) to form acetyl Co A, and in this process one molecule of NAD^+ is reduced to NADH. Acetyl Co A then enters the Krebs cycle.

The anaerobic lactic system

Although what we have said so far is quite true, the actual fate of pyruvic acid depends upon the availability of oxygen at the time it is produced. If oxygen is not available then aerobic cellular respiration (the Krebs cycle and the electron transport chain) cannot proceed and the reduced coenzyme ($NADH + H^+$) is unable to deliver its hydrogen atoms. Actually the hydrogen atoms are 'off-loaded', but not to the electron transport chain. They are instead added to pyruvic acid, which, as a consequence, is reduced to lactic acid. This is illustrated in Figure 7.6.

In what conditions might oxygen be unavailable in sufficient amounts?

One example is in muscle cells during vigorous exercise.

And what are the effects?

You may already have worked out that, in the dissociation of lactic acid, hydrogen ions (H^+) are liberated and the intracellular pH falls, with the consequences of fatigue and a burning or aching pain. You might also be interested to know that one component of the training effect (taking vigorous exercise regularly) is an improvement in blood supply so that the individual becomes able to endure longer periods of

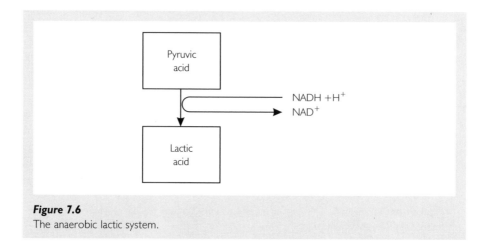

Figure 7.6
The anaerobic lactic system.

vigorous exercise. Eventually, when oxygen becomes available once again, lactic acid is converted back to pyruvic acid, which then enters the aerobic pathway. Let us continue with this by looking at the Krebs cycle.

The Krebs cycle

The Krebs cycle is shown in Figure 7.7. It looks quite complicated doesn't it? Don't worry though; we shall not look at the detail of every stage. Upon entering the Krebs cycle, Co A is lost from the acetyl group and recycled. The acetyl group then combines with oxaloacetic acid to produce citric acid.

How many carbon atoms does citric acid possess?

It has six – two from the acetyl group and four from oxaloacetic acid. Citric acid enters a series of reactions at the end of which the oxaloacetic acid is released and becomes free to pick up another acetyl group from acetyl Co A and re-enter the cycle. In fact, that is the whole reason the process is called a cycle – the Krebs

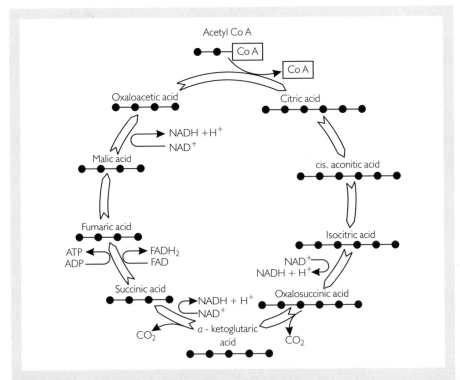

Figure 7.7
The Krebs cycle.
(Adapted from Sackheim G. I. (1996) *An Introduction to Chemistry for Biology Students* 5th edn. Belmon, CA: Benjamin/Cummings.)

cycle, after Hans Krebs; the citric acid cycle, after citric acid of course; or the tricar-boxylic acid cycle, since the acids involved have three carboxyl (COOH) groups.

Let us consider some of the important points of the cycle. Note that while citric acid has six carbon atoms, a-ketoglutaric acid has only five and succinic acid has four. Once again these lost carbon atoms can be accounted for by the production of CO_2. Note also that some important molecules have been produced – ATP, NADH + H^+ and $FADH_2$.

How many molecules of each of the above have been produced?

You should have noted that for each molecule of acetyl Co A which enters the Krebs cycle one molecule of ATP, three molecules of NADH ($+H^+$) and one of $FADH_2$ have been produced. But remember that for each molecule of glucose which enters glycolysis two acetyl Co A molecules are produced, so we have to double the Krebs cycle figures!

The electron transport chain

The electron transport chain is a series of reactions which occur in the inner mito-chondrial membrane and which lead to the production of ATP. It is illustrated in Figure 7.8. We do not intend to explain the details of the chain, so Figure 7.8 is identified as a diagrammatic representation of the principle involved. It does not accurately illustrate a specific stage of the chain.

Why is it referred to as a chain?

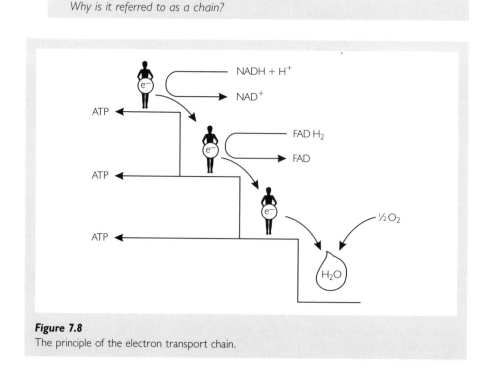

Figure 7.8
The principle of the electron transport chain.

A series of reactions is involved in which electrons are passed from one carrier to another. These electron carriers (some of which are called cytochromes) thus form a chain in a similar way to a line of people passing objects to one another. Indeed, in Figure 7.8, the electron carriers are represented by figures. Each electron carrier has a greater affinity for electrons than the previous one and less affinity than the next one. Consequently, electrons are passed along the chain from the first to the last and the carriers are alternately reduced (by gaining electrons) and oxidised (by losing electrons). Furthermore, the energy possessed by the carriers decreases from the beginning to the end of the chain. This is represented by the figures standing on successively lower steps.

> *What happens to the energy liberated by this process?*

It is used to create further ATP.

> *What is the source of the electrons?*

It is the hydrogens supplied by NADH + H^+ and by $FADH_2$, which are oxidised to NAD^+ and FAD respectively. Note that $FADH_2$ delivers its electrons at a lower energy level than does NADH + H^+, so it generates fewer ATP molecules – two instead of three.

> *And what happens to the electrons at the end of the chain?*

Another good question. They are picked up by half an oxygen molecule ($\frac{1}{2}O_2$) which thus becomes negatively charged and which in turn picks up two hydrogen ions (H^+) to form water. This is the only step in cellular respiration where O_2 is actually required.

That is almost everything as far as the catabolism of glucose goes. All we need to do now is add up the net ATP gain.

> *Do this now, but do not forget that, in glycolysis, one molecule of glucose produces two molecules of pyruvic acid and that in the electron transport chain NADH + H^+ produces three molecules of ATP while $FADH_2$ produces only two.*

Now have a look at Table 7.6. Did you get the right answer? Perhaps we should also point out that this is the theoretical maximum number of ATP molecules which can be produced from one glucose molecule. This maximum number is not always produced but we avoid additional detail here.

Fat catabolism

Thus far we have concentrated upon carbohydrate catabolism and the release of energy from the oxidation of glucose. However, although brain cells use glucose

Table 7.6 The net ATP production following the aerobic oxidation of glucose.

Process	Products per glucose unit	Net production of ATP after electron transport chain
Glycolysis	2 ATP	2 ATP
	2 NADH + 2 H$^+$	6 ATP
Formation of acetyl Co A	2 NADH + 2 H$^+$	6 ATP
Krebs cycle	2 ATP	2 ATP
	6 NADH + 6 H$^+$	18 ATP
	2 FADH$_2$	4 ATP
Total		38 ATP

exclusively, most cells are able to metabolise other molecules, including fats. The first step is the hydrolysis of fat in the fat cells themselves.

> *What are the products of the hydrolysis of fats?*

Fatty acids and **glycerol**. Many body cells are able to add a phosphate molecule to glycerol to produce glyceraldehyde 3-phosphate (G3P).

> *Where have we heard of G3P before?*

Have a look at glycolysis once again and you will soon find G3P. So glycerol can enter glycolysis via G3P. Since G3P is equivalent to half a glucose molecule, 19 ATP molecules are produced from glycerol.

> *What about fatty acids?*

The liver is able to metabolise them in a process called **beta oxidation** (β-oxidation) – the result is acetyl Co A. You should have no difficulty in remembering that acetyl Co A enters the Krebs cycle in the mitochondria.

> *How much energy is available from the metabolism of fat?*

The answer to this question is that it depends upon the length of the fatty acid chain. All have more carbon atoms than glucose and consequently yield more acetyl groups than does glucose. In fact, this may create something of a problem. Remember that after acetyl Co A enters the mitochondria Co A is recycled and the acetyl group combines with oxaloacetic acid to produce citric acid – the first step in the Krebs cycle. If there are insufficient oxaloacetic acid molecules to pick up the acetyl groups and form citric acid acetyl Co A accumulates and the liver converts acetyl Co A molecules to **ketones**.

> *Do you remember what ketones are?*

They are molecules which contain hydrocarbon groups (CH) attached to a carbonyl group (CO). One example, acetone, is shown below:

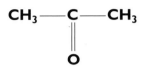

Other examples also possess a carboxyl group (COOH) and are referred to as **ketoacids**. One example is acetoacetic acid. Ketones may be metabolised in muscle cells to produce energy, but an accumulation in excessive quantities (ketosis) is detrimental.

> *Can you work out what some of the effects of ketosis are?*

Have a look at the following practice point.

Practice point 7.2: ketosis

When the availability of carbohydrates as a source of energy is reduced the body begins to draw on its stores of fat. The process of mobilising fatty acids and glycerol as sources of energy involves a number of hormones, including glucagon, from the *a* cells of the Islets of Langerhans in the pancreas; adrenaline (epinephrine) from the adrenal medulla; glucocorticoids, from the adrenal cortex; and growth hormone from the anterior pituitary. You can find out more about these hormones from a physiology text.

> *What circumstances might lead to fat being used as the principal source of energy?*

We have already noted that it is when there is a reduced availability of carbohydrates, and this occurs during prolonged fasting or starvation and also in diabetes mellitus. In the most familiar form of diabetes mellitus (type I), the hormone insulin is deficient, and since it is required, in most cells, for the transport of glucose across cell membranes, there is a cellular deficiency of carbohydrate.

> *What are the effects of ketosis?*

Obviously whole books and journals are written about diabetes mellitus, so we're just skimming the surface here. Firstly, remember that since ketones such as acetoacetic acid liberate hydrogen ions and lower blood pH (see Chapter 4 (Acids, bases and pH balance)). Consequently, the term **ketoacidosis** is also used. A falling pH is the principal stimulation of respiration and leads to a characteristic deep and fast respiratory pattern called Kussmaul's respirations. Secondly, acetone is volatile and has the characteristic smell of pear drops which can be detected on the breath. Thirdly, ketones are eliminated in the urine and can be detected using a simple reagent strip. Finally, high levels of ketones in the blood may eventually lead to coma. Clearly, understanding ketosis has important practical implications. For details of nursing care you will need to consult a nursing text.

Protein catabolism

When other sources of energy are unavailable, for example, in starvation, proteins may be catabolised and used to provide energy. In the liver the amino group (NH_2) is removed from amino acids in a process called **deamination**. This NH_2 group is then converted to ammonia (NH_3) and finally to urea, which is excreted in the urine. The fate of the remaining part of the molecule depends upon which amino acid is being catabolised, but it is converted to one of the substances in the Krebs cycle. For example, following deamination, the amino acid alanine is converted to pyruvic acid, while leucine is converted to acetyl Co A. Further detail is not given here, but we are sure you get the idea! The important thing to note is that the body's primary source of energy is glucose, but fats and proteins may both be catabolised in certain circumstances.

Basal metabolic rate

Metabolic rate is the rate at which energy is expended by the body. It is measured in kilojoules per hour (kJ/h). Metabolic rate is affected by a range of different conditions.

> *Can you think of some of them?*

Your list might include the sleep–wake cycle, exercise, hormones, the feeding–fasting state, stress and illness. Clearly if we are to compare the metabolic rates of different individuals recordings must be made under identical conditions. In fact, basal conditions are chosen – hence the term **basal metabolic rate** (BMR). *Basal* refers to the state in which the individual is fasted, rested and inactive but awake. It is not the lowest metabolic rate that the body experiences, since that occurs during sleep. We do not describe the measurement of metabolic rate here, but it is important to understand the concept of metabolic rate because it is referred to in clinical practice. For example, BMR is influenced by the hormone thyroxine, which is secreted by the thyroid gland in the neck. Consequently, abnormal variations in BMR may reflect under- or over-secretion of thyroxine. You might like to find out more about this hormone in a physiology text. You could follow this up by considering nursing care in thyroid dysfunction.

Summary

I Energy is defined as the ability to do work.

2 Energy exists in a number of forms.

3 It can not be created or destroyed, but it can be converted from one form to another.

4 The body derives its energy from the oxidation of nutrient molecules.

5 The energy value of food is measured in joules (J).

6 The body's chemical form of energy is adenosine triphosphate (ATP).

7 There are four main stages of glucose catabolism:

  a glycolysis

  b the conversion of pyruvic acid to acetyl Co A

  c the Krebs cycle

  d the electron transport chain

8 The breakdown of glucose to liberate energy may be summarised in the following way:

$$C_6H_{12}O_6 + 6O_2 \longrightarrow 6CO_2 + 6H_2O$$

38 ADP   38 ATP

9 Fats and proteins may, at times, be catabolised to liberate energy.

Self-test questions

7.1 Which of the following are examples of kinetic energy?

  a A glucose molecule

  b Flowing blood

  c A compressed spring

  d Oxygen molecules diffusing in to blood

7.2 Which of the following statements about energy are true?

  a It is impossible to change one form of energy into another.

  b In ordinary chemical reactions, energy cannot be destroyed.

  c One form of energy may be transformed into another.

  d In ordinary chemical reactions, energy cannot be created.

7.3 Which of the following statements are true?

  a The calorie is the amount of energy required to raise the temperature of 1 kg of water by 1 °C.

  b 1 Calorie = 1000 calories.

  c 1 calorie = 4.2 joules.

  d 1 Calorie = 42 joules.

7.4 Which of the following statements are true?

  a Pyruvic acid is the end product of glycolysis.

  b The Krebs cycle occurs in the cytoplasm.

  c Glycolysis occurs in the cytoplasm.

  d The maximum number of ATP molecules generated from one molecule of glucose is 38.

7.5 Match the substance on the left with the appropriate description on the right.

a Pyruvic acid

b $FADH_2$
c Oxygen
d Oxaloacetic acid

i The final recipient of electrons in the electron transport chain
ii Converted to acetyl Co A
iii Recycled in the Krebs cycle
iv Transports hydrogen to the electron transport chain

7.6 Which of the following statements are true?
    a The electron transport chain occurs in the cytoplasm.
    b Two molecules of ATP are needed for glycolysis to proceed.
    c Glycolysis may proceed in the absence of oxygen.
    d One molecule of ATP is generated in the conversion of pyruvic acid to acetyl Co A.

7.7 Match the substance on the left with the appropriate description on the right.
    a Acetyl Co A   **i** Part of the electron transport chain
    b Citric acid   **ii** Enters the Krebs cycle
    c NAD   **iii** Has six carbon atoms
    d Cytochrome   **iv** A derivative of niacin

7.8 Match the substance on the left with the number of carbon atoms on the right.
    a Pyruvic acid   **i** 2
    b Glucose   **ii** 3
    c Oxaloacetic acid   **iii** 4
    d An acetyl group   **iv** 6

7.9 Match the reaction on the left with the appropriate description on the right.
    a ADP $\longrightarrow$ ATP   **i** Hydrolysis
    b $NAD^+ \longrightarrow NADH + H^+$ **ii** Oxidation
    c $FADH_2 \longrightarrow FAD$ **iii** Reduction
    d Fats $\longrightarrow$ Fatty Acids + Glycerol   **iv** Phosphorylation

10 Regarding the reaction shown below, which one of the following substances has been oxidised?

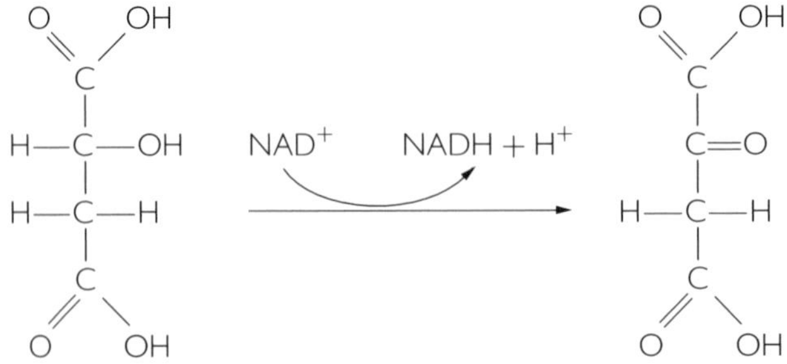

Malic acid   Oxaloacetic acid

    a Malic acid
    b NAD^+
    c $NADH + H^+$
    d Oxaloacetic acid

Answers to self-test questions

7.1 b and d	**7.8** a ii
7.2 b, c and d	b iv
7.3 b and c	c iii
7.4 a, c and d	d i
7.5 a ii	**7.9** a iv
b iv	b iii
c i	c ii
d iii	d i
7.6 b and c	**7.10** a
7.7 a ii	
b iii	
c iv	
d i	

Further study/exercises

7.1 We have looked at the energy liberating reactions of the body, but how do the energy requirements of individuals vary? To answer this question, compare the energy requirements of the following groups:

a A baby up to 6 months

b A child of 5 years

c Male and female adolescents

d A pregnant woman

e A lactating woman

d An elderly person

Barker H. M. (1991). *Beck's Nutrition and Dietetics for Nurses*, 8th edn. Edinburgh: Churchill Livingstone, Chapters 13, 14, 16

Garrow J. S. and James W. P. T., eds. (1993). *Human Nutrition and Dietetics*. Edinburgh: Churchill Livingstone, Chapters 25–27

MAFF (1995). *Manual of Nutrition*, 10th edn. London: The Stationery Office, Chapters 5, 13

Truswell A. S. (1992). *ABC of Nutrition*, 2nd edn. London: British Medical Association, Chapters 4–7

7.2 Clearly the energy requirements of individuals depend not only upon their age and lifestyle, but also on the presence of illness. How do the following influence the energy requirements of the body?

a Elective surgery

b Major trauma

c Extensive burns

Barker H. M. (1991). *Beck's Nutrition and Dietetics for Nurses*, 8th edn. Edinburgh: Churchill Livingstone, Chapter 21

Dudek S. G. (1993). *Nutrition Handbook for Nursing Practice*, 2nd edn. Philadelphia: J. B. Lippincott Company, Chapter 14

8 Heat and body temperature

Learning outcomes

1 Distinguish between heat and temperature.

2 Define the terms *latent heat of fusion* and *latent heat of vaporisation*.

3 Outline the Celsius and Kelvin scales for the measurement of temperature.

4 Describe various temperature measurement devices, including the mercury-in-glass thermometer, the thermistor and the infrared tympanic membrane thermometer.

5 Distinguish between core and peripheral body temperature.

6 Discuss some of the factors which influence body temperature in the healthy.

7 Describe the procedure for taking body temperature with a clinical thermometer, a thermistor and an infrared tympanic membrane thermometer and discuss the effects of site and duration of measurement upon the value obtained.

8 Describe four physical processes by which heat is transferred and relate these to the body.

9 Describe the mechanisms of temperature homeostasis.

10 Discuss the causes, effects and management of pyrexia, heat exhaustion, heat stroke and hypothermia.

Introduction

Heat and temperature are important topics of study for health professionals. Indeed, one of the first clinical skills acquired by student nurses is the measurement of body temperature. Until recently this was an uncomplicated task, there being only one device, the clinical (mercury-in-glass) thermometer, available for the purpose. However, today you will also need to be familiar with thermistors and infrared tympanic membrane thermometers too. Of course, many patient problems remain the same as ever – the elderly with hypothermia (low body temperature) or the feverish child with an infection, for example. On the other hand, situations such as the infant in whom profound hypothermia has been induced purposefully during a heart operation have only become more common in recent years.

The cases noted above illustrate the fact that a greater understanding of the science of heat and of temperature and its measurement is currently demanded of health care professionals than in the past. Consequently, the contents of this chapter may prove very useful. Let us begin our study of these topics by asking a question.

What is the difference between heat and temperature?

Heat and temperature

If you have read the previous chapter you will already know that heat is a form of energy. Consequently, it might surprise you to learn that, until the end of the 18th century, heat was thought of as a material substance. Although we now know that this is not the case, we still speak of heat being transferred from one object to another. It is fine to talk this way so long as we remember that what is being transferred, heat, is a form of energy. Temperature is not the same as heat; instead it may be regarded as the degree of hotness.

But don't two objects with the same temperature have the same amount of heat energy?

If you think about this for a moment you will quickly realise that the answer to this question is no. For example, imagine two containers of water at room temperature – let us say 20 °C. Suppose one contains 1 kg and the other 2 kg. Now imagine that the same amount of energy is delivered to them both – 4.2 kJ perhaps.

Will the temperature of the water in both containers rise by the same degree?

Intuitively we know that the answer is no – the smaller volume will be hotter. We might also remember that 4.2 kJ is sufficient energy to raise the temperature of 1 kg of water by 1 °C and so conclude that it would take twice as much energy to raise 2 kg of water to the same degree. Actually we have not here taken into account the amount of energy required to heat the container and the energy losses while heating takes place. However, the point of this discussion is not to work out the actual temperature rise but simply to show that heat and temperature are not the same thing at all. This difference is further illustrated when we consider the concept of **latent heat**.

Latent heat and changes in state

Let us think again about the experiment involving the heating of ice which we introduced at the beginning of Chapter 3 (Water, electrolytes and body fluids). In this experiment we noted that, as the ice is heated, the temperature initially rises but remains stable for a time as the ice melts. Once melted the temperature rises again but becomes stable as the water boils. These changes are illustrated by the graph in Figure 8.1.

What is happening between points b and c?

The ice is melting and there is no rise in temperature, despite the fact that heating continues.

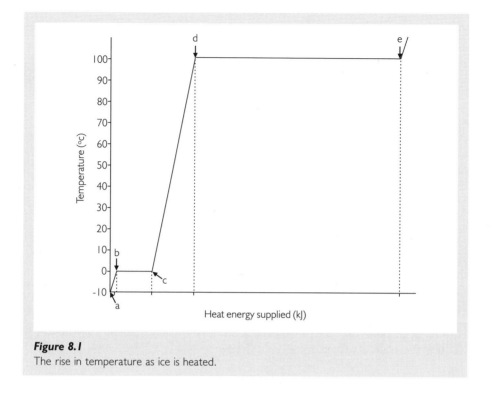

Figure 8.1
The rise in temperature as ice is heated.

What is happening between points d and e?

The water is vaporising (boiling), and once again there is no temperature rise despite the fact that heating continues.

So how much energy does it take to get from a to b, compared with b to c, c to d and d to e?

Suppose we start with 1 kg of ice at −10 °C. It takes only 42 kJ of energy to raise the temperature of the ice from −10 °C to 0 °C (a to b). Surprisingly, it then takes a further 330 kJ to melt the ice (b to c). To raise the temperature of 1 kg of water from 0 °C to 100 °C then requires 420 kJ (c to d), but very much more (2260 kJ) to vaporise the water.

You may recall that the heat supplied during the melting of ice is used to enable water molecules to partially escape the strong attraction which they have for each other in solid state. The heat required for this process of melting is referred to as the *latent heat of fusion*. Here the term 'latent' may be interpreted as 'hidden', since the heat cannot be detected with a thermometer. Similarly, the heat supplied during boiling enables water molecules to escape the weak attraction they have for each other in the liquid state and so enter the gaseous state. The heat required to vaporise the water is, not surprisingly, referred to as the *latent heat of vaporisation*.

Can we make use of latent heat?

Practice point 8.1: latent heat and ice packs

One example of the use we make of latent heat of fusion is the application of ice packs to the body for the purpose of cooling.

What advantage does ice have over water for cooling?

It isn't just a matter of temperature. Remember those figures quoted earlier? It takes relatively little energy to raise the temperature of 1 kg of ice by 10 °C (42 kJ) compared with the latent heat of fusion (330 kJ). When ice packs are applied to the body there is initially no change in the pack temperature as the ice melts, but there is a considerable cooling of tissues to which the ice is applied.

What use might this be put to?

One application is the reduction of swelling and pain following injury, of joints for example, through the mechanism of *vasoconstriction* (constriction of blood vessels). Our bodies also make use of the latent heat of vaporisation, but we shall consider this later under sweating.

Temperature scales

The measurement of energy, including heat, in joules has already been described, so this chapter will focus upon temperature. One temperature scale has already been mentioned, and it is probably the one which is most familiar to you – the Celsius scale. Fortunately, it is also the scale used in clinical practice. It is named after Anders Celsius who, in the 18th century, proposed a scale based upon one hundred divisions between the melting point of ice and the boiling point of water. Consequently, ice melts at 0 °C and water boils at 100 °C.

What is normal body temperature in degrees Celsius?

You should have little trouble remembering that it is 37 °C. Of course there are other scales too – indeed, although we have stressed the need to use SI (Système International) units the degree Celsius is not the SI unit of temperature, which is the Kelvin (K *not* °K). The Kelvin scale is so named because it was proposed by William Thomson who subsequently became Lord Kelvin. In this scale each division is equal to a division on the Celsius scale; to convert from Celsius to Kelvin simply add 273.

What are the melting and boiling points of water in Kelvin?

They are 273 K and 373 K respectively.

How did this scale come about?

We shall not say too much about this but it relates to the predicted temperature at which particles of matter have the lowest kinetic energy. This is $-273\,°C$, which is taken to be the zero point on the Kelvin scale – sometimes called absolute zero.

Finally, you may have heard of the Fahrenheit scale, but it is now rarely used and certainly not in clinical practice. Its use should be avoided, but for the sake of completeness we note that ice melts at $32\,°F$ and water boils at $212\,°F$. One degree Fahrenheit is not therefore equal to one degree Celsius. The following shows how the two units are converted:

$$°C = °F - 32 \times 5/9$$

$$°F = °C \times 9/5 + 32$$

Temperature measurement devices

If we wanted to devise a means of measuring changes in temperature we might think of making use of a substance some physical property of which itself changes with temperature. A good choice would be mercury, which expands uniformly as temperature rises, is a good conductor of heat (see later) and is opaque and therefore visible. The mercury-in-glass thermometer makes use of these properties of mercury and it is such an effective device that it still finds use today – 350 years after it was invented!

The mercury-in-glass thermometer

This type of thermometer is illustrated in Figure 8.2. In this device the mercury is located in a bulb, but may expand up a fine tube which runs the length of the thermometer. In clinical thermometers, which are used to measure body temperature, this tube is very fine indeed. It has to be since the degree of expansion of mercury

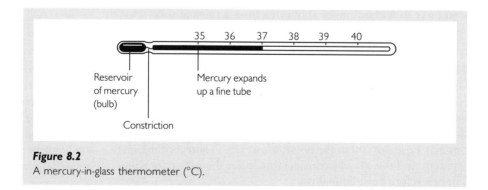

Figure 8.2
A mercury-in-glass thermometer (°C).

over the range of variation in body temperature is slight. The fineness of the tube makes the mercury column within it difficult to see, and students initially complain that they find clinical thermometers difficult to read. The trick is to rotate the thermometer between the fingers and thumb while holding it horizontally at eye level. At a certain point the fine column will be magnified by the glass of the thermometer and a reading can be taken from the scale.

> While I am doing this, will the mercury not contract again? Do I have to rush to get a reliable reading?

You are quite right to ask these questions, but clinical thermometers have a special feature which ordinary mercury-in-glass thermometers do not possess – a constriction just above the bulb. This does not impede the mercury as it expands but it does prevent it from receding once the thermometer has been taken from the body. It is therefore an example of a maximum reading thermometer.

> So how do I get the mercury column down again?

It has to be shaken down – perhaps you have seen nurses doing this. Some care needs to be taken, since glass thermometers are not difficult to break and mercury is toxic. *In the event of a mercury spillage you should follow the relevant policy of the clinical area in which you are working. Special kits are available for collecting spilt mercury.*

The thermistor

This is an example of an energy-converting device – thermistors convert heat into electrical energy and give a digital read-out of temperature. They consist of a heat-sensitive probe which is connected to a recording device. The probe is flexible and blunt and can be inserted into a body orifice. It may be passed into the rectum via the anus or into the nasopharynx or oesophagus via the nostril. It has to be said that they are not entirely comfortable! Consequently, they are generally used in unconscious and more seriously ill patients. You will certainly find them in critical care units. While it is not difficult to pass a probe into the rectum, it does require some skill to pass one into the oesophagus, especially if the patient is unconscious. The technique is essentially the same as for passing a nasogastric tube, and you might like to find out more about this from a nursing text.

Liquid crystal thermometers

The term *liquid crystal* sounds like a contradiction in terms, but actually refers to liquids which possess some of the properties of certain crystals. One of these is colour change associated with temperature variation. When plastic strips are impregnated with liquid crystals they can be used to indicate temperature by placing the strip on the forehead. They have the advantage of requiring no skill in use, and this may be important in the home. However, they record the *surface* skin temperature only and this may be considerably lower than the *core* temperature (see later). They therefore have limited use.

Infrared tympanic membrane thermometers

These devices do not measure temperature directly, but indirectly from the intensity of the infrared radiation emitted from the tympanic membrane (ear drum). Infrared radiation consists of energy in the form of rays which form part of the electromagnetic spectrum. Perhaps you are already familiar with infrared, but if not it may help you to know that visible light is also part of this spectrum. We shall consider it in more detail in Chapter 11 (Light and vision). An example of an infrared tympanic membrane thermometer (such as the Braun Thermoscan) is illustrated in Figure 8.3. Such devices are quick, simple to operate and accurate. Consequently, they are gaining use in a variety of care settings and in homes. We shall describe the procedure for taking temperatures in a practice point later, but first let us consider body temperature in a little more detail.

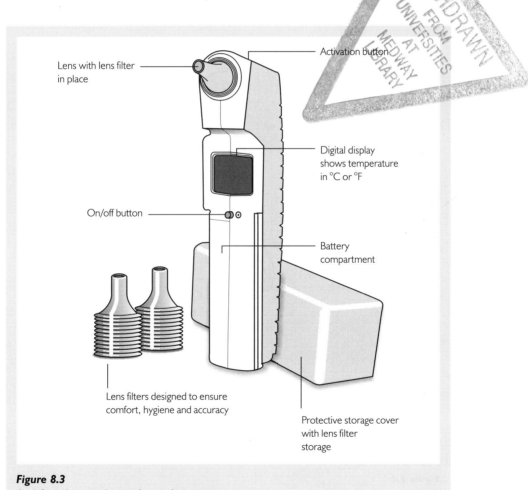

Lens with lens filter in place

Activation button

Digital display shows temperature in °C or °F

On/off button

Battery compartment

Lens filters designed to ensure comfort, hygiene and accuracy

Protective storage cover with lens filter storage

Figure 8.3
An infrared tympanic membrane thermometer.

Body temperature

The temperature of the body at any one time depends upon the amount of heat generated by cellular metabolism and heat exchanged with the environment. At first sight the term *body temperature* appears self-explanatory, but if you study Figure 8.4 for a moment you will see that the value recorded varies according to the site that a measurement is taken. The implication is that we have to clarify what we mean by the term.

Core temperature is that of the organs of the body cavities – the brain in the cranium, the heart and lungs in the thorax and the liver in the abdomen. In normal health the resting core temperature varies from 37 °C by less than one degree Celsius and it is this value which is most useful in clinical practice. Core temperature does vary with activity level and it may rise to 40 °C during vigorous exercise. Table 8.1 shows the normal range of body temperature for various activity

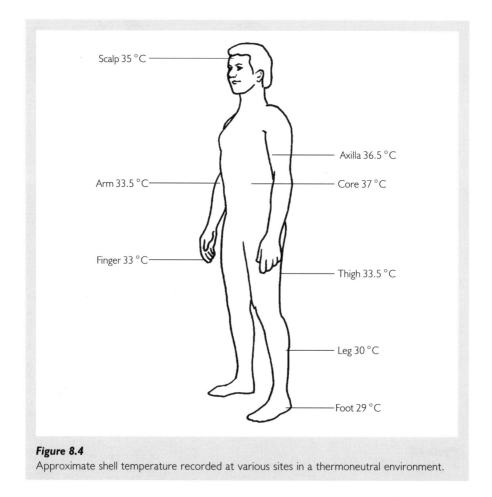

Figure 8.4
Approximate shell temperature recorded at various sites in a thermoneutral environment.

Table 8.1 Approximate core temperature at different activity levels.

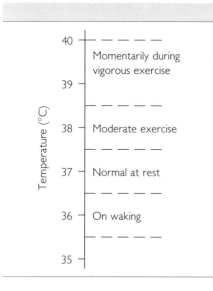

Temperature (°C)

40 — Momentarily during vigorous exercise

39 —

38 — Moderate exercise

37 — Normal at rest

36 — On waking

35 —

levels. *Peripheral* body temperature, sometimes called *shell* temperature, is the temperature of the surface of the body and it has a lower value than core temperature. At certain sites, for example the feet, peripheral temperature is considerably below the core value.

Actually the progressive fall in temperature as blood flows through the limbs, illustrated in Figure 8.4, indicates that an important heat-conserving mechanism is in operation. As blood flows away from the torso in the arteries of the limbs, heat is lost to the blood flowing in the opposite direction in adjacent and parallel veins. This means that blood returning to the torso is re-warmed, as is shown in Figure 8.5. It is an example of a counter-current

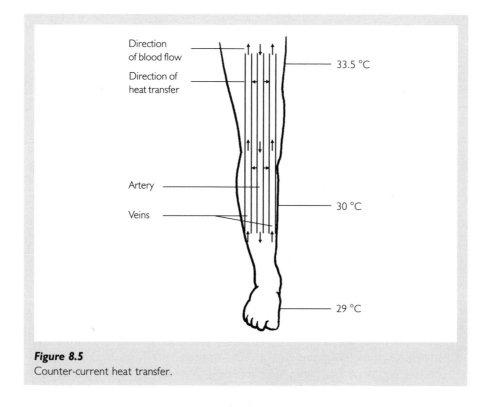

Direction of blood flow

Direction of heat transfer

Artery

Veins

33.5 °C

30 °C

29 °C

Figure 8.5
Counter-current heat transfer.

mechanism which results in relatively stable temperatures at each point along both vessels. Were this mechanism not in operation blood would cool as it passed through the arterial system and not be re-warmed as it returned in the venous system. This would tend towards a reduction in core temperature.

Practice point 8.2: taking a temperature

The ability to measure body temperature with a clinical (mercury-in-glass) thermometer remains an important skill even when 'high-tech' devices are available. Arrive on duty to find that no one has placed the tympanic membrane thermometers on charge and you will know exactly what I mean! In some care situations such equipment may simply be unavailable.

How do I ensure that an accurate measurement of core temperature is obtained?

The answer is to chose an appropriate site and leave the thermometer in place for long enough. The oral cavity is an obvious choice, but the bulb has to be placed adjacent to the sub-lingual artery at the base of the tongue. This ensures that the temperature of blood which has travelled only a short distance from the heart is recorded. Even so, the value obtained will be a little below that of the actual core temperature since the sublingual pocket is not completely isolated from environmental influences. The difference may be about 0.4 °C.

Are there any problems with the use of the oral cavity?

Yes, but they are not difficult to perceive. The patient has to be conscious and co-operative. Do not use the oral cavity in infants or young children; many clients with learning disability are also unsuitable. In addition, no foods or fluids should have been consumed or cigarettes smoked in the 15 minutes prior to a measurement being taken since these significantly affect accuracy.

Cleanliness is also important if cross-infection is to be avoided, and thermometers should be cleansed with a 70% isopropyl alcohol swab before use. In addition, a fresh disposable plastic sheath may be used for each patient.

An alternative to the oral cavity is the axilla (armpit), but the depth of this cavity in the elderly and emaciated may mean that it is difficult to keep the thermometer in place adjacent to the axillary artery. Even when this is achieved the value of the measurement may be 1 °C lower than the value recorded in the oral cavity. Clearly it is important to record the site of measurement as well as the value itself. In addition, the axilla is unreliable when hypothermia or shock are present. Unfortunately, in these cases, the oral site may be also be contra-indicated since the patient may not be fully conscious. Instead, a rectal measurement may be taken with a clinical thermometer reserved for this purpose. It is inserted up to 4 cm into the rectum and held in place while a recording is made. The isolation of the rectum from the external environment means that a rectal measurement is usually higher than one taken orally, although this does not necessarily mean that it is more accurate.

The rectum is some distance from the central circulation and changes in core temperature may not be immediately reflected in rectal temperature. In addition, insertion of the thermometer into soft stool will separate the thermometer from the rectal wall and thus give an unreliable result.

You will also have realised that it is important to avoid exposing hypothermic patients to further cooling when taking rectal temperature measurements and the use of a rectal probe, which may be left in place, may be more desirable in this case. An even better solution is to use an infrared tympanic membrane thermometer, as accurate results will be obtained without disturbance to the patient. Rectal measurements used to be taken in infants, but this is no longer recommended due to the risk of rectal ulceration or even perforation. Once again the tympanic membrane thermometer is the best alternative

The second consideration as far as accuracy goes is the duration of measurement. Mercury-in-glass thermometers require a period of time for the temperature of the mercury to equilibrate with that of the body. In the case of a thermometer placed within the oral cavity three minutes should be allowed, but longer may actually be required in some instances, especially in the case of an axillary measurement.

What about the infrared tympanic membrane thermometer?

It has a clear advantage here. The tip of the detector head is inserted a little way into the external auditory canal and an instant digital read-out is obtained. One important precaution which is taken is the use of a fresh disposable sheath over the tip of the detector head with each patient so as to avoid any risk of transmission of ear infection.

Cyclical variations in body temperature

Body temperature is not constant throughout the day but varies in a rhythmical fashion. Such body rhythms are referred to as circadian rhythms and those which occur on a daily basis are referred to as diurnal. The highest body temperature is recorded between 16:00 and 20:00 and the lowest value between 02:00 and 06:00. The difference is in the range 0.5–1.0 °C and should be borne in mind when interpreting a temperature record.

Practice point 8.3: body temperature and natural birth control

In addition to the diurnal variation of body temperature noted above, that of the female also varies with the menstrual cycle. Daily measurements of body temperature may be used to indicate the onset of the absolutely infertile period – that is the part of the menstrual cycle when intercourse will not lead to conception.

The ovary secretes the hormone progesterone following ovulation (the release of the oocyte/egg cell). This hormone that is responsible for the peak in body temperature that remains until one or two days before the next period. The absolutely infertile phase is considered to have begun once three high temperatures have been recorded. Once progesterone levels begin to fall the temperature also drops, but it is worth noting that the variation of body temperature attributable to progesterone is generally less than one half of one degree Celsius. Consequently, as accurate as possible a measurement should be made. It is a good idea to use the same thermometer for each measurement and to take the temperature at the same time of day – ideally the first thing in the morning. The measurement of body temperature in an attempt to determine when intercourse will not lead to conception is one aspect of a natural means of birth control. Other observations are also made by women who practice natural birth control, and you might like to find out more about what is involved and the reliability of this method.

The transmission of heat

We have already noted that heat may be transmitted from one object to another and we are so familiar with this that we think little about it. After all, when we are cold it is the most natural thing in the world to stand by a heater!

What are the physical processes involved in the transmission of heat?

Conduction

Conduction is the process whereby heat is transmitted from one object to another in contact with it. If we go to pick up a spoon which has been resting for some time in a hot drink we should not be surprised to discover that it too is hot.

What has actually occurred to make the spoon hot?

Remember that we described how the kinetic energy of water molecules increase as water is heated? The high kinetic energy water molecules of the hot drink collide with, and so lose energy to, the atoms of the metal spoon, and as a consequence the temperature of the spoon rises. When we try to pick up the spoon some of this energy is also transmitted to us and our fingers also become hot. The important thing to note with regard to this form of heat transmission is contact between objects.

Does all matter conduct heat equally well?

You clearly know that it does not. Indeed, we make decisions about conductivity all the time.

For example, what do we really mean when we say that we need a warm coat for winter?

Well for a start we do not mean that the coat is warm! The coat will be warm or cold depending upon where it is at the time. What we mean is that it keeps us warm by not allowing heat to escape readily from our bodies. To put it another way, we could say that it is a poor conductor or a good insulator. In everyday language we often make the mistake of referring to temperature when we are actually commenting upon a material's conductivity. For example, we may use the term 'cold steel', but actually we are indicating that metals are good conductors of heat. Consequently, even when metal is at room temperature, if we touch it with bare skin heat is readily conducted away from our body and we detect this as a peripheral cooling. Similarly, if we walk barefoot on a tiled floor we might describe it as cold, when in reality it is simply a better conductor of heat than carpet. The one occasion that we do use the proper terminology is when we talk of insulating the loft. Here we acknowledge that we are using a poor conductor – we do not usually say that we are warming the loft!

Convection

In our definition of conduction we spoke of heat being transferred from one object to another. Here the word *object* stands for any form of matter – living organisms, solids, liquids or gases. In fact, heat is conducted from hot objects to a cooler atmosphere around them.

> Now work out what happens to the density of the air surrounding a hot object.

Perhaps you have figured out that as air is heated the kinetic energy of the gas molecules increases and they move further apart. Consequently, the air becomes less dense and rises. Its place around the hot object is taken by denser, cooler air, which then becomes heated, and the process, called convection, is repeated. In this way, a hot object placed in a room in which the air is still begins to produce currents of air – we call them convection currents. Indeed, the expression is used to describe heaters – convector heaters or convectors. The important thing to note is that convection is a further physical process by which hot objects lose heat.

Radiation

You will no doubt have realised that one thing which conduction and convection have in common is that heat is lost from matter to matter. Now think about the warming of the Earth by the Sun.

> Is matter involved in the transmission of heat in this case?

Clearly not, since the Earth is separated from the Sun by the vacuum of space. Another process must be involved here, and it is this which we refer to as radiation. Once again the language used here should be familiar – we do after all refer to some heating devices as radiators. The word radiation actually refers to something diverging from a point. In this case that something is energy in the form of rays – infrared rays in fact. Incidentally, dark objects absorb infrared radiation more readily than light ones, and silvered surfaces reflect infrared just as they do light. Consequently, you may have noticed that black cars become very hot in summer, but this is less of a problem in white ones. You will also have no doubt worked out that there is a good reason why Mediterranean houses are often painted white.

Evaporation

Heat is lost by wet objects as water changes from the liquid to the gaseous state – remember the latent heat of vaporisation? This occurs at much lower temperatures than the boiling point, since all water molecules do not have the same kinetic energy and some higher energy molecules will escape the attraction of others at temperatures well below 100 °C. We call this process evaporation and we are quite familiar with it in everyday life. It is of course the reason that puddles dry after a shower of rain. We should also note that while all the four processes which we have described depend upon a temperature gradient (heat is lost from hot objects to cooler ones) evaporation also depends upon a vapour pressure gradient. This means that, for evaporation to be effective, the atmosphere must be dry. If the air around a wet object is saturated with water vapour, so that it is impossible for it to take up any more water, evaporation will not occur. You may have experienced this if you have visited a tropical country and found that wet clothing takes a long time to dry.

Is there anything which promotes the evaporation of water?

You might have figured out that an air current does, since air with a higher water vapour pressure (that is, containing more water vapour) around the wet object is replaced by drier air, which can then take up more water by evaporation. This is, of course, the reason that washing dries quicker on a windy day. We shall note the importance of this to the body later.

Preventing heat losses: the vacuum flask

If we understand the mechanisms by which heat is transmitted we can also work out how heat losses can be reduced. Take the example of the vacuum flask illustrated in Figure 8.6. Heat exchange between liquids placed in the flask and the environment is kept to a minimum by a number of features. Firstly, the inner part of the flask consists of a double-walled glass bottle, the two walls of which are separated by a vacuum. This prevents heat losses by conduction and convection. In addition, the silvered surface of the glass bottle is reflective, and this reduces heat losses by radiation.

What about heat exchanges and the body?

Heat exchanges and the body

The exchange of heat between the body and the environment occurs by the same four physical processes which we described earlier and which affect inanimate

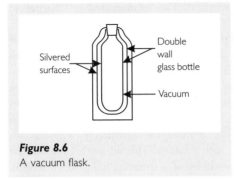

Figure 8.6
A vacuum flask.

objects. They are illustrated in Figure 8.7. Ordinarily some 60% of the heat loss of the body occurs by the process of radiation, although it should be borne in mind that the body may also gain heat by this process. An example is the warming produced by sunbathing. It is of course possible that this will eventually lead to hyperthermia (elevated body temperature), but do not confuse this with

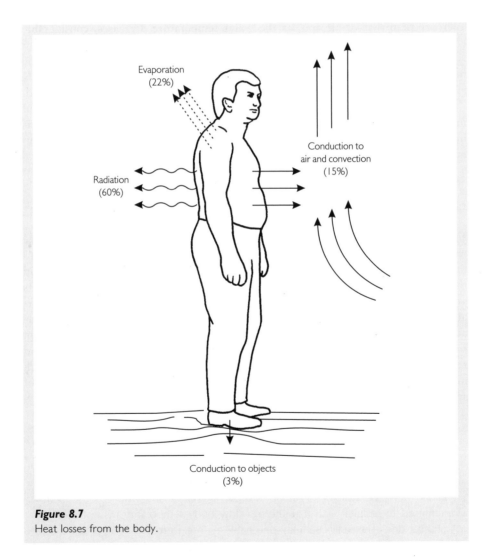

Figure 8.7
Heat losses from the body.

sunburn which is caused by ultraviolet radiation. This is discussed in Chapter 11 (Light and vision).

Much less body heat is lost by conduction and convection, since, for most of our lives, we are fairly well insulated, either by our clothing or the bedclothes when asleep. When these are removed it is usually in circumstances where the environmental temperature is quite high; a sunny day, or after the bathroom heater has been on for a little while! Once again it is possible for the body to gain heat by conduction and a hot bath is a good example. There are some obvious practical applications here. The temperature of any clinical environment in which patients have to undress should take the increased heat loss into account. This invariably means that it is too hot for the staff, and uniforms should be of a suitable, lightweight design. In addition, patients who require assistance to get out of baths should not be unattended for significant periods of time. Note that water has a much greater conductivity than that of air and, as the water temperature decreases, considerable cooling may result.

Taken together, conduction and convection account for about 18% of the body's heat losses, but this figure can be increased when air is moved rapidly across the surface of the body, such as occurs in draughts and when fans are used.

Heat is lost from the body by evaporation at a basal level affecting the lungs and mucous membranes of the mouth and nose. However, during exercise, when sweating occurs, the latent heat of vaporisation accounts for a considerable increase in heat loss by this process.

Can you think of other, perhaps less obvious, examples of increased heat loss through evaporation?

You may have worked out that preventing water loss from the body is one of the functions of the skin; so what about people who have lost a good deal of the integument? You should realise that this is exactly what happens in extensive burns and this knowledge will help you to predict what some of the problems of such patients are. In addition, surgery in which body cavities are opened also results in considerable cooling and, unless hypothermia is to be induced as part of the procedure, operating theatres are usually kept quite warm. Indeed this, together with gowning-up, sometimes presents a problem for students, who have to get used to the heat and restrictive clothing. It is probably this, rather than the sight of blood, which makes some feel unwell at the operating table! On average, the evaporation of sweat accounts for about 22% of the heat lost from the body.

Practice point 8.4: heat loss in the fallen elderly

Let us now put to use some of our knowledge of the physical processes by which heat is lost from the body. The case history of an elderly person who falls in the home, is unable to get up without assistance and who is subsequently admitted to hospital with hypothermia (low body temperature) is not at all an unusual one. The individual may be unable to rise from the floor because of a cerebrovascular accident (stroke) or a fractured neck of femur (more common in elderly females). If any lengthy period is spent in inactivity metabolic rate will fall and along with it body temperature. Hypothermia is dealt with a little later, but for now let us concentrate on the processes by which heat is lost.

Heat loss by radiation will continue much as before, but losses by conduction may be considerably increased due to contact between the body and the floor. Carpeting will provide some degree of insulation, but heat loss through tiled floors or those covered simply with linoleum will be very great. In houses without double glazing there may be draughts and these are often worse at floor level because of gaps under doors. Losses through evaporation will be increased if the individual is wet; examples include falling while getting out of the bath and slipping on a floor which is already wet.

Temperature homeostasis

The body strives to maintain core temperature by keeping heat production and heat loss in balance. There are essentially three components to the mechanism by which this is achieved – detection, interpretation and response. Temperature detection is performed by structures of the nervous system called thermoreceptors, some of which are located in the skin and mucous membranes (peripheral), while others are found deep within the body (central). Impulses from thermoreceptors travel along nerves to the brain and a structure called the hypothalamus interprets them and initiates a response in various body systems. The hypothalamus strives to maintain a set-point of core temperature and for this reason it has sometimes been described as the body's thermostat.

Can you work out which body systems play a part in temperature homeostasis? How does the body respond to a disturbance in temperature?

After you have thought about the answer to the questions above, have a look at Figure 8.8.

The hypothalamus possess both a heat promoting and a heat losing centre. It is these areas which determine the response of the body to a disturbance in temperature. The body systems mainly affected are the skin, skeletal muscle, adrenal glands (situated above the kidneys) and the thyroid gland (situated in the neck).

Let us first of all consider a decrease in temperature. Nervous impulses to the skin induce cutaneous vasoconstriction (blood vessel constriction), which results in a decrease in blood flow to the periphery and, as a consequence, less heat is lost to the environment by conduction, convection and radiation. This is of course the reason why we look pale when cold. In addition, vasoconstrictor nerves are concentrated in the distal ends of limbs, so the greatest effect is in the hands and feet. Other responses are concerned with increasing heat production. For example, the involuntary contraction and relaxation of skeletal muscle,

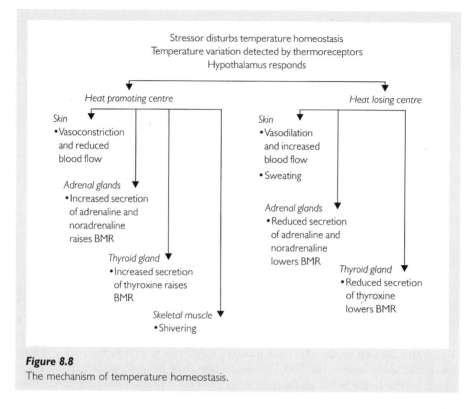

Stressor disturbs temperature homeostasis
Temperature variation detected by thermoreceptors
Hypothalamus responds

Heat promoting centre   *Heat losing centre*

Skin
• Vasoconstriction
  and reduced
  blood flow

Skin
• Vasodilation
  and increased
  blood flow

• Sweating

Adrenal glands
• Increased secretion
  of adrenaline and
  noradrenaline
  raises BMR

Adrenal glands
• Reduced secretion
  of adrenaline and
  noradrenaline
  lowers BMR

Thyroid gland
• Increased secretion
  of thyroxine raises
  BMR

Thyroid gland
• Reduced secretion
  of thyroxine
  lowers BMR

Skeletal muscle
• Shivering

Figure 8.8
The mechanism of temperature homeostasis.

which we call shivering, generates considerable additional heat. Furthermore, the effect of the increased secretion of adrenaline (epinephrine) and noradrenaline (norepinephrine) from the adrenal medulla and thyroxine from the thyroid gland is to raise metabolic rate.

In contrast, when temperature is increased the response of the body is essentially the opposite of that described above. Blood vessels in the skin dilate (cutaneous vasodilation) and heat losses by conduction, convection and radiation are increased. At the same time, the evaporation of sweat results in considerable heat loss through the latent heat of vaporisation, while a reduction in metabolic rate leads to decreased heat production.

One aspect of temperature regulation which we have not mentioned is that of a behavioural response. We change our behaviour with temperature fluctuations, since we find these uncomfortable. Perhaps this sounds like stating the obvious, but there is an important point here which is that it is not always possible for patients to respond appropriately to temperature fluctuations. We are sure that you can think of many examples. Patients may be immobile, unable to talk and express their needs or may even be unconscious. Of course infants require special consideration, as do those with learning disabilities and even some with mental health problems. A profoundly depressed patient may simply lack the motivation to respond to cold.

Thermoregulation abnormalities

Thermoregulation abnormalities involve either a raised body temperature (hyperthermia) or a reduced body temperature (hypothermia). In addition, the terms *pyrexia* and *fever* are also used when body temperature is elevated. However, they should not be considered to be synonymous with hyperthermia, as you will see below.

Conditions in which body temperature is raised.

The term *elevated body temperature* appears self-explanatory, but at what point do we regard a temperature rise to be abnormal? Commonly it is when there is a persistent elevation above 37.5 °C. The key word here is *persistent*, since, although the normal diurnal variation of body temperature is less than one degree Celsius, a healthy individual may have a recorded value of body temperature of between 36 °C and 40 °C depending upon activity level. Consequently, we would not describe a body temperature of 40 °C as abnormal if it occurred immediately after vigorous exercise. On the other hand, a temperature of 37.5 °C which persisted for several hours would be considered abnormal.

Are there problems associated with an elevated body temperature?

Firstly, the individual feels unwell, and secondly, metabolic rate rises by 10% for every one degree Celsius rise in body temperature. The consequent increase in heat production then contributes to a further rise in metabolic rate and so on. If body temperature exceeds 41 °C physiological regulatory mechanisms are overwhelmed and it will then continue to rise to the upper lethal limit of 44 °C.

Conditions in which body temperature is raised fall broadly into two groups – impaired heat dissipation and increased heat production. Impaired heat dissipation occurs when environmental temperature is high, especially if this is combined with high humidity. Possible consequences are heat exhaustion or the more serious heat stroke (see later).

Increased heat production may have one of a number of causes. It may be induced by drugs or it may be due to excessive secretion of the hormone thyroxine (thyrotoxicosis). Another cause is damage to the heat regulating mechanism itself, which may occur in head injury or cerebro-vascular accident (stroke). However, perhaps the most common cause of increased heat production is the release of pyrogens into the bloodstream. These may be infective organisms, toxins or substances released from damaged cells which have the effect of raising the set-point of temperature that the hypothalamus strives to maintain. Since infections are commonly accompanied by increased heat production, body temperature is often taken when infection is suspected or when it is a possible complication of an intervention such as surgery. However, since pyrogens may be released from damaged cells when no infection is present, temperature may be elevated in disease processes which are not infective. Myocardial infarction (heart attack) is a good example of this.

Table 8.2 **Methods employed to cool the body.**

Means by which heat losses due to conduction, convection and radiation may be increased	Means by which heat losses due to evaporation may be increased
Reducing environmental temperature Removing excessive clothing Creating gentle currents of air with an electric fan	Washing with tepid water (27–30 °C)

Hyperthermia

The use of this term should be confined to conditions in which temperature is elevated because of impaired heat dissipation or when there is an increased metabolic heat production. The affected individual commonly complains of feeling hot and will sweat and appear flushed (cutaneous vasodilation). Suitable interventions involve measures to increase heat loss from the body.

> *Can you work out what these are?*

Have a look at Table 8.2.

> *But if we want to increase heat loss, why do we not direct electric fans towards the patient, use cold water in washing and apply ice packs?*

The answer to this is that such aggressive measures induce cutaneous vasoconstriction and shivering and would be counter-productive.

Heat exhaustion

We have noted that the response of the body to a hot environment includes sweating. If sweating is excessive or prolonged (perhaps when the individual is working or playing sport) fluid and electrolyte losses can be considerable. This may lead to heat cramps. In addition, cutaneous vasodilation predisposes the individual to orthostatic hypotension (low blood pressure when standing) because blood becomes pooled in the lower extremities. This may result in fainting (heat syncope). Note that, in such an individual, the skin feels cool and wet because of the evaporation of sweat and this is important in preventing the temperature from becoming dangerously high. In fact, body temperature may be only a little above normal in this case. Treatment includes allowing the individual to rest in a cool place, gentle cooling measures and cool drinks. Tomato juice replaces salt as well as fluid and may be considered to have an advantage over fruit juices. In addition, sachets of electrolyte powders which are diluted in water are available for the purpose of rehydration and treatment of electrolyte depletion.

Heat stroke

This is a much more serious condition than heat exhaustion although it shares the same causes. In heat stroke the individual's ability to respond to a high environmental temperature is impaired and, as a consequence, core temperature rises dramatically. In this case the individual does not sweat and the skin is hot and dry. Predisposing factors include dehydration, lack of acclimatisation, obesity and excess intake of alcohol. It is also more common in the elderly. Cooling measures should be implemented and cool drinks given, but a rapid temperature rise to the upper lethal limit may occur before treatment can be instituted.

Pyrexia

This term is synonymous with fever and its use should be confined to temperature increases which result from the re-setting of the set-point of the hypothalamus by pyrogens. It is important to understand the mechanism of action of pyrogens so that appropriate nursing interventions can be made. When the set-point of temperature which the hypothalamus strives to maintain is raised, heat conserving and generating responses come into play.

> *What does the patient feel like at this stage?*

He feels cold and generally unwell, he will appear pale, due to cutaneous vasoconstriction, he will shiver and his metabolic rate will be increased. As a consequence his core temperature will rise until it reaches the new set-point.

> *Are there any benefits to developing a pyrexia?*

There may be some. The immune system may be more effective when temperature is raised and an elevated metabolic rate may be helpful in the healing process. You can find out more about this in a physiology text.

> *What can the nurse do to help the patient with pyrexia? Should measures be taken to cool him?*

The answer to this question is not quite as obvious at it may first seem. Initially, when the patient feels cold and unwell he will almost certainly seek a warm bed. Even though his core temperature is raised, attempts to cool him will be counter-productive since they will initiate further shivering and vasoconstriction. Antipyretic drugs may be given and these include aspirin, other non-steroidal anti-inflammatory drugs (NSAID), and paracetamol. Since 1986 the use of aspirin in children under twelve years has ceased because of its association with a serious condition called Reye's syndrome. If you are to be caring for children you might like to find out more about this. Needless to say, a paediatric paracetamol preparation should be used instead.

After a period of time the hypothalamic set-point returns to normal. This may be part of a natural recovery process, follow the administration of an antipyretic or, in the case of a bacterial infection, follow the use of antibiotics. At this point the core temperature will now be above the set-point and heat losing responses will be initiated. The patient will feel hot, will appear flushed (cutaneous vasodilation) and will perspire. Interventions aimed at promoting heat loss such as discussed before are now appropriate.

Malignant hyperpyrexia

This is the rapid and extreme rise in temperature associated with drugs such as the muscle relaxant suxamethonium chloride and the general anaesthetic Halothane. A full discussion of the cause and management of this condition is beyond the scope of this book, but if you have a clinical placement in operating theatres you might like to find out more about it.

Hypothermia

Hypothermia is the state of an abnormally low body temperature. It may be described as mild, moderate or severe, but different authors ascribe different values to these descriptions. Consequently, we identify clinical signs for different values of core temperature for adults in Table 8.3.

The principal cause of hypothermia is a low environmental temperature, so the condition is more common in winter months.

> *Who is most at risk of accidental hypothermia?*

You should have been able to work out that those at risk include individuals taking part in outdoor activities such as hill-walking, the elderly and infants. The elderly are at risk of hypothermia for physiological and social reasons. Physiological reasons include reduced adipose (fat) tissue and an impaired shivering response. The elderly also tend to be less active than the young. Social reasons include poor housing and low incomes. Infants are at risk of hypothermia because of a high surface area to volume ratio, which results in greater heat losses. They are also obviously reliant upon adults to provide suitable clothing and shelter.

Table 8.3 **Clinical signs for different core body temperatures.**

Core temperature (°C)	Clinical signs
36–35	Cutaneous vasoconstriction
	Shivering
	Increased blood pressure
34–33	Poor coordination
	Confusion
	Normal blood pressure
32–31	Consciousness is clouded
	Reduced shivering
	Low blood pressure
	Reduced heart and respiratory rates
30–29	Loss of consciousness
	Muscular rigidity
below 28	Risk of cardiac arrest

How should the patient with hypothermia be helped?

Clearly the interventions made in hypothermia are aimed at re-warming the patient. These may be divided into passive techniques (preventing further heat loss) and active techniques (involving the transfer of heat to the body).

Passive techniques include insulating the patient with blankets and increasing the environmental temperature. Ambient room temperature is taken to be 21 °C, so the temperature of the environment in which the hypothermic individual is treated should be maintained above this – up to 30 °C. Reflective blankets (foil or space blankets) should be avoided if a suitable environmental temperature can be maintained, since they will reflect infrared radiation from the patient. They do, however, have a use in preventing hypothermia in a low environmental temperature such as occurs when hill-walkers are stranded on a mountainside. In this case they function to reduce heat loss from the body by radiation. Passive techniques are suitable for use in all hypothermic patients and are the main treatment when the temperature is above 32 °C.

However, if the patient's temperature is below 32 °C, active re-warming techniques may be employed. These include the administration of warmed intravenous fluids and gastric lavage with warmed saline (passing warmed saline into the stomach via a nasogastric tube). More dramatic is the establishment of an extracorporeal circulation – that is, an artificial circulation is established outside the body: blood is re-warmed by passage through a heat exchanger and then pumped back to the patient.

Why not simply apply heat to the surface of the body?

This could indeed be done, but there are a number of problems. Firstly, surface heating may prevent shivering, since the muscles are warmed before the core. Secondly, if the skin is warmed first, blood flow to the extremities will increase and this may result in a fall in blood pressure. Such a fall is poorly tolerated in the elderly, who may have pre-existing heart disease. Thirdly, as peripheral circulation is increased cold blood is returned to the heart, and this may cause a drop in core temperature which may precipitate ventricular fibrillation (a form of cardiac arrest). This may be the explanation for sudden death which occurs after treatment has begun.

There is obviously much more that might be said on the subject and you might look at this topic in more detail.

Other uses of heat and cold

Sterilisation

Sterilisation refers to the elimination of all living organisms from an object, and heat, in one form or another, is often used to achieve this. It has a number of advantages, including controllability and the rapidity and certainty of its effects. In addition, unlike chemical methods, it leaves behind no potentially harmful substances.

Heat sterilisation may involve dry or moist heat. Examples of dry heat include the incineration of contaminated disposables, such as dressings; passing wire loops through the flame of a Bunsen burner flame in the microbiology laboratory; and the drying of glassware in hot air ovens.

Moist heat sterilisation includes pasteurisation – a process named after Louis Pasteur, who developed it. A temperature of 65 °C for 30 minutes or 72 °C for 15 seconds kills all bacterial pathogens which might normally be present in milk. Boiling an object will kill most organisms, although some spores may survive even prolonged treatment. Washing in boiling water is a suitable treatment for cutlery, crockery and linen, but it is inadequate for surgical instruments, dressings and theatre gowns. In these cases steam under pressure is used in a device called the autoclave.

Clearly heat treatment does have one disadvantage – the article being treated must be able to survive the temperature used.

If it cannot, what other methods could be used?

Perhaps you might like to find out more about alternative methods for yourself.

Induced hypothermia

In heart surgery, hypothermia is induced using topical and systemic techniques as a means of protecting the heart from a lack of oxygen – oxygen requirements are much lower in cold tissues. Only a small number of students who read this book will visit units in which such techniques are used, but you will realise that one aspect of the post-operative care of such individuals is the re-warming of the patient.

Summary

1 Temperature is not the same as heat. Heat is a form of energy, while temperature may be regarded as the degree of hotness.
2 The SI unit of temperature is the Kelvin, but degrees Celsius (°C) are used in clinical practice.
3 The amount of heat required to melt a solid is referred to as the latent heat of fusion and the amount of heat required to vaporise a liquid is referred to as the latent heat of vaporisation.
4 The two most important temperature measurement devices in clinical practice are the mercury-in-glass thermometer and the infrared tympanic membrane thermometer.
5 The temperature of organs within the body is referred to as the core temperature, while the temperature at the surface of the body is called the peripheral temperature.
6 Normal core temperature is taken to be 37 °C.
7 The most reliable site for recording body temperature with a mercury-in-glass thermometer is the sub-lingual pocket of the oral cavity.

8 The mercury-in-glass thermometer must be left in the sub-lingual pocket for at least three minutes, but the infrared tympanic membrane thermometer gives an immediate and accurate value of core temperature.

9 The transfer of heat occurs by conduction, convection, radiation and evaporation.

10 Temperature homeostasis involves the nervous, endocrine and cardiovascular systems and the skin.

11 Thermoregulation abnormalities include pyrexia, heat exhaustion, heat stroke and hypothermia.

Self-test questions

8.1 Match the value on the left with the appropriate description on the right.

a 330 kJ	**i** Amount of energy required to raise the temperature of 1 kg of ice from $-10\,°C$ to $0\,°C$
b 2260 kJ	**ii** Amount of energy required to raise the temperature of 1 kg of water from $0\,°C$ to $100\,°C$
c 42 kJ	**iii** Amount of energy required to vaporise 1 kg of water
d 420 kJ	**iv** Amount of energy required to melt 1 kg of ice

8.2 Which one of the following is considered to be the normal body temperature in degrees Celsius?

a 35

b 43

c 37

d 40

8.3 Which one of the following sites is considered to give the most accurate value of core body temperature?

a Sub-lingual pocket

b Axilla

c Rectum

d Tympanic membrane

8.4 What is the minimum period of time that a clinical thermometer should be placed in the sub-lingual pocket if an accurate reading is to be obtained?

a 1 minute

b 3 minutes

c 10 minutes

d 15 minutes

8.5 Which of the following is considered to be the normal diurnal variation of body temperature in degrees Celsius?

a Less than 0.5

b 0.5–1.0

c 1.0–2.0

d Greater than 2

8.6 By which one of the following processes is most heat lost from the body of an individual at rest at normal room temperature?

    a Conduction

    b Convection

    c Radiation

    d Evaporation

8.7 Which one of the following is considered to be the best definition of a raised body temperature?

    a A temperature of 39 °C or above

    b A persistent elevation of 37.5 °C or above

    c A temperature of 41 °C or above

    d A persistent elevation of 40 °C or above

8.8 You observe that an individual suspected of a pyrogenic infection looks pale, shivers, is cold to touch and has an oral temperature 39 °C. He complains of feeling cold and takes to his bed asking for another blanket. Which one or more of the following interventions would be appropriate?

    a Administer prescribed antipyretic drugs

    b Tell the patient that he has a fever and that he should not have an extra blanket

    c Give him an extra blanket

    d Turn on an electric fan

8.9 It is a hot day, and after a game of beach volley-ball a friend complains of feeling unwell. He appears flushed and is sweating profusely. His oral temperature is 38.5 °C. Which one or more of the following interventions would be appropriate?

    a Take your friend to a shaded place

    b Suggest that he recline and rest

    c Provide cool drinks and fan him gently

    d Suggest that he take some aspirin

8.10 An elderly person is admitted to hospital after neighbours find him sitting in a chair in a cold flat. He is conscious but his coordination is poor and he appears confused. His temperature taken by an infrared tympanic membrane thermometer is 33 °C. Which one or more of the following interventions would be appropriate?

    a Check his temperature frequently

    b Cover him with plenty of blankets

    c Try to raise his temperature with hot water bottles

    d Check his pulse and blood pressure

Answers to self-test questions

8.1	a iv	**8.5**	b
	b iii	**8.6**	c
	c i	**8.7**	b
	d ii	**8.8**	a and c
8.2	c	**8.9**	a, b and c
8.3	d	**8.10**	a, b and d
8.4	b		

Further study/exercises

8.1 Suppose that you are a practice nurse involved in visiting the housebound elderly. What practical advice could you give regarding keeping warm in the winter?

Otty C. and Roland M. O. (1987). Hypothermia in the elderly: scope for prevention. *British Medical Journal*, **295**(6595), 419–420

Thomas L. (1989). Insulating the elderly: what measures can be taken to prevent the onset of hypothermia. *Nursing the Elderly*, **1**(5), 8–10

Wright J. (1991). Accidental hypothermia. *Professional Nurse*, **6**(4), 197–199

Vyvyan M. Y. T. (1992). Making sense of hypothermia. *Nursing Times*, **88**(49), 38–40

8.2 Pyrexia in children may give rise to concern because of the possibility of febrile convulsions. What are febrile convulsions and how should the child with pyrexia be cared for?

McCance K. L. and Huether S. E. (1994). *Pathophysiology the Biological Basis for Disease in Adults and Children*. 2nd int. edn. St Louis: Mosby, p. 605

Whaley L. F. and Wong D. L. (1985). *Essentials of Pediatric Nursing*, 2nd edn. St Louis: Mosby, p. 846

Further reading

Mosse J. and Heaton J. (1990). *The Fertility and Contraception Book*. London: Faber & Faber, Chapter 4.

9 Force, mechanics and biomechanics

Learning outcomes

After reading the following chapter and undertaking personal study you should be able to:

1 Distinguish between scalar and vector quantities.
2 Define the terms *speed*, *velocity*, *acceleration*, *mass*, *weight* and *momentum*.
3 Define the term *force* and distinguish between balanced and unbalanced forces.
4 Define the term *friction* and relate it to the structure of synovial joints and to lubrication.
5 State Newton's laws of motion and relate these to the body.
6 Describe what is meant by the force of gravity and relate this to prolonged standing and the effects of immobility.
7 State what is meant by the term *centre of gravity* and relate this to the stability of the body and to lifting.
8 Describe three forms of lever and give examples of each within the body.
9 Relate an understanding of levers to lifting technique and back injury.
10 Distinguish between fixed and moveable pulleys and describe their operation.
11 Draw vector diagrams in order to resolve two forces acting simultaneously upon an object.
12 Relate pulleys and vectors to orthopaedic traction.

Introduction

In this chapter we will consider how things move (mechanics) and in particular how our bodies move (biomechanics). We begin with an explanation of terms and note that the quantities to which we refer may be described as either **scalar** or **vector** quantities. The difference between them is simple: scalar quantities have size only, while vector quantities have both size and direction. The significance of this distinction will become clear as you work through the chapter.

Speed and velocity

We know that speed is a measure of how fast something is travelling, and it is expressed as the distance travelled per unit of time as shown below:

$$\text{speed} = \frac{\text{distance}}{\text{time}}$$

The SI (Système Internationale) unit of distance (length) is of course the metre, and we commonly express speed in terms of the number of metres travelled in one second (m s^{-1} or m/s). For example, nervous impulses travel as fast as $120\,\text{m s}^{-1}$ in some myelinated neurons (nerve cells). However, when referring to objects such as cars and trains it is obviously more convenient to use kilometres per hour (kph), or in the UK miles per hour (mph). The same units are also used to measure velocity.

How does velocity differ from speed?

The answer is that the term *velocity* is only used when an object is travelling in a straight line – that is, velocity is a vector quantity while speed is a scalar quantity. Consequently, we cannot describe an electron orbiting the nucleus of an atom at constant speed as having constant velocity, since its direction of movement is constantly changing – it is circling round the nucleus.

Acceleration

When the velocity of an object is changing we might describe it as accelerating (getting faster) or decelerating (getting slower). Acceleration is defined as the rate of change of velocity:

$$\text{acceleration} = \frac{\text{velocity } (\text{m s}^{-1})}{\text{time } (\text{s})}$$

Consequently, the units of acceleration are metres per second per second (m s^{-2}).

Think about this for a moment.

Suppose an object at rest started to accelerate at a rate of 1m s^{-2}.

What would be its velocity after three seconds?

If you have answered $3\,\text{m s}^{-1}$ you would be right. This increase in velocity is referred to as a positive acceleration. If an object is slowing down scientists tend to refer to negative acceleration rather than deceleration. The term *acceleration* is also used of an object whose speed is constant but whose direction of travel is changing. Consequently, in the example of the atom used above, electrons are constantly accelerating, since their direction of travel is not linear.

What would cause a moving object to change its direction or speed?

Clearly it must have been acted upon by some **force**. We shall consider the concept of force a little later, but first let us deal with mass and weight.

Mass and weight

Most people use these terms interchangeably, but to the scientist they mean two quite different things. **Mass** is the amount of matter contained in an object, and it is measured in kilograms. It is therefore a scalar quantity, and also a fundamental quantity. This latter term refers to the fact that the mass of a stationary object does not change with its position. A 1 kg bag of flour has the same mass on Earth, on the Moon or in a rocket travelling to the Moon.

> *What about weight?*

Weight is not the same as mass; rather, it is a force, and the weight of our 1 kg bag of flour is certainly not the same in all circumstances. For example, it weighs less on the Moon than on the Earth, and even less in space, where it becomes weightless. Clearly we need to say more about this concept of force, and (since we have mentioned planets, space and weightlessness) more about **gravity** too. Before we do so, we should note that, unlike mass, weight is a vector quantity, since a force always acts in a particular direction. For example, weight on Earth always acts towards the centre of the Earth.

Momentum

This term is used of the product of the mass and velocity of an object.

> *So if a lorry and a car were both travelling at the same speed which would have the greater momentum?*

If you think it is the lorry you are right, since although both vehicles have the same velocity the lorry has the greater mass. On the other hand, a moving car has greater momentum than a stationary lorry, which, because it has zero velocity, has zero momentum. You will also no doubt realise that objects with the greatest momentum are most difficult to stop.

Force

We commonly use this word in everyday language to refer to some form of compulsion. For example, if our car develops a fault we might say 'I was forced to stop'. When applied to physical objects we recognise that a force involves some kind of push or a pull. The SI unit of force is named the newton (N) after Sir Isaac

Newton. One newton is the force which, when applied to a mass of 1 kg produces, in the absence of friction, an acceleration of $1 \, \mathrm{m\,s^{-2}}$. This means that if a force of 1 N were applied to a stationary mass of 1 kg it would be travelling at $1 \, \mathrm{m\,s^{-1}}$ after 1 second, $2 \, \mathrm{m\,s^{-1}}$ after 2 seconds and so on. This definition specifies 'in the absence of friction', and friction makes a very great difference to the force required to move an object.

Friction

When two surfaces which are in contact are moving over each other there is a force which resists this movement, and it is this which we call **friction**. The amount of friction depends upon the nature of the two surfaces. For example, friction is greatest between rough surfaces and less between polished surfaces. However, even highly polished surfaces appear rough when viewed under a microscope, so friction is always present between surfaces in contact.

> *Is friction a good or a bad thing?*

As you might expect, it depends. The ability to walk depends upon friction between our shoes and the floor, and a walking frame without friction-producing rubber pads on the legs would be positively dangerous! On the other hand, friction between an object which I want to move and the floor increases the effort which I will need to move it. Other problems with friction are that it produces heat and wears away surfaces which are rubbing against each other.

> *If we wanted to reduce the friction between two surfaces what methods could we use?*

One obvious thing that we could do is to fit one of the objects with wheels. Hospital beds are an example, since these frequently have to be moved, often with patients still on them. Another solution would be to use a lubricant – car engine oil is a good example. However, it might be more useful to think instead about friction and the body.

Practice point 9.1: friction and the body

One obvious site where friction occurs is within a joint. Figure 9.1 is an illustration of the structure of a synovial joint. Examples of synovial joints include the knee, elbow, shoulder and hip.

> *So are there any wheels here?*

Obviously not, but synovial joints do contain very smooth surfaces which pass across each other. Upon examination one might almost imagine that the ends of

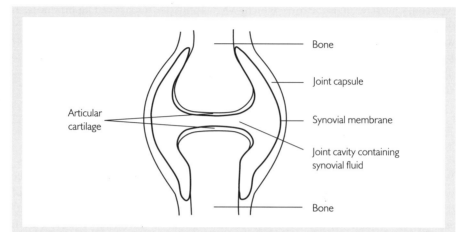

Figure 9.1
The generalised structure of a synovial joint.

articulating bones have been polished! In actual fact this 'polished surface' effect is produced by hyaline cartilage, which covers the ends of long bones where a joint is formed. In addition, the joint space between the bones contains a viscous fluid – synovial fluid – and this lubricant reduces friction still further. Of course, joints may eventually be damaged by friction, and this is essentially the cause of osteoarthrosis (osteoarthritis). In this condition the cartilage has become worn away and there is inflammation of the joint. Osteoarthrosis is more common in the elderly because of a greater period of wear and tear, but it is also associated with causes of increased wear and tear such as obesity and playing sport.

Friction must also be considered when inserting instruments or catheters into the body. One example is the catheterisation of the urinary bladder – a procedure which nurses commonly perform. In this procedure a sterile catheter is passed into the urinary bladder via the urethra. If trauma to the urethra is to be avoided a lubricant is required, but this must be non-toxic. K-Y™ jelly is the lubricant of choice.

The concept of balanced and unbalanced forces

Suppose we were to imagine a man trying to push a very heavy object which is resting on the floor. Despite applying a great deal of force to the object it just does not move.

Why not?

Because it is too heavy? That is what we would say in everyday language, but we might also explain that the man cannot overcome the force of friction. When he pushes against the object the force which he applies is opposed by friction. We might also say that the forces are *balanced*, since there is no movement of the object. Now imagine that the object has wheels. It is the same weight as before, but now when the man pushes against it friction has been very much reduced and there is much less opposition to the force he applies. The forces are *unbalanced* and the object moves. Forces are unbalanced when there is a net force acting in a particular direction which produces movement.

Newton's laws of motion

We have already mentioned Sir Isaac Newton, and now we are going to consider three laws which he formulated in order to describe objects in motion.

Newton's first law (law of inertia)

This may be stated in the following way:

All objects remain at rest or in a state of uniform motion unless acted upon by an unbalanced force.

This law simply states that objects continue to do the same thing unless a force compels them to do otherwise. For example, if they are stationary they do not move unless an unbalanced force compels them to do so and if in motion they do not change speed or direction unless an unbalanced force acts upon them. This is exactly what the expression uniform motion refers to – constant speed and direction. This first law of Newton's is sometimes called the law of **inertia**. Inertia is essentially the tendency of a body to resist a force which is applied to it. Let us think a little more about this in the context of rapid deceleration injuries.

Practice point 9.2: rapid deceleration injuries

It may be stating the obvious to note that these are injuries which result from the rapid deceleration of the body without impact with another object. One possible cause is a car crash. Among the most common are whiplash injuries to the neck, and in order to work out what happens imagine that you are driving at modest speed when your car is in a head-on collision with another.

What happens next?

Obviously, your car decelerates very quickly because of the impact.

But what about you? Remember Newton's first law?

Your body continues to travel forward, and if you are not wearing a seat belt you may strike on the steering wheel or be thrown through the windscreen. However, let us assume that you are wearing a seat belt – your chances of survival are now much better.

But what about your head? It isn't restrained is it?

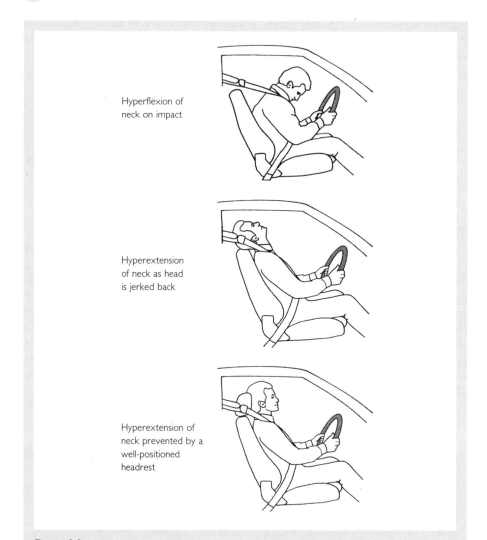

Hyperflexion of
neck on impact

Hyperextension
of neck as head
is jerked back

Hyperextension of
neck prevented by a
well-positioned
headrest

Figure 9.2
How a whiplash injury occurs.

Not initially, and it does indeed continue to travel forwards so that your neck becomes bent forward to an extreme degree (hyperflexion). Remember though that your body is held in place in the seat, so that your head is now jerked backwards and becomes hyperextended – ouch! The description of this type of injury as whiplash is quite graphic, and you will no doubt understand that the alternate hyperflexion and hyperextension of the neck may result in painful muscular and ligamentous injury to the neck. Whiplash injury is illustrated in Figure 9.2.

Are there any features of car design which help to prevent whiplash?

Yes. The main problem is the hyperextension of the neck as the head is jerked back, and you will realise that a properly adjusted headrest limits this movement and helps to reduce neck injury. This is also illustrated in Figure 9.2. However, it is worth noting that a headrest which is positioned too low may actually make the neck injury worse by acting as a point about which the head pivots. Consequently, head-rests should be positioned so that the pad is behind the head and not the neck.

Whiplash is not the only injury to be caused by rapid deceleration. Another structure which may be affected is the aorta – the artery which leaves the heart to carry oxygenated blood to the body. The heart is located in the centre of the chest and, like the head in the above example, it too swings forwards then back-wards when our bodies are stopped suddenly. In contrast, the aorta is held more rigidly and the force of the swinging action of the heart may be sufficient to cause the aorta to rupture. Needless to say, the affected individual usually rapidly haemorrhages to death. The crash position shown on airline safety cards is an attempt to reduce this type of injury, and it has been suggested that lives could be saved in air crashes if aircraft seats faced backwards. In this position the swinging action of the heart within the chest is prevented.

Newton's second law

Since an unbalanced force produces a change in motion, the degree of change may be used to determine the quantity of the force as illustrated below:

$$\text{force (N)} = \text{mass (kg)} \times \text{acceleration (m s}^{-2}\text{)}.$$

Alternatively, we could also state this law in the following way:

The acceleration of a fixed mass is proportional to the force applied to it.
One obvious application of the law is that the acceleration of a wheelchair which we are pushing is proportional to the force we apply to it. This example may appear too obvious, but science does not have to be difficult all the time!

Newton's third law

This law is often stated in the following way:
To every action there is an equal and opposite reaction.
It certainly is a simple formulation, but perhaps a better one would be:
When an object exerts a force upon another, the second object exerts an equal force on the first but in the opposite direction.
Stated in this way it becomes clear that two objects are involved. Consider walking as an example. In this case the two objects are the individual and the Earth. The act of walking involves the application of a force to the Earth but this is opposed by equal force exerted by the Earth against the individual. Of course in this case the two objects have different masses, so it is the individual who is pushed forward and not the reverse! It is quite often the case, as in that above, that the opposing force is friction, and we have thought about this already.

What happens when friction is removed?

In the case of walking our feet would not grip, the force we exert would not be opposed and we would slip over. Remember that many patient-related objects such as wheelchairs, commodes, beds and bed tables have wheels which reduce friction and remove opposing force. It is of course important to lock these wheels when patients are using these devices, otherwise an accident may result. Once again this is a rather obvious point, but it is not at all uncommon to find accident reports which cite failure to lock wheels or ineffective brakes as a cause of injury to the patient or carer. The next time you are involved in assisting a patient to move, perhaps a simple bed to chair transfer, make a point of checking that the brakes hold under modest force. If they do not, you and the patient may be at risk of injury. Needless to say, such defective equipment should be labelled as such and not used until repaired.

The force of gravity

I suppose that everyone knows the story of the apple which Newton saw fall and which led him to describe gravity. Consequently, the idea that the Earth attracts all objects to itself is a familiar one. In actual fact, experiments have demonstrated that every object exerts a force of attraction on every other object, and that this attraction is proportional to the mass of the two objects added together and inversely proportional to the square of the distance between them. This may at first be difficult to believe. When a brick is dropped from a building it is quite obvious that it is attracted to the Earth since it accelerates downwards. What is not obvious is that the Earth is attracted towards the brick with equal force, but this is actually what is happening. Of course, the Earth is massive compared to the brick, so its acceleration is effectively zero.

Does gravity affect our bodies?

Yes, as the following practice point makes clear.

Practice point 9.3: gravity and the body

We have noted that all objects are attracted towards the Earth, and in this context the word *object* includes the blood in our bodies. When we move from a recumbent to an upright position there is a tendency for blood to pool in the legs. This is not normally a problem since a reflex vasoconstriction (blood vessel constriction) affecting the legs reduces this tendency. However, if someone has to stand for a long time, especially if it is hot, then there may be a reduction in the volume of blood returning to the upper body from the legs and, as a consequence, of cerebral (brain) blood flow too. The effect of this is that they experience a momentary loss of consciousness, commonly called a faint (syncope).

So what is the first-aid measure in fainting?

Quite simply we make gravity work in favour of cerebral blood flow and lay the individual flat with the legs slightly elevated.

It is also worth noting that the reflex vasoconstriction noted above becomes impaired following prolonged recumbency. In the case of patients standing after a period of bed rest it would be unwise to expect them to rise from the bed and stand in one movement – they may faint. Instead, ask them first to sit on the edge of the bed and become used to the legs hanging down for a moment. This way if they feel faint they can simply lie back on the bed. However, if they feel no ill effects they can stand, but they should not walk away from the bed immediately.

Actually, bed rest has a number of unhelpful effects upon the body and you may be interested to know that research in this field has been conducted by American and Russian space agencies. Researchers used bed rest to simulate a loss of gravitational stress on the upright body. Consequently, we now know a great deal about the effects of bed rest, one of which is a change in the balance of activity between bone-forming cells (osteoblasts) and bone-resorbing cells (osteoclasts). As a consequence of the loss of force exerted longitudinally through the bones, bone resorption is accelerated, bone density decreases and renal excretion of the important bone mineral calcium increases. Actually, few illnesses require complete bed rest and loss of bone density may be prevented by encouraging the patient to stand for periods during the day.

Centre of gravity

The **centre of gravity** of an object of uniform density and simple shape, such as a cube, is the geometric centre of the space which that object occupies. This is illustrated in Figure 9.3. The idea of the centre of gravity may be a familiar one.

Of what significance is this point?

The centre of gravity is a point within an object at which the entire weight of the object can be thought of acting for the purpose of considering **torque**. Torque is the tendency of a force to produce rotation about a pivot point (**fulcrum**). If an object is supported under its centre of gravity it can be balanced so that it is not subject to any turning force. This is best illustrated in Figure 9.4, in which the object is a plank.

The plank is a simple shape. Let us assume that it has uniform density. Consequently, its centre of gravity is its geometric centre, marked by X. When the plank is supported below the centre of gravity it is not subject

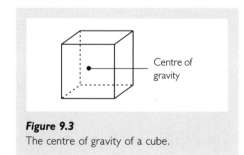

Figure 9.3
The centre of gravity of a cube.

Centre of gravity

to a net turning force and the condition of static equilibrium exists. A static equilibrium exists when an object is at rest and remains in a fixed position. In everyday language we say that the plank is balanced.

In contrast, the same plank supported at any other point will not be in a static equilibrium. It will be subject to a turning force and will therefore tip over.

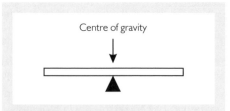
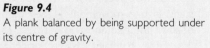

Figure 9.4
A plank balanced by being supported under its centre of gravity.

How do we relate an understanding of the centre of gravity to our bodies?

This is what we consider next.

Practice point 9.4: centre of gravity and our bodies

The first thing to note is that our bodies do not have uniform shape and density, so the position of the centre of gravity is not immediately obvious. The centre of gravity in the individual standing with his arms at his sides is illustrated in Figure 9.5.

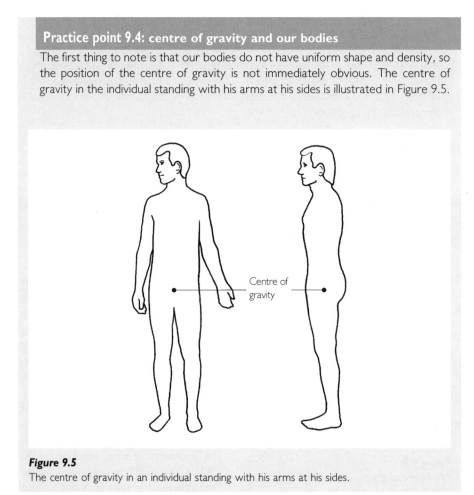

Figure 9.5
The centre of gravity in an individual standing with his arms at his sides.

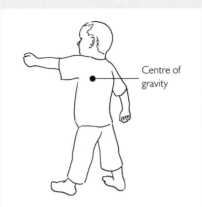

Figure 9.6
The position of the centre of gravity in an infant.

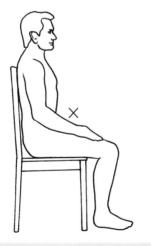

Figure 9.7
The position of the centre of gravity when sitting.

In what structure is the centre of gravity located?

If you have noted that it lies within the pelvis you are correct. In fact the exact point is in line with the third sacral vertebra. The point to note here is that the centre of gravity is in the lower torso, and this produces stability. Consequently, standing requires little muscular effort.

What about infants – is the centre of gravity in the same place?

You have probably realised that it is not (see Figure 9.6). Take the example of the child who is beginning to walk. Because the head of an infant is large in proportion to the body, the the centre of gravity is above the pelvis and this produces instability – just at the time the child is learning to walk! Adults can simulate this instability by raising the arms above the head, which raises the centre of gravity. Walking also requires a movement of the centre of gravity – this time to one side of the pelvis in order to allow the non-weight-bearing leg to be swung forward.

So what about the centre of gravity in the sitting and lying positions?

Figure 9.7 shows that when sitting the centre of gravity lies outside the body. This makes getting up from this position difficult, and the way in which we normally deal with this is to bend forward towards the edge of the seat, thus moving the centre of gravity closer to the body.

Nonetheless the action of standing still involves some effort and contraction of the muscles of the abdomen too.

Can you work out which individuals might find this difficult?

Perhaps you have noted the frail elderly, pregnant women, the obese and those with abdominal wounds.

What could be done to help?

It is not difficult to work out that upright but comfortable chairs with armrests are generally easier to rise from, and there are models with electric motors which lift the seat for those with special difficulties. The next time that you visit a placement where the frail elderly are cared for, examine the types of seating available and assess its suitability. Ask the patients what they think about it too.

When lying recumbent the centre of gravity is once again within the pelvis and moving the pelvis will turn the body over. We make use of this fact when turning the unconscious patient in to the recovery position. You might like to find out about this if you are not familiar with this aspect of first aid. Of course, when the patient is

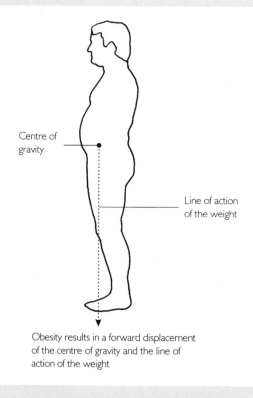

Centre of gravity

Line of action of the weight

Obesity results in a forward displacement of the centre of gravity and the line of action of the weight

Figure 9.8
The forward displacement of the centre of gravity in obesity.

sitting in bed the centre of gravity is once again outside the body, and this makes the patient more difficult to move.

The centre of gravity is also affected by body weight, as Figure 9.8 shows. In obesity the additional body weight is not evenly distributed throughout the body – the abdomen is one important site for the deposition of fat. Consequently, obesity results in a forward displacement of the centre of gravity, and this may contribute to back strain as will be explained later.

Stability

We have noted the meaning of the term *centre of gravity,* and now we turn to the topic of stability. When we think about stability it is helpful to imagine a vertical line passing through the centre of gravity. Consider Figure 9.9 for a moment.

Which of the four shapes is in a stable equilibrium and which will fall over?

If you have noted that (a) is in stable equilibrium and will not fall over and that (c) is not in equilibrium and will fall over you are right.

Why will (c) fall over?

The answer is that the line of action of its weight falls outside the base.

What about (b)?

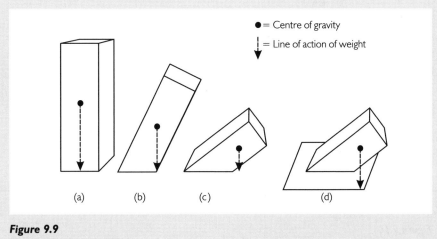

● = Centre of gravity

╎ = Line of action of weight

(a)   (b)   (c)   (d)

Figure 9.9
The stability of four shapes.

In this case the line of action of its weight falls on the edge of the base. The object will not fall over – it is in equilibrium but it is an unstable equilibrium. Only a little force will cause it to fall over.

> *What about (d)? Will (d) fall over?*

The answer is no, since although (d) looks similar to (c) it has an extended base, and the line of action of the weight remains within this base. So the key to stability is the size of the base. The relevance of this is seen when we think about someone who is mildly intoxicated and finding it difficult to keep his balance.

> *How does he increase his stability?*

Perhaps you have realised that walking with the feet further apart, thus widening the base, is an automatic attempt to increase stability.

> *Can we make use of this principle in patients who are unstable? What artificial means are there of increasing the size of the base?*

You will have no difficulty realising that walking sticks, crutches and walking frames are all attempts to increase stability by widening the base, as Figure 9.10 shows. The important thing to note here is that all these aids have to be fitted for the individual and their use taught. For example, if crutches are to be effective they have to be the right length for the patient. In addition, if they are held vertically

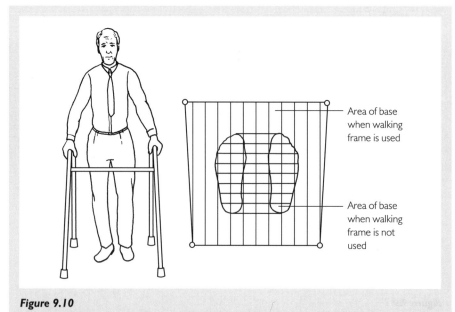

Figure 9.10
A walking frame increases the size of the base.

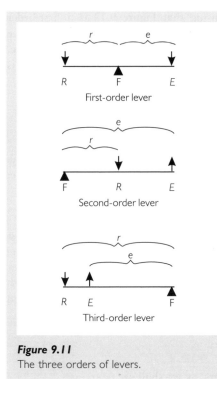

Figure 9.11
The three orders of levers.

downwards little effect on the size of the base is achieved – they have to be held at an angle away from the body.

Levers

A **lever** consists of a bar or a plank arranged so that it can pivot about a point referred to as the fulcrum (F). Actually, the plank used previously to illustrate the principle of the centre of gravity is one example of a lever. Levers also involve two forces – an effort force (E) and a resistance force (R). The latter is sometimes referred to as the load. Three types of lever are distinguished on the basis of the relationship between the fulcrum, the effort and the resistance. These three types of lever are referred to as first-, second- and third-order levers. Consider the three illustrations in Figure 9.11.

> *What is the relationship between the fulcrum, effort and resistance in each of the types of lever?*

You should have noted that in a first-order lever the fulcrum is between the two forces, in a second-order lever the resistance is between the fulcrum and the effort, and in third-order levers the effort is between the fulcrum and the resistance. Note too that the perpendicular (at right angles) distance between the fulcrum and the line of action of the effort force is called the effort arm (*e*) and the perpendicular distance between the fulcrum and the line of action of the resistance force is called the resistance arm (*r*).

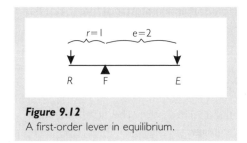

Figure 9.12
A first-order lever in equilibrium.

You should also be able to see that in second-order levers the effort arm is always longer than the resistance arm, while in third-order levers the resistance arm is always longer than the effort arm. In first-order levers either arm may be longer or both may be the same length. Just look over that last sentence again. Now consider Figure 9.12.

No doubt you had no trouble recognising that Figure 9.12 is a first-order lever, but note that in this case the effort arm is twice as long as the resistance arm. You might think 'so what'; after all, we did say that this could be the case. Indeed, but imagine that our first-order lever is in a stable equilibrium – that it is balanced.

> Is the effort force (E) greater than, less than or equal to the resistance force (R)?

Perhaps you have figured out that the effort force is less than the resistance force, since the effort force has a longer lever arm. In fact the following equation is true:

$$Ee = Rr$$

Consequently, we can work out that in this case the value of the effort force (E) must be exactly half that of the resistance force (R), since the effort arm (e) is twice the length of the resistance arm (r). In fact, that is the whole point about first-order lever arrangements such as this – they enable us to oppose a resistance force with a lesser effort force by using a longer lever arm. In essence we have just explained the use of the crowbar!

> Are levers important in the body?

They are indeed. Actually, bones act as levers, as we shall see next.

Practice point 9.5: levers and the body

If you remember the three types of lever you will be able to recognise them when you see examples in the body. Three are shown in Figure 9.13.

An example of a first-order lever is the flexion and extension of the neck, such as occurs when we nod the head. Standing on the toes, as one might to look over an obstacle, is an example of a second-order lever, while contraction of the biceps muscle of the upper arm is an example of a third-order lever. When the biceps muscle contracts it becomes shorter and wider and we can feel this increase in girth as it bulges in the upper arm. In this case the biceps shortens to a relatively small extent compared to the distance through which the hand moves. Remember that in third-order levers the effort arm is always shorter than the resistance arm. Consequently, the effort force must be greater than the resistance force since this acts through a longer lever. Thus in third-order levers the resistance force is moved through a greater distance than the effort force, but the cost is that a greater effort force is required.

Levers and lifting

Figure 9.14 illustrates someone attempting to lift a sack. The first-order lever system involved is superimposed upon the image.

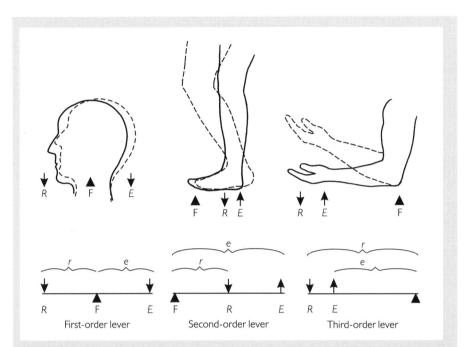

First-order lever   Second-order lever   Third-order lever

Figure 9.13
Three examples of levers in the body.

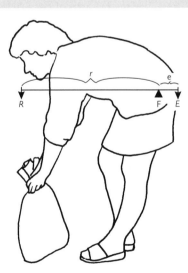

Figure 9.14
The lever system involved in lifting.

Once again the effort force is indicated by E, the effort arm by e, the resistance force by R, the resistance arm by r and the fulcrum by F. In this diagram it is clear that the resistance force is provided by the sack and the upper body.

What is providing the effort force?

It is provided by the contraction of the muscles of the back. The point to note here is that, because the fulcrum is close to the back, the effort force must be considerably larger than the resistance force since the resistance force acts through the longer lever arm. The application of this understanding to practice is to note that the posture adopted by the individual in this illustration is much more likely to result in back injury than one in which the resistance arm is reduced in length.

How could the resistance arm be reduced in length?

Quite simply by bringing the resistance force, the sack, closer to the body. In addition, flexing of the spine should be avoided and the sack lifted by straightening the knees. The reason for this is explained in the next section. Finally, spreading the feet apart will improve stability by widening the base.

Levers and back injury

How are back injuries sustained?

As you might expect, there are many forms of back injury. One injury which you might have heard about is commonly called a slipped disc. The disc being referred to is an intervertebral disc. Intervertebral discs are located, as their name implies, between adjacent vertebrae of the spine. Each disc consists of an inner semi-fluid nucleus pulposus which confers elasticity and compressibility. External to this is a strong outer ring of cartilage called the annulus fibrosus which contains the nucleus pulposus and limits its expansion. This structure makes the discs highly suitable for their role in cushioning the force transmitted up the spine when walking, running or jumping. In addition, the discs allow the spine to flex (bend forwards) and to a lesser extent to bend from side to side too.

In the case of a slipped disc the nucleus pulposus protrudes through a rupture in the annulus fibrosus.

Can you work out what the effects of this might be and how the condition might be treated?

The effects result from compression of the spinal cord or a spinal nerve by the herniated disc, and this produces pain and possibly numbness too. Initial treatment includes bed rest and analgesics (painkillers), but surgical removal of a damaged disc is sometimes indicated.

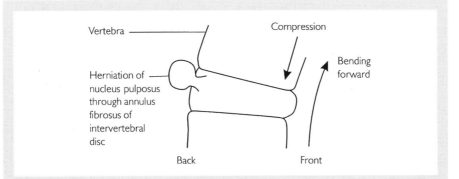

Figure 9.15
A slipped disc.

So how are slipped discs caused?

Bending forward to lift a heavy object, as previously described, is the cause, and the effect of this on the disc is illustrated in Figure 9.15.

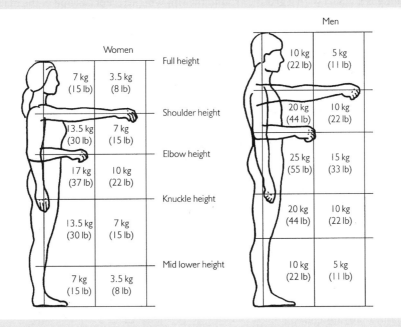

Figure 9.16
Numerical guidance on maximum lifting capacities. (Taken from The National Back Pain Association (1997). *The Guide to the Handling of Patients. Introducing a Safer Handling Policy.* 4th edn. Middlesex: The National Back Pain Association.)

In this diagram the forces applied to a disc while bending forward and lifting a heavy object are shown. Note that this compression of the disc is not uniform along the whole breadth of the disc. The anterior portion is compressed, while posteriorly the disc is stretched, and here the annulus fibrosus is thinned and ruptured.

What is a safe weight to lift?

This is a difficult question to answer. Guidance is provided in Figure 9.16. Note that the numerical values given differ for men and women and also according to the position of the weight to be lifted. The maximum values are 25 kg for men and 16.6 kg for women, but these figures only apply to loads close to the lower body. If heavier weights are involved a detailed risk assessment is required. In addition, patient handling not only involves heavier weights but also positions outside the ideal. For further information on risk assessment and patient handling the student is referred to the source of the illustrations in Figure 9.16.

Pulleys

A **pulley** is a wheel with a groove around the rim which accepts a rope in the manner shown in Figure 9.17. If the pulley wheel rotates about a fixed axle, that is the axle does not move up or down, it is referred to as a fixed pulley. Figure 9.17 is an example of a fixed pulley.

What is the purpose of a fixed pulley?

Fixed pulleys are used to change the direction of a force but they do not produce any mechanical advantage. That is, an effort force of 100 N must be exerted to raise a resistance force (load) of the same value. The expression *ideal mechanical advantage* (IMA) is sometimes used, and this is given by the equation below:

$$IMA = \frac{load}{effort}$$

In the fixed pulley system the IMA is therefore 1 – that is, there is no mechanical advantage.

But are there pulley systems which do produce a mechanical advantage?

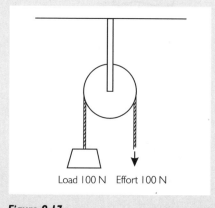

Load 100 N   Effort 100 N

Figure 9.17
A fixed pulley.

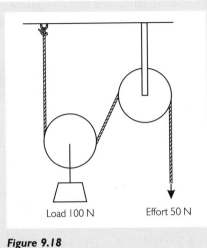

Figure 9.18
A single moveable pulley system.

Yes there are. These involve move-able pulleys, as shown in Figure 9.18. A moveable pulley is one whose axle is not fixed and which can therefore move up and down as well as rotate.

In the case of a single moveable pulley system a load of 100 N requires an effort of only 50 N to lift it.

So what is the IMA?

You are right if you think it is 2 (100/50). You may also have realised that the IMA is also equal to the number of ropes supporting the load. More complicated pulley systems, such as a block and tackle, produce greater mechanical advantages, but the actual mechanical advantage is always less than the IMA because of the increased friction involved.

Vectors

Remember that vector quantities are those which possess both size and direction. A force is one example of a vector quantity.

What happens if an object is subject to two forces simultaneously? What is the net force and direction of its action?

To work out the answer to this question we have to add the two vectors together in order to discover a single vector, called the resultant, which would have the same result. We can do this by drawing a vector diagram on a piece of graph paper. Consider Figure 9.19 one as an example.

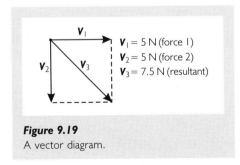

$V_1 = 5$ N (force 1)
$V_2 = 5$ N (force 2)
$V_3 = 7.5$ N (resultant)

Figure 9.19
A vector diagram.

In this diagram an object, repre-sented simply by a dot, is subject to two forces acting simultaneously and at right angles to each other. These forces are represented by lines the length of which is proportional to the magnitude of the force. In this case both forces are 5 N, and so both lines are of equal length. Note that the two lines form part of a parallelogram (a shape with parallel sides). If we now complete the shape we can resolve the direction of the resultant by drawing a line from the object to the opposite corner of the parallelogram. The magnitude of the resultant can be calculated simply by measuring this line.

Practice point: 9.6 pulleys, vectors and orthopaedic traction

Traction simply refers to a pulling force, so the term orthopaedic traction means the application of a pulling force to the bones. It is sometimes used to immobilise broken bones and to keep the broken ends in alignment, especially where muscle contraction tends to cause the broken ends to override each other, thus shortening the limb. Traction may be applied directly to the skeleton through pins inserted through a bone (skeletal traction) or indirectly through the skin using bandages (skin traction). Figure 9.20 shows one type of traction. Note that both fixed and moveable pulleys are used.

Which of the pulleys are fixed and which is moveable?

You should have noted that pulleys (a), (b) and (d) are fixed but (c) is moveable. The presence of pulleys means that the force (weight) acts in a number of directions and that the leg is subject to two vectors. To work out the resultant we have to draw a vector diagram – in fact this is already done for you.

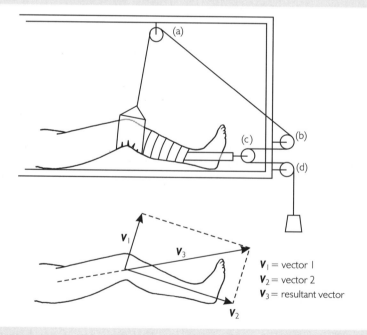

V_1 = vector 1
V_2 = vector 2
V_3 = resultant vector

Figure 9.20
An example of orthopaedic traction.

So what is the direction of the resultant?

It is in line with the shaft of the femur – the broken bone in this case.

It is especially important to note that since a single moveable pulley is involved the force applied to the leg is actually twice that of the weight applied to the rope. Clearly if a mistake were to be made, and too great a weight attached, the resultant traction on the leg would be considerable due to this mechanical advantage.

It is possible to see examples of traction in both paediatric and adult orthopaedic units. If you are involved in the care of a patient in traction, find out more about the system involved and the nursing care of the patient. However, it is worthwhile pointing out that advances in internal and external fixation techniques have meant that the use of traction has declined in recent years.

Summary

1 Scalar quantities, such as mass and speed, have size only, while vector quantities, such as weight, velocity, acceleration and force, have both size and direction.

2 The behaviour of objects in motion is described in Newton's three laws.

3 Fainting caused by the pooling of blood in the legs is one example of the effect of gravity on the body.

4 The centre of gravity of the body varies with position and obesity.

5 An object does not fall over if the line of action of its weight passes through the base.

6 Levers are described as first-, second- or third-order depending upon the relative positions of the fulcrum, resistance force and effort force.

7 A first-order lever diagram can be used to examine the forces involved when lifting.

8 Appropriate lifting technique involves avoiding heavier weights than recommended, holding the weight close to the body and avoiding flexing of the spine.

9 Pulleys are devices used for changing the direction of a force; some pulley systems also produce a mechanical advantage.

10 The effect of two forces upon an object can be resolved using a vector diagram and the parallelogram method.

Self-test questions

9.1 Which of the following statements are true?

a Mass is an example of a vector quantity.

b An electron orbiting the nucleus of an atom is constantly accelerating.

c Force is an example of a vector quantity.

d Momentum is the product of mass and velocity.

9.2 Which of the following statements are true?

a When a person is sitting the centre of gravity lies within the pelvis.

b When an individual is supine the centre of gravity lies within the pelvis.

    c Obesity causes an anterior displacement of the centre of gravity.

    d The centre of gravity of a child lies outside the pelvis.

9.3 Which of the following statements are true?

    a Whiplash injury can be explained by reference to Newton's second law.

    b The statement 'to every action there is an equal and opposite reaction' is an expression of Newton's third law.

    c Newton's first law is sometimes referred to as the law of inertia.

    d Force = Mass × Acceleration.

9.4 Which of the following statements are true?

    a Speed = Distance/Time

    b Acceleration = Velocity/Time

    c The acceleration of a fixed mass is proportional to the force applied to it.

    d When two vehicles are stationary the one with the greatest momentum is that with the greatest mass.

9.5 Which of the following statements are true?

    a A seesaw is an example of a first-order lever.

    b Flexing the elbow is an example of a third-order lever.

    c Flexing the neck is an example of a third-order lever.

    d Standing on the toes is an example of a second-order lever.

9.6 In the case of a first-order lever in equilibrium in which the resistance force is 10 N, the resistance arm 3 m and the effort arm 1 m, what is the value of the effort force?

    a 120 N

    b 60 N

    c 30 N

    d 15 N

9.7 In the case of a first-order lever in equilibrium in which the resistance force is 50 N, the resistance arm 1 m and the effort arm 10 m, what is the value of the effort force?

    a 500 N

    b 250 N

    c 50 N

    d 5 N

9.8 In the case of a fixed pulley a load of 50 N could be lifted by which one of the following?

    a 100 N

    b 50 N

    c 25 N

    d 15 N

9.9 In the case of a single moveable pulley system a load of 50 N could be lifted by which one of the following?

    a 100 N

    b 50 N

    c 25 N

    d 15 N

9.10 Which of the following statements are true?

    a IMA = effort/load

    b The IMA of a fixed pulley system is 1.

c All pulleys change the direction of a force.
d The IMA of a single moveable pulley is 2.

Answers to self-test questions

9.1 b, c and d   **9.6** c
9.2 b, c and d   **9.7** d
9.3 b, c and d   **9.8** b
9.4 a, b and c   **9.9** c
9.5 a, b and d   **9.10** b, c and d

Further study/exercises

9.1 When you next visit a clinical area in which patients need help to move and position themselves, identify the moving and handling aids available and find out how to use them.

9.2 Suppose an immobile patient needs to be moved from a hospital trolley to a bed. How should this be done? What should you do to ensure the safety of the patient and the staff involved?

The National Back Pain Association (1997). *The Guide to the Handling of Patients. Introducing a Safer Handling Policy*, 4th edn. Middlesex: The National Back Pain Association

10 Electricity, magnetism and medical equipment

Learning outcomes

After reading the following chapter and undertaking personal study you should be able to:

1 Describe magnetism and distinguish between permanent magnets and electromagnets.
2 Distinguish between static and dynamic (current) electricity.
3 Explain the terms *potential difference*, *current*, *resistance* and *power* and identify the units used to measure each.
4 Briefly describe the effects of electrocution and distinguish between macroshock and microshock.
5 Identify some of the important aspects of electrical safety.
6 Describe the action potential and the conduction system of the heart.
7 Briefly describe important medical uses of electricity including the ECG monitor, the defibrillator, diathermy and the artificial pacemaker.

Introduction

It is probably not until a power cut that we come to realise how heavily dependent we are upon electricity and not only for light. An average home usually contains a large number of items of electrical equipment including cookers, heaters, television sets, HiFi units and so on. It is not then surprising to discover that modern healthcare also shares this dependency upon electricity. In the course of your careers you will encounter medical equipment of many different kinds, some of which are described in this chapter. In addition, you will also learn about the importance of electricity in the proper functioning of the body. Good examples are the electrical impulses transmitted in nerves and these are also described here. However, firstly we spend some time thinking about a related phenomenon – magnetism.

Magnetism

Of all the phenomena described in this book, **magnetism** must be one of the most familiar.

Think about the occasions when you have used a magnet.

Most of us, at some point in our childhood, owned a magnet, and as adults we may have used the magnetic pointer of a compass to help us find our way. Magnetism is not of course a newly observed phenomenon. The ancient Greeks noticed that certain stones would attract objects made of iron, and the description of these rocks as magnetic is derived from the name of the city of their discovery – Magnesia.

Permanent magnets

The description of a magnet as *permanent* is probably self-explanatory, but this expression is used to distinguish magnets in which the **magnetic field** is not dependent on the flow of an electric current from **electromagnets**, which are described later. The earliest permanent magnets were created by rubbing ferrous metals with magnetic rock. Subsequently, when small permanent magnets were floated on water, it was noticed that one end, or **pole**, always pointed north, so the ends of bar magnets came to be described as the north and south poles. Of course this observation led to the development of the compass, in which the magnetic bar is shaped like a pointer. Such devices were in use by the Middle Ages. The need to float the pointer on water was subsequently eliminated by balancing the pointer on a pivot.

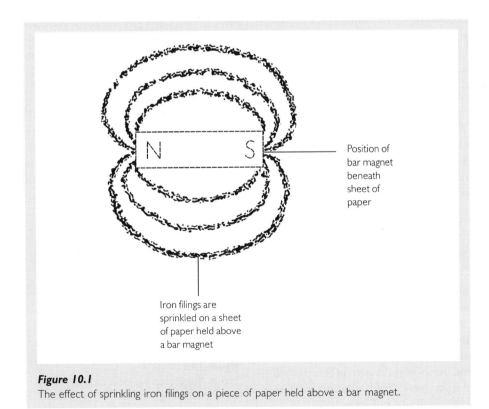

Position of
bar magnet
beneath
sheet of
paper

Iron filings are
sprinkled on a sheet
of paper held above
a bar magnet

Figure 10.1
The effect of sprinkling iron filings on a piece of paper held above a bar magnet.

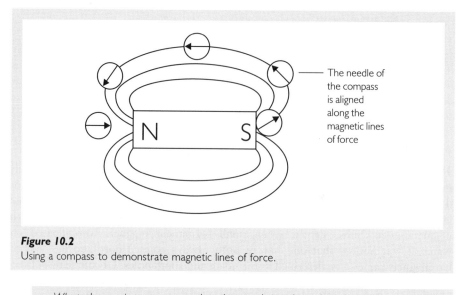

Figure 10.2
Using a compass to demonstrate magnetic lines of force.

What observations can we make when we bring the poles of two bar magnets together?

Like poles repel; unlike poles attract. Actually, if as a child you owned a train-set with magnetic couplings you may have worked this out long ago! You may also have performed an experiment at school in which iron filings are sprinkled upon a sheet of paper which is held above a bar magnet.

When this is done, what do we observe?

The iron filings do not remain scattered in a diffuse pattern, but instead they fall in a regular arrangement corresponding to magnetic lines of force, as shown in Figure 10.1. These lines of force are collectively referred to as the **magnetic field**, which is the region in which magnetic force can be demonstrated. If we were to take a small compass we could also demonstrate these lines of force by the movement of the needle. This is shown in Figure 10.2.

Electromagnets

A connection between electricity and magnetism was confirmed when, in the early 19th century, it was shown that a magnetic field is present around a wire in which current is flowing. As soon as the current is turned off the magnetic field disappears. This effect is shown in Figure 10.3.

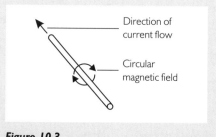

Figure 10.3
The magnetic field around a wire in which a current is flowing.

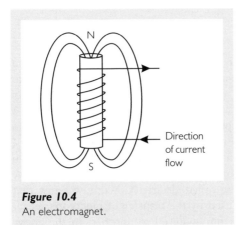

Figure 10.4
An electromagnet.

If the wire is formed into a coil which surrounds a cylinder, a magnetic field similar to that of the bar magnet is formed and the device is called an electromagnet. This is illustrated in Figure 10.4. One use of very large electromagnets is in cranes which are used to lift cars in scrapyards.

Are there medical uses of electromagnets too?

Yes there are. For example, you may witness an electromagnet being used to remove small metal objects which have penetrated the eye.

Practice point 10.1: magnetic fields and the body

At present, the beneficial or harmful effects of magnetic fields upon the body remain largely unconfirmed. Some believe that magnetic fields may be used to alleviate inflammation and improve blood supply to injured areas. Others allege that magnetic fields associated with high-voltage power supplies may, in some way, contribute to illness. It will be interesting to see what research in this field reveals.

However, it is certainly the case that artificial pacemakers may be inhibited by magnetic fields, and for this reason patients who have such devices implanted should not lean over car engines which are running. In addition, they should ask for security devices, such as those used in libraries and airports, to be turned off before they pass through them.

The solenoid

The magnetic field of an electromagnet is stronger if it contains an iron core, but if the core is not fixed it will move forward, that is to the north pole of the electromagnet, when the current is flowing. This effect is the basis of the *solenoid* or electrical switch. It is the kind of switch used when security doors are fitted to apartments, since the solenoid can be operated from a distance. It also finds use in items of medical equipment such as artificial ventilators.

Electricity

Like magnetism, electricity is a commonly experienced phenomenon, but one which is probably poorly understood. We commonly distinguish between two forms of electricity – *static* electricity and *dynamic* electricity.

Static electricity

Once again the origins of this aspect of science go back to the ancient Greeks, who noticed that the semi-precious stone amber (fossilised tree sap) could be made to attract filaments of a cloth with which it was rubbed. In a similar way, you may have noticed that hair is attracted to the comb following vigorous combing. As a child you may even have torn up small pieces of paper in order to watch them 'jump' onto the comb as it is brought near to them. Centuries after the Greeks, any material with which this phenomenon could be demonstrated became known as an *electric*, a name derived from the Greek *elektron*, meaning amber, and the mysterious power of attraction came to be called electricity. We now know that the phenomenon of static electricity is caused by the transfer of **electrons** (negatively charged sub-atomic particles) from one material to another. In the above cases this is achieved by rubbing.

Remember that only two types of charge exist – positive and negative. When electrons are transferred from one object to another the recipient object becomes negatively charged and the object which has lost electrons becomes positively charged.

What do we know about the behaviour of charged objects when they are brought together?

Like charges attract, while unlike charges repel. Thus static electricity is the phenomenon of a build-up of electrons. Since the electrons are not moving the use of the description 'static' is quite appropriate.

Does the build-up of electrons remain forever?

No. Eventually something will happen to enable the electrons to be discharged. For example, they may leak away slowly through the atmosphere. This is especially so in humid conditions. Alternatively, they may be suddenly discharged by contact with a **conductor**. You may find this occurring when making a hospital bed. These have wheels with rubber tyres, so that a build-up of charge is only dissipated when someone who is earthed touches the bed.

Is static electricity dangerous?

Not normally, although the discharge of static electricity during bed-making can smart! Nonetheless, there are occasions when static electricity is potentially dangerous. For example, flammable gases may be ignited by the heat of a spark caused by the discharge of static electricity. In addition, during a storm, charged regions may develop in the atmosphere and the discharge of electrons from one region to another results in a very large spark which we call lightning. The danger presented here is when a discharge is made to earth.

Insulators and conductors

Since we have already used the term **conductor** it is probably a good idea to define this term and compare conductors with **insulators**. Note that some materials do not allow the movement of electrons, so when rubbing transfers electrons to them the charge is not lost. Such materials are called insulators or poor conductors. Examples are amber, plastic, rubber, ceramics and wood. Other materials, such as metals, are referred to as conductors since electrons readily pass through them. Distinguishing between conductors and insulators is important when we move on to consider dynamic electricity.

Dynamic electricity

If static electricity involves the build-up of electrons in one place, then dynamic electricity obviously involves the flow of electrons. A number of concepts have to be introduced if dynamic electricity is to be understood. Do not worry though: we can introduce them by using a simple illustration – that of a water tank and tap, as shown in Figure 10.5.

You may wonder what a water tank and tap have to do with electricity, but stick with this illustration for a moment.

What will happen if the tap is opened?

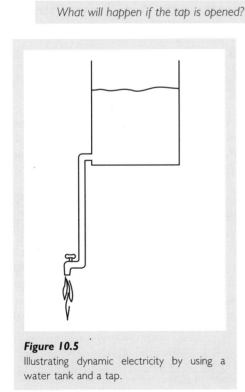

An obvious question – water will indeed flow out of it and we could measure this flow in litres per minute (l/min).

But why does the water flow?

Once again the answer is fairly obvious. It is because the tank is high up, perhaps in the loft, and the tap is low down, perhaps in the kitchen. In consequence, there is a pressure difference between the two points, and it is this pressure difference which causes the water to flow.

And if the water in the tank was not replaced, what would happen to the flow rate as the volume in the tank became depleted?

Figure 10.5
Illustrating dynamic electricity by using a water tank and a tap.

We are sure that you have the right answer once again. As the volume of water in the tank became depleted the pressure difference between the tank and the tap would become reduced and the flow rate would fall.

Could any other factor affect the flow rate?

Yes, the diameter of the pipes could. Provided the pressure does not change, a greater flow rate could be maintained through a wider pipe.

So what has this illustration got to do with electricity?

First of all, note that electrons also flow 'downhill' from a point of high concentration to a point of low concentration. That is, dynamic electricity consists of a flow of electrons from negative to positive. Actually, when electrons are flowing in this way we say that there is a *current* of electricity, so perhaps the illustration of the tank of water and tap is not such a strange one after all. Incidentally, for this reason we also speak of dynamic electricity as current electricity.

Can we measure electrical current just as we measure the flow of water?

Yes we can, although we do not use l/min! The SI (Système Internationale) unit of current is the **ampere** (or amp), named after the scientist André Ampère. There is a current of one amp when one *coulomb* of charge is flowing each second:

electrical current (amps) = number of coulombs/second

What's that – a coulomb of charge?

The charge of a single electron is far too small to be of practical use in the measurement of electrical current. Consequently, the SI unit of charge, the coulomb (named after Charles Coulomb), is used, and this is much larger. In fact, one coulomb is equal to the charge of 1.6×10^{19} electrons.

What pushes these electrons along? Is there an electrical equivalent of the pressure difference between the tank and the tap?

Yes there is. The term *electrical potential* is the amount of energy per coulomb that a charge possesses. It is measured in *joules per coulomb* – a unit renamed the **volt** after the scientist Allessandro Volta. The difference in the electrical potential difference between two points is called the **potential difference**, and it too is measured in volts. Current only flows between two points when there is an electrical potential difference between them.

Does the diameter of the wire through which a current is passing affect the current in the same way that the diameter of a pipe affects the flow of water?

Once again the answer is yes. Just as only the widest pipes can carry the greatest flow of water, so only the thickest wires can carry the greatest currents. You might have learned as much from experience. For example, a low current is conveyed from a transformer to the track of a toy train-set by thin wires, but the greater current which is carried to the transformer from the mains requires thicker wires.

Actually, the concept being introduced here is that of **resistance** – a property that impedes the flow of current through a conductor. Less resistance is offered by thicker wires. Resistance is not only dependent upon the diameter of the wire but also upon the material from which it is made, the least resistance being offered by the best conductors. The length of a wire also influences resistance – that is, the resistance increases with length. The SI unit of resistance is the *ohm* (Ω), named after Georg Ohm, and the relationship between current, electrical potential difference and resistance is given in the following equation (Ohm's law):

$$I \text{ (amps)} = V(\text{volts})/R(\text{ohms})$$

This means that when an electrical potential difference of 1 volt drives a current of 1 amp the resistance is 1 ohm.

> *What would happen if too great a current were to be passed through a wire?*

The answer is that some of the energy would be lost as heat and light. If the current is great enough the heat generated will actually melt the wire and we might say that the wire has burned out. In order to avoid this, very thick wires supply X-ray tubes, which may require a potential difference of 200 000 V. However, sometimes the emission of heat and light is actually desired – in electric heaters and light bulbs for example. In these cases the diameter of the wire is chosen carefully so that the desired effect is produced without the wire melting. We might also select a wire of a certain thickness so that if too great a current is drawn through it it does melt and the flow of current ceases.

> *Have you worked out that this is what a fuse is?*

The problem with fuses is that, even with a low current rating 3 amp fuse, the current required to burn out the fuse is much greater than that required to kill. Consequently, all electrical devices should be used with earth leakage circuit breakers (see later).

Current flow

While the illustration of the tank of water and tap works quite well, it does have one deficiency. It illustrates best the flow of current in one direction. However, for much of the time we have to deal with current flow which changes direction.

Direct current

When electrons flow in one direction through a circuit the current is described as *direct current* (d.c.). This is the type of current flow produced by batteries. Batteries may be of a number of types, but all make use of the fact that the atoms of certain metals dissociate (break up) when placed in an **electrolyte** solution (solutions of ions). For example, if a rod of zinc is partially immersed in a dilute solution of sulphuric acid (H_2SO_4), the zinc atoms dissociate into zinc ions (Zn^{2+}), which enter solution, and electrons, which remain in the rod. The dissociation of zinc is much more rapid than that of copper. Consequently, if a similar copper rod is inserted into the same solution a potential difference will exist between the two rods. If the two rods are now joined by a wire to form a circuit, electrons will flow from the zinc rod to the copper rod – that is, there is a flow of current. Batteries which employ electrolyte solutions like this are referred to as wet cells. A car battery is an obvious example. Other batteries are called dry cells, since the electrolyte is in the form of a paste.

Direct current is most often employed in small electrical devices such as torches, toys and radios. Artificial pacemakers which are implanted in to the body are also powered by batteries.

Alternating current

In this type of current the electrons flow first in one direction and then in the other. Consequently, it is described as alternating current (a.c.). The domestic power supply is alternating current and in this case the electrons flow 50 times in each direction every second. Consequently, we say that the current has a frequency of 50 Hz.

> *Does a.c. have advantages over d.c.?*

It has some. The energy losses in the power cables of electricity generating companies are less when a.c. is used and transformers can be used with a.c. (A transformer is a device which is used to change one voltage in to another).

> *But if a.c. simply involves the rapid oscillation of electrons, what do we mean when we speak about using electricity?*

Good question; we are not, after all, consuming electrons but the energy transferred by their oscillation.

> *Do you remember that energy is measured in joules?*

We are sure that you do, although the information that is really useful to have about electrical devices is the rate at which they consume energy – that is **power**. Power is expressed in joules/second:

$$\text{power (J/s)} = \frac{\text{energy used (J)}}{\text{time (s)}}$$

You should also be able to see that the following is also true:

$$\text{power (J/s)} = \frac{\text{potential difference (V)}}{\text{current (amps)}}$$

since

$$\text{potential difference (V)} = \text{joules/coulomb}$$

and

$$\text{current} = \text{coulomb/second}$$

so

$$\text{power (J/s)} = \frac{\text{joules}}{\text{coulomb}} \times \frac{\text{coulomb}}{\text{second}}$$

Actually, J/s is a further unit which has been renamed – this time after James Watt. The **watt** (W) is, however, a relatively small unit of power for most purposes, so the kilowatt (kW) is used. This is equal to 1000 watts.

Electrocution

In this section we are going to consider what happens when an electric current is applied to the body.

> *What two things are required for current to flow?*

An electrical potential difference is required to drive the current and a circuit is needed through which the current can flow. It is important to keep these two factors in mind, since an understanding of them informs many aspects of electrical safety.

> *Why is electrocution potentially dangerous?*

There are a number of reasons. An electric current may produce painful muscular contractions and affected individuals may not be able to release themselves from the source of the current. Affected respiratory muscles experience sustained contraction and the individual may not be able to breathe. Electrocution may also induce cardiac arrest – that is, the heart ceases to pump effectively. This does not necessarily mean that the heart is still. For example, the *arrhythmia* (abnormal rhythm) called ventricular fibrillation is characterised by rapid uncoordinated contractions. The appearance of the heart in this state has been likened to a bag of worms! This description does at least convey the impression of a quivering but ineffective rhythm. Finally, burns may be produced at the site of entry of the current to the body and at its site of exit too. These range from minor to very severe indeed. Skin grafting may be required, or in the case of a severely burned limb, amputation may be performed.

Is it possible to predict the kind of effects electrocution will produce in a particular case?

There are many variables which affect the severity of the damage done by an electric current passing through the body, and some of these are described below.

Duration of current: this is perhaps the most obvious. The longer the duration, the worse the effects. Consequently, earth leakage circuit breakers are designed to interrupt the current in a fraction of a second. These devices compare the current flowing in the live wire with that flowing in the neutral. Under perfect conditions the current in each should be identical. If not it means that some of the current is leaking away – perhaps to an individual who is being electrocuted! When this occurs we say that the device is tripped and the current flow is stopped.

The amount of current: the amount of current flowing is the most significant factor in determining the effects of electrocution. Currents of 1–5 mA will produce a definite shock, while death becomes possible at currents above about 10 mA.

What factors determine the amount of current flowing?

Remember Ohm's law $(I = V/R)$? Both potential difference and resistance influence the amount of current flowing.

Potential difference: although electrical safety warnings often provide information about the electrical potential difference in volts, it is not the potential difference *per se* which is important but the current that it can drive. From Ohm's law we can see that, provided the resistance remains unaltered, a greater potential difference will drive a greater current.

Resistance: electrocution by the domestic supply of 240 V sometimes proves fatal and other times does not.

Why is this?

There could be a number of reasons, but one has to do with resistance. For example, the resistance offered by dry skin may be up to 1000 times that of wet skin, so the current driven by the same potential difference through wet skin will be higher than through dry skin. Thus far we have assumed that the current has to pass through the skin and other soft tissues in order to reach the heart. This is, of course, usually the case.

Under what circumstances could a current be passed directly to the heart?

One example is when a transvenous pacing wire is in place. In this case, a pacemaker wire is passed through the venous circulation to the right ventricle of the heart while the other end remains outside the body and is attached to an external impulse generator (pacemaker box).

In such a case as this even a very low current inadvertently passed down the wire to the heart may induce ventricular fibrillation, since the high resistance of the skin and other body tissues has been bypassed. Indeed, the current required may even be below the level of perception of the individual through whom it is conveyed to the patient. Consequently, this form of electrocution is termed *microshock* in order to distinguish it from the more common form, which we call *macroshock*.

Frequency: the heart is more susceptible to some frequencies of alternating current than others. For example, it is susceptible to the frequency of the domestic supply (50 Hz), while the very high frequency of **diathermy** (see later) causes burns without inducing muscle contraction or ventricular fibrillation.

Practice point 10.2: electrical safety

In a textbook such as this, the advice given about electrical safety must be of a rather general nature. In addition, the hospital setting is sometimes treated differently from your own home. For example, in the home, an electrician would not usually be called to replace a light bulb or fuse. However, in the hospital there will be suitably qualified staff to carry out such tasks. A hospital will have a health and safety policy which covers electrical safety. You should familiarise yourself with the policies of the institutions in which you gain experience, and you should, of course, follow them. Each department will also have a health and safety officer who will also be able to offer advice when required. Nonetheless, the following are some general points worthy of note:

1 Plugs should be properly fitted. In EC countries new items of electrical equipment will already have plugs fitted, but they can subsequently be broken in use and require replacement by suitably qualified staff.

2 Plugs should be fitted with a fuse of the correct current rating for the equipment to which they are attached.

3 Hospitals will have a policy about the testing and checking of electrical equipment and it should be adhered to. Testing is usually performed annually and the date of the test recorded on a sticker placed on the equipment.

4 All equipment should be operated in accordance with the manufacturer's instructions.

5 Do not place objects containing fluids on top of items of electrical equipment.

6 Do not operate electrical equipment with wet hands.

7 Check equipment for signs of obvious damage such as broken plugs or frayed flex. Do not use damaged equipment, but label it as such and have it removed for repair by suitably qualified staff.

8 Prevent microshock by avoiding the creation of a circuit involving the patient. For example, do not touch the patient while at the same time touching an item of electrical equipment or some other conducting object.

> ### Practice point 10.3: first aid in electric shock
>
> For a full description of what to do in electric shock you should consult a first-aid manual. However, the following are points for general guidance.
>   1 Do not touch someone who is being electrocuted; tell others to keep away too.
>   2 Do not make contact with them via any conducting object.
>   3 Turn off the current at the mains.
>   4 Once you have ensured that it is safe to do so, check that the patient is breathing and that he has a heartbeat.
>   5 If breathing has ceased and a heartbeat is absent, commence cardio-pulmonary resuscitation (mouth-to-mouth breathing and closed chest cardiac compression).
>   6 Use others to call for help.
>   7 Burns should only be given attention once breathing and a heart beat has been restored.

Electricity in the body

The action potential

Neurons (sometimes spelled *neurones* and otherwise referred to as nerve cells or nerve fibres) are able to produce electrical impulses, which are referred to as **action potentials**. They are able to do this because of two features identified below:
  1 *A resting membrane potential* – there is a potential difference (electrical gradient) across the cell membrane.
  2 *Ion channels* – pores which allow the movement of ions in or out of the cell. The channels may be open or closed and they are therefore described as being gated.
The neuron actively pumps sodium ions (Na^+) out of the cell in exchange for potassium ions (K^+). This means that the concentration of Na^+ is greatest outside the cell, while the concentration of K^+ is greatest inside the cell. We might describe this effect by saying that there is a ion concentration gradient across the cell membrane. The mechanism by which this is achieved is called simply the *sodium pump*. The ions are not exchanged one-for-one. Instead two K^+ are exchanged for three Na^+. This results in an electrical potential difference across the cell membrane and it can, of course, be measured.

> *What is the SI unit of electrical potential difference?*

You will no doubt have remembered that it is the volt (V), although this is too large a unit to use here. Instead we use the millivolt (mV). Remember that one volt equals 1000 millivolts. The potential difference across the membrane of a nerve cell in which an impulse is not being conducted is referred to as the resting membrane potential and its value is in the range of $^-40$–$^-90$ mV. This means that the inside of the cell (axoplasm) is charged negatively compared to the outside. We describe the cell in this state as being polarised.

> *So what happens for an impulse to be generated?*

The first thing to note is that certain sodium channels open, and these allow Na^+ to diffuse through the cell membrane.

> *In which direction does Na^+ move?*

You may have been able to work out that Na^+ diffuses into the cell. This is because the concentration of Na^+ is greater outside the cell than inside. Diffusion is fully explained in Chapter 3 (Water, electrolytes and body fluids).

> *And what effect does this influx of Na^+ have on the membrane potential?*

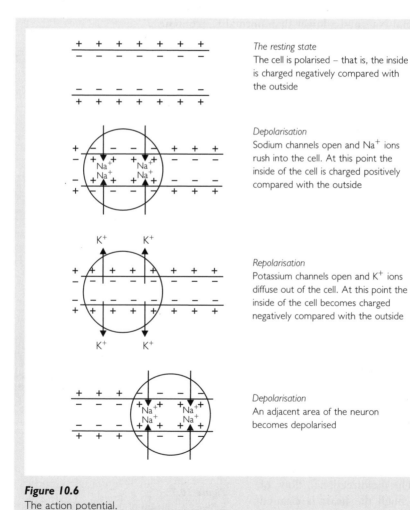

The resting state
The cell is polarised – that is, the inside is charged negatively compared with the outside

Depolarisation
Sodium channels open and Na^+ ions rush into the cell. At this point the inside of the cell is charged positively compared with the outside

Repolarisation
Potassium channels open and K^+ ions diffuse out of the cell. At this point the inside of the cell becomes charged negatively compared with the outside

Depolarisation
An adjacent area of the neuron becomes depolarised

Figure 10.6
The action potential.

Perhaps you have worked out that the axoplasm becomes more positive. When this occurs we say that the cell has *depolarised*. Actually, the membrane potential may now be in the region of $^+30\,mV$. Following the entry of Na^+, the sodium channels close and potassium channels open.

> In which direction does K^+ diffuse?

It diffuses out of the cell.

> And what effect does this efflux of K^+ have?

The axoplasm becomes negative once again. Consequently, we say that the cell is repolarised. Following repolarisation the resting membrane potential is restored but the cellular contents of Na^+ and K^+ are different. The cell still has a higher than normal Na^+ and a lower than normal K^+ content.

> What mechanism is responsible for restoring the normal intracellular ion concentration?

The sodium pump. You should also note that the changes described above bring about the same sequence of events in adjacent areas of the neuron. Consequently, an action potential consists of a wave of depolarisation which passes down the length of the neuron and which is then followed by a wave of repolarisation. This sequence of events is illustrated in Figure 10.6. All nervous impulses are brought about in this manner, so an understanding of this aspect of science is fundamental to an understanding of the nervous system as a whole.

The conduction system of the heart

The heart is a hollow muscular pump with four chambers. The two upper chambers are called *atria* and the two lower chambers are called *ventricles*. The unidirectional flow of blood through the heart is ensured by the presence of *valves*. Actually,

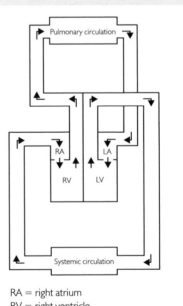

RA = right atrium
RV = right ventricle
LA = left atrium
LV = left ventricle

Figure 10.7
Schematic diagram of the circulation.

you might say that there are two blood circulations in the body. The right side of the heart pumps blood to the lungs (*pulmonary circulation*), while the left side of the heart pumps blood to the rest of the body (*systemic circulation*). Figure 10.7 is a schematic representation of the circulation.

Clearly, if the heart is to be effective its contractions must be coordinated. To this end the heart possess a conduction system through which the action potential, which brings about contraction, passes. However, the conduction system of the heart is not comprised of neurons but of modified cardiac muscle cells. These cells are organised in one of two ways. They either form small masses called *nodes* or they form extended structures called *bundles* or *fibres*.

The rate of contraction of the heart is determined by a small mass of tissue called the *sino-atrial node*, which is located in the wall of the right atrium. Consequently, this node is referred to as the *pacemaker*. The wave of depolarisation initiated by the sino-atrial node passes over the atria in a downwards direction. As a consequence the atria contract in a downwards direction and so empty into the ventricles. Next, the wave of depolarisation is slowed down as is passes through a second node – the *atrioventricular node*.

What purpose does the slowing of the impulse serve?

Perhaps you have realised that this slowing of the impulse allows time for blood to flow into the ventricles.

The wave of depolarisation now passes quickly to the apex of the heart through the *bundle of His* and the *bundle branches*. At the apex each bundle branch gives rise to *Purkinje fibres*.

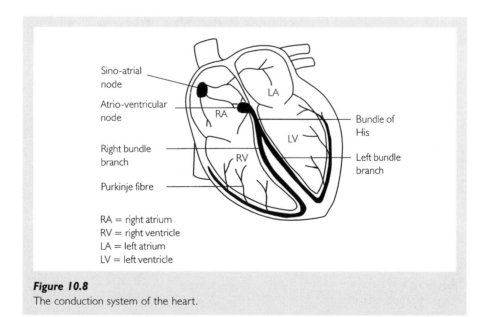

Figure 10.8
The conduction system of the heart.

In which direction is the wave of depolarisation now passing?

Upwards. As a consequence the ventricles contract upwards and blood is pumped into the body's two main arteries – from the right ventricle into the *pulmonary artery* and from the left ventricle into the *aorta*. The conduction system of the heart is shown in Figure 10.8.

Important items of medical equipment

The ECG monitor

Have you ever heard of cathode rays?

It is a rather old-fashioned term, but **cathode rays** are actually streams of electrons produced by a heated filament which forms the negative electrode (**cathode**) of a device called the cathode ray tube. Since the cathode ray tube is the basis of the oscilloscope (hospital monitor), and of television sets too, it is important that we say something about how it works. The oscilloscope is illustrated in Figure 10.9. The electrons produced at the cathode are accelerated towards the **anode** (positive electrode) because of the large potential difference between the two electrodes. After passing through a hole in the anode, the stream of electrons strikes a display screen which, as a consequence, fluoresces.

So what would an observer see on the screen?

An observer would see a fluorescent spot corresponding to the point at which the stream of electrons strikes the screen. However, note that before striking the

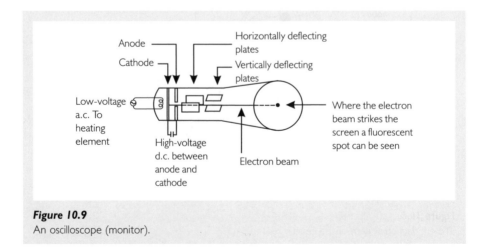

Figure 10.9
An oscilloscope (monitor).

screen the beam has passed between two pairs of plates which are orientated at right angles to each other and across which a potential difference can be applied. By this means the beam can be deflected vertically or horizontally. In addition, the potential difference applied to the horizontally deflecting plates (x plates) steadily increases before suddenly changing polarity and steadily increasing once again.

> *What effect does this have?*

To an observer, the fluorescent spot moves horizontally from the left of the screen to the right as the potential difference increases. It is then deflected to the left by the sudden change in polarity, but this happens too quickly to be seen. What the observer actually sees is what appears to be another spot moving from left to right. Of course the process continues so long as the power is switched on.

> *What are the vertically deflecting plates (y plates) used for?*

This question is best answered by considering the ECG (electrocardiograph) monitor. In this case electrodes are applied to the patient's chest and these detect the small currents generated by the heart. These currents are then amplified and applied to the plates in order to produce a vertical deflection of the electron beam. The resultant wave pattern would be quite difficult to interpret were it not for the fact that the fluorescence of the screen persists for long enough for the eye to see the path taken by the spot.

> *What does a typical ECG trace look like?*

This is what we consider in the following practice point.

Practice point 10.4: the normal ECG

Figure 10.10 shows what a normal ECG trace looks like. Note that several features are labelled. The first positive (upwards) deflection is labelled a *P wave*. This results from atrial depolarisation. Following this there is a complex of waves called the *QRS complex* which results from ventricular depolarisation. The *T wave* results from ventricular repolarisation. This normal pattern is referred to as *sinus rhythm* because the wave of depolarisation originates in the sino-atrial node. Abnormalities of rhythm are referred to as *arrhythmias*, and you will learn about these during your studies. When you are next involved in the care of a patient who is attached to an ECG monitor have a look at the trace and see if you can identify the waves. Ask the staff to explain any arrhythmia to you.

The defibrillator

Heart disease is the leading cause of death in industrial countries. Sudden death due to heart disease is usually the result of an abnormal heart rhythm called *ventricular fibrillation.* The termination of ventricular fibrillation by the administration of a strong electric shock is called **defibrillation** and the electrical device which generates the current is called a **defibrillator**.

> *Have you seen a defibrillator?*

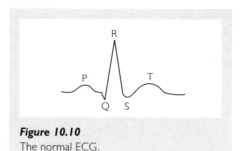

Figure 10.10
The normal ECG.

You may not yet have seen a defibrillator in use, but if you have undertaken a practice placement in a hospital you should know where the nearest one is kept – you might be asked to get it urgently!

Defibrillators may be operated from the mains, but most also contain rechargeable batteries and can be operated when an external source of current is unavailable. This is only the case when the batteries are charged, so, when not in use, defibrillators should still be plugged in. The electrical component common to all defibrillators is the **capacitor** – a device responsible for storing charge. The capacitor may be charged either from the mains or from the internal batteries. This is only done immediately prior to using the defibrillator when the charge button is pressed. The amount of energy to be used can be regulated and this is administered to a patient via two paddles which are placed on the chest (external defibrillation). Both paddles have 'fire' buttons and both buttons must be pressed if a charge is to be delivered. This feature reduces the likelihood of accidental discharge. A simplified circuit diagram for a defibrillator is given in Figure 10.11. If the chest has been surgically opened, as is the case during a heart operation, sterile paddles may be applied directly to the heart (internal defibrillation). However, it should

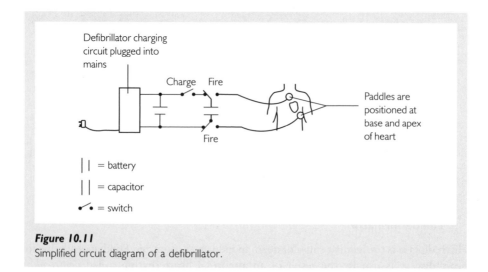

Figure 10.11
Simplified circuit diagram of a defibrillator.

be noted that much less energy is required for internal defibrillation than for external defibrillation.

Defibrillators are potentially dangerous pieces of equipment and the proper procedure for their use must be followed. However, general advice about defibrillator safety is given in the following practice point.

Practice point 10.5: defibrillator safety

1 Defibrillators should only be operated by staff who have undergone training in the procedure to be followed.
2 Defibrillators should only be used in accordance with the relevant hospital policy and the manufacturer's instructions.
3 Defibrillators should never be discharged with the paddles held in the air.
4 The capacitor should be charged to the correct energy level prior to defibrillation.
5 Burns to the patient's skin should be avoided by using conductive jelly pads beneath the paddles and by pressing the paddles firmly against the patient's body. The jelly pads should not touch, otherwise a conductive pathway over the patient's chest may be formed.
6 Prior to defibrillation the operator should give a warning in a clear voice and allow time for the resuscitation team time to stand back from the patient's bed.
7 During defibrillation the resuscitation team should not touch the patient, the bed or any conductive object in contact with the patient.
8 During defibrillation the patient's body must not be in contact with any conducting object through which the current may pass to earth.

Diathermy

Diathermy involves the use of high-frequency alternating current for the purpose of cutting tissue (*electrosurgery*) or sealing blood vessels (*electrocautery*). The frequency of the current is indeed very high −400 kHz to 3 MHz (up to 3 million Hz!).

Is diathermy not dangerous for the patient?

You might think so, since a current of 0.5 amps may be used. Indeed certain precautions do need to be taken, but remember that nerves and muscles are most sensitive to low frequencies, such as that of the domestic supply (50 Hz). They are unaffected by the high frequencies of diathermy. Note too that in *monopolar* diathermy the current enters the body through the narrow diathermy probe – the *active* electrode, which is held by the surgeon – but it leaves the body through a large electrode placed beneath the patient's buttocks or thigh (*indifferent* electrode). This arrangement is shown in Figure 10.12. The resistance to current flow offered by the tissues of the body results in heating, but the current density is obviously very much greater at the active electrode than at the very much larger indifferent electrode. Consequently, the temperature at the tip of the active electrode rises to several hundred degrees Celsius.

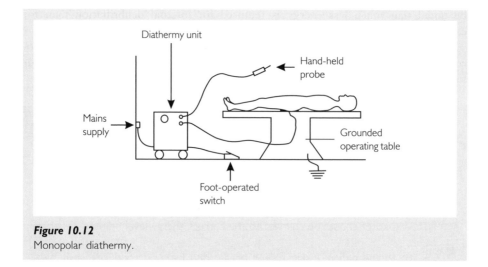

Figure 10.12
Monopolar diathermy.

In electrocautery a bleeding vessel is gripped with artery forceps and these are touched with the active electrode. The heating effect dries out the cells of the vessel wall and they shrink. This causes the vessel to contract and a clot forms within it. In this way bleeding is arrested. In electrosurgery a temperature of up to 1000 °C is created immediately beneath a narrow active electrode and the cells disintegrate instantly.

In bipolar diathermy both electrodes are located within the probe held by the surgeon. Since the damage produced by bipolar diathermy is less than with monopolar diathermy it is favoured by plastic surgeons and paediatric surgeons.

The artificial pacemaker

Heart disease may involve the conduction system as well as the heart muscle itself. The result is a disorder of conduction called a conduction defect. Conduction defects result in a slowed pulse rate (bradycardia). In fact, the pulse may be too slow to maintain cerebral perfusion and the patient loses consciousness (Stokes–Adams attack). Sick sinus syndrome and atrioventricular block are two examples of conduction defects. In sick sinus syndrome there is a slowing of the rate of depolarisation of the sino-atrial node. However, impulses are not blocked as they travel through the conduction system and therefore each ECG trace consists of a P wave, a QRS complex and a T wave. Consequently, we describe this as sinus bradycardia. In atrioventricular block some of the impulses originating from the sino-atrial node are not transmitted by the atrioventricular node and this is seen on the ECG as P waves which are not followed by QRS complexes.

What could be done to help the patient with a conduction defect?

No doubt you realise that he may be treated with an artificial pacemaker (Figure 10.13). Artificial pacemakers consist of external or implantable pulse generators

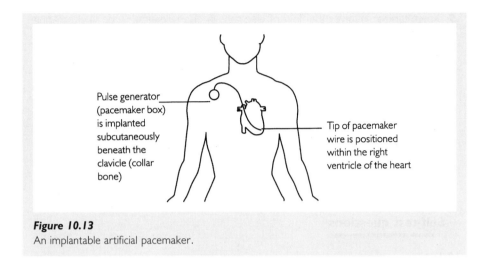

Figure 10.13
An implantable artificial pacemaker.

which produce pulses of current. It is these pulses which initiate a wave of depolarisation and as a consequence contraction of the heart. The current reaches the heart through a wire passed transvenously (through the venous circulation) to the right ventricle.

Transcutaneous electrical nerve stimulation (TENS)

TENS is an electrical method of providing relief from both chronic and acute pain. You might witness its use by women to alleviate the pain of early labour. As its name implies, TENS involves applying an electrical stimulus to the skin. Unpleasant effects such as muscle contraction are avoided by using moderately high frequencies of up to 5000 Hz. The TENS apparatus is portable and it therefore enables users to continue with their daily activites while wearing it.

How does TENS work?

This is not an easy question to answer. It may be that the action potentials in sensory nerves which TENS stimulates act to block other sensory information which the brain would interpret as pain. It appears that part of the spinal cord acts like a gate and the stimulation provided by TENS serves to 'close the gate' to other impulses. The gate control theory of pain is commonly described in physiology and nursing texts and you might like to find out more about it.

Summary
1 Electromagnets may be used to remove objects from the eye.
2 Electricity involves the transfer of electrons.
3 A build-up of electrons which are not flowing is referred to as static electricity.
4 Dynamic (current) electricity involves the flow of electrons.

5 Direct current involves the flow of electrons in one direction, while in alternating current the direction of current flow changes in a cyclical manner.

6 The relationship between potential difference, current and resistance is given in Ohm's law.

7 Power is the amount of energy consumed in one second and it is measured in watts (joules/second).

8 The action potential is one example of electricity in the body.

9 Important medical uses of electricity include the electrocardiogram (ECG), the defibrillator, diathermy, the artificial pacemaker and transcutaneous electrical nerve stimulation (TENS).

Self-test questions

10.1 Which of the following statements are true?

    a Magnets in which magnetic force is dependent on the flow of an electric current are described as permanent.

    b The region in which magnetic force can be demonstrated is referred to as the magnetic field.

    c Like poles of bar magnets attract.

    d The magnetic lines of force which surround a wire through which a current is passing lie parallel to the wire.

10.2 Which of the following statements are true?

    a When objects become charged by rubbing we find that opposite charges attract.

    b A build-up of electrons is referred to as static electricity.

    c Static electricity is also known as current electricity.

    d Objects through which electrons readily pass are known as insulators.

10.3 Which of the following statements are true?

    a The current produced by a battery flows in one direction.

    b Batteries may be constructed by placing two rods of the same metal in an electrolyte solution.

    c Transformers can be used with direct current.

    d It is more economical to transmit alternating current over long distances.

10.4 Which of the following statements are true?

    a Resistance increases with the length of a wire.

    b Resistance increases with the diameter of a wire.

    c The greatest resistance is offered by the best conductors.

    d The resistance offered by dry skin is greater than that offered by wet skin.

10.5 Which of the following statements are true?

    a The amount of current flowing is the most significant factor in determining the effects of electrocution.

    b The heart is relatively insensitive to alternating current at 50 Hz.

    c Direct currents cannot produce a fatal electric shock.

    d The action of an earth leakage circuit breaker is to rapidly interrupt the flow of current.

10.6 According to Ohm's law, what would be the current flowing in amps when the potential difference is 5 volts and the resistance is 2 ohms?

a 10

b 5

c 2.5

d 2

10.7 Which of the following statements are true?

a The resting membrane potential of a neuron is approximately $-30\,mV$.

b The resting membrane potential arises because Na^+ is pumped into the cell.

c The sodium pump exchanges two K^+ for each Na^+.

d At rest, the inside of the cell membrane of a neuron is charged positively compared with the outside.

10.8 Which of the following statements are true?

a Depolarisation is brought about by an influx of Na^+.

b The membrane potential becomes more positive during depolarisation.

c Repolarisation is brought about by an influx of K^+.

d The membrane potential becomes more negative during repolarisation.

10.9 Concerning the ECG, which one of the following events results in a P wave?

a Ventricular depolarisation.

b Atrial depolarisation.

c Atrial repolarisation.

d Ventricular repolarisation.

10.10 Which one of the following statements correctly describes why diathermy does not result in ventricular fibrillation?

a The current used is too small.

b The resistance of the soft tissues is too high.

c The heart is insensitive to high frequencies.

d The potential difference used is too small.

Answers to self-test questions

10.1 b

10.2 a and b

10.3 a and d

10.4 a and d

10.5 a and d

10.6 c

10.7 a

10.8 a, b and d

10.9 b

10.10 c

Further study/exercises

10.1 We have mentioned the ECG and ventricular fibrillation, but what other common arrhythmia are there? Find out about the following:

a Atrial fibrillation

b Heart-block (there are a number of different types)

    c Ventricular tachycardia

    Hampton J. R. (1997). *The ECG Made Easy*, 5th edn. Edinburgh: Churchill Livingstone

10.2 In addition to the ECG, a number of other electrical tests may be performed in hospital. Find out about the following:

    a Electroencephalography (EEG)

    b Electromyography (EMG)

    c Electrochochleography (ECOcG)

    d Electrical response audiometry (ERA)

    Evans D. M. D. (1994). *Special Tests. The Procedure and Meaning of the Commoner Tests in Hospital*. London: Mosby

Light and vision

Introduction

In this chapter we are going to consider light and describe the process of vision. So much of our normal everyday activity depends upon our ability to see that loss of sight is always contemplated fearfully.

What exactly is light and how do we see?

Mankind has theorised about this throughout history. One belief current at the time of the ancient Greeks was that vision was the result of something being emitted from the eye itself – an idea which now sounds quite absurd. In the more recent past Newton thought of light in terms of a stream of particles, while for some considerable time the behaviour of light as a wave has been accepted. However, some of the properties of light cannot be adequately explained by wave theory and the notion of light as particles re-emerged. Clearly light is a complex phenomenon. For our purposes much of what we observe of light can be explained on the basis of wave theory, so this is where we begin.

Transverse waves

Imagine that we have a tank of water in which a cork is floating. We then create a wave on the surface of the water.

What happens to the cork?

Quite obviously it moves up and down when the wave passes, but note that it is not pushed forward by the wave. The wave in our tank is an example of the *transverse* kind. Transverse waves are those in which a particle of the medium oscillates in a direction which is perpendicular to the direction of the propagation of the wave. Note too that waves involve the transfer of energy from one place to another without the transfer of matter. The particles of the medium in which the wave is travelling oscillate, in this case up and down, while the energy of the wave is transmitted the length of travel of the wave. The nature of a transverse wave is shown in the diagram in Figure 11.1.

A number of measurements can be made of transverse waves and some of these are described below.

Wavelength

The **wavelength** (λ) of a transverse wave is the distance between two crests or two troughs on successive waves.

Amplitude and displacement

Displacement (d) is the distance travelled by a particle when moving from a peak to a trough (or vice versa). **Amplitude** (a) is the maximum displacement of a particle from its rest or mean position and it is therefore equal to half the displacement.

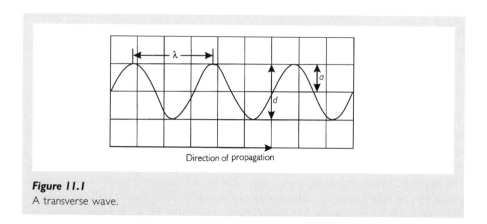

Direction of propagation

Figure 11.1
A transverse wave.

Frequency

Frequency (f) is the number of complete waves which pass a particular point in one second. You might hear frequency described in terms of cycles per second, but the proper name for this unit is the hertz (Hz).

Period

The period (T) is the length of time it takes for one wavelength to pass a particular point. Period and frequency are inversely related, so that the following equations are true:

$$T = 1/f \quad \text{and} \quad f = 1/T$$

Velocity

The velocity of a wave (v) is how fast the wave is moving in the direction of propagation. It is not a constant quantity but depends upon the medium through which the wave is passing. For example, the velocity of light in air is $3 \times 10^8 \, \text{m s}^{-1}$, but in glass it is $2 \times 10^8 \, \text{m s}^{-1}$. The relationship between velocity, wavelength and frequency is given in the wave equation below:

$$v = \lambda/T \quad \text{or} \quad v = f\lambda$$

Phase

Imagine that a particle at the rest position is displaced upwards, firstly to a peak and then downwards to a trough before returning to the rest position again. It has completed one cycle of oscillation. Now imagine two particles at the same part of the cycle but on adjacent waves. They might both be at the rest position, both at peaks or troughs or both be at an intermediate position. Such particles are said to be in phase. Obviously two particles at different positions are then said to be out of phase. The extent to which two particles are out of phase is related to 360° of a circle. For example, if a particle in the rest position is compared with one at a peak or trough of an adjacent wave they are described as being 90° out of phase. Similarly, a particle in the rest position and one half-way up a crest are 45° out of phase.

Intensity

Intensity is a measure of the amount of energy being carried by a wave and it is expressed as the amount of energy passing through an area of one square metre (which is perpendicular to the direction of propagation of the wave) in one second. The units of the measurement of intensity are therefore $\text{J s}^{-1} \, \text{m}^{-2}$.

The electromagnetic spectrum

We have taken time to describe transverse waves because light may be thought of as the transfer of energy in the form of such waves. However, there is a problem here.

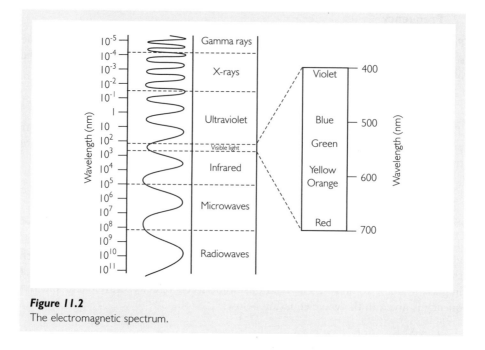

Figure 11.2
The electromagnetic spectrum.

In addition to being described as transverse, waves in a tank of water may also be referred to as mechanical – a reference to the fact that they require a material medium through which to pass.

> *What about light? Does it require a material medium through which to pass?*

You know that it does not, since it crosses the vacuum of space.

> *So in the case of light, just what is it that is oscillating like our cork in the water tank?*

The answer to this question is that light is an *electromagnetic* wave and that the properties which are undergoing oscillation are the direction of magnetic and electric fields. Electromagnetic waves are difficult to visualise because, unlike the water in our tank, these fields cannot be seen.

Light is not an isolated electromagnetic phenomenon. Rather it forms a small part of a spectrum of waves called, unsurprisingly, the *electromagnetic spectrum*. This spectrum is illustrated in Figure 11.2. Have a look at this diagram for a moment. The electromagnetic spectrum includes radio waves, microwaves, infrared, ultraviolet, X-rays and gamma rays. Each form of electromagnetic radiation differs from another by having a different wavelength and frequency. For example, radio waves have a longer wavelength and lower frequency than do gamma rays. In this chapter we shall concentrate upon visible light, but we shall also mention ultraviolet. Other forms of electromagnetic radiation, such as X-rays and gamma rays, are dealt with in Chapter 13, while still other forms are not considered at all in this book. However, it is important to realise that electromagnetic radiation forms a continuous spectrum of waves and that the wavelength at which the name given to a particular form of radiation changes is somewhat arbitrary. For example, there is no rigid boundary between ultraviolet and X-rays.

White light

In view of what we have noted above regarding the continuous nature of the electromagnetic spectrum, you will not be surprised to learn that visible light is not itself compromised of waves of a single wavelength either. Instead, it is formed from a spectrum of visible wavelengths of between 400–700 nm, each of which we perceive as a different colour. These colours become separated in a prism. We shall say more about this later; what is important to know now is that the result of adding these separate wavelengths together is the mixture of wavelengths which we call white light.

The properties of light

What happens to light when it hits an object?

It depends upon the substance from which the object is made. For example, the object may transmit the light – that is, light passes through it. In this case the substance is described as being *transparent*. Air, glass and pure water are examples of transparent substances. Other substances transmit light, but in doing so scatter it, so that objects viewed through the substance cannot be seen clearly. In this case the substance is described as being *translucent*, frosted glass is an obvious example. Still other substances do not transmit light at all and are described as being *opaque*. Examples include wood and metal.

Reflection

If light is not transmitted what then happens to it?

It might be *absorbed*, as is the case with black objects. Alternatively, the light will bounce off the substance; to be more correct, we say that it has been *reflected*. Shiny opaque substances such as polished metals are good at reflecting light, but even transparent substances reflect some light.

Does a substance reflect all the wavelengths of light equally?

Think about this for a moment and then answer a further question.

Why does an object appear red?

An object appears red because it reflects only red light, while other frequencies are transmitted or absorbed.

How would a red object appear in pure blue light?

In pure blue light there is obviously no red to be reflected, so the object would appear black.

Refraction

We have already noted that light readily passes through transparent substances, but now we are going to deal with what happens when it passes from one transparent substance to another.

Do you remember what we said about the velocity of light in glass and air?

Perhaps you recall that light travels more slowly in glass than in air – in fact, the denser the substance the slower light travels through it. Thus light passing from air to glass slows down, but in addition it bends towards the *normal*. The normal is a line drawn at right angles to the surface of the glass at the point at which the light strikes the glass. The bending of light in this way is called **refraction**. In contrast, when light leaves the glass and re-enters the air it speeds up once again and bends away from the normal. This process is illustrated in Figure 11.3.

Do we experience the effects of refraction in everyday life?

We do: one example is the distortion of the position of objects in water. Any child who has tried to catch fish using only a net knows that it is not as easy as it might

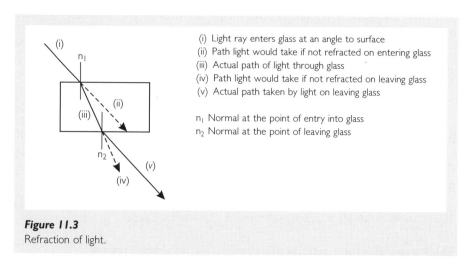

(i) Light ray enters glass at an angle to surface
(ii) Path light would take if not refracted on entering glass
(iii) Actual path of light through glass
(iv) Path light would take if not refracted on leaving glass
(v) Actual path taken by light on leaving glass

n_1 Normal at the point of entry into glass
n_2 Normal at the point of leaving glass

Figure 11.3
Refraction of light.

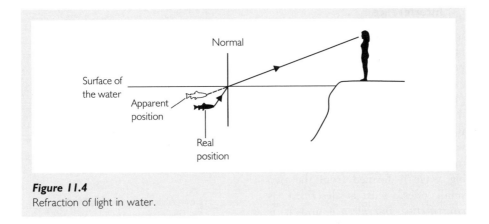

Figure 11.4
Refraction of light in water.

first seem – this is not only because fish can be very quick when they want to be! Have a look at Figure 11.4 for a moment.

> *What happens to a ray of light reflected off the body of the fish as it leaves the water?*

You should remember that as light moves to a less dense medium it accelerates and bends away from the normal. This creates a problem in the interpretation of the visual image for the brain of an observer, since it is 'used' to light travelling in straight lines. Consequently, the apparent position of the fish is closer to the surface of the water than the actual position. There is clearly an important point here – especially if you are involved in caring for children; bodies of water always appear shallower than they really are. Consequently, parents who warn their children 'it's deeper than you think' are, perhaps unwittingly, displaying a knowledge of refraction.

Refraction also accounts for the splitting of white light into its separate wavelengths when it enters a prism, as shown in Figure 11.5. This separation of colours is due to the fact that the different wavelengths of visible light are refracted at different angles. This effect is clearly an important consideration for manufacturers of optical instruments if colour aberrations are to be avoided. Of course, not all prisms are artificial. Droplets of water act as prisms, and the refraction of sunlight shining through rain droplets accounts for the formation of rainbows.

Once you have understood refraction you will find it much easier to comprehend the mechanism of vision and the behaviour of lenses, which we deal with later.

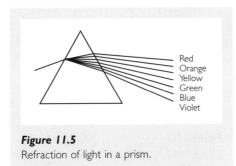

Figure 11.5
Refraction of light in a prism.

Interference

Interference is simply the effect of the meeting of two waves such as can be created in a tank of water. For example, if two waves which are in phase meet (crest meets crest and trough meets trough) the resultant wave formed has an amplitude which is equal to the sum of that of the two original waves. That is, one big wave is formed from the two small ones. This is described as *constructive interference* and it is illustrated in Figure 11.6. In contrast, if two waves which are 180° out of phase meet (crests and troughs meet) the result is *destructive interference*, and in effect the two waves cancel each other out, as Figure 11.6 also shows.

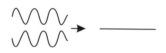

Constructive interference results from the meeting of two waves which are in phase

Destructive interference results from the meeting of two waves which are out of phase (in the above case by 180°)

Figure 11.6
Diagram showing interference.

Experiments have shown that light is capable of interference too, and this of course gives credence to wave theory. In fact, the colours seen in a film of oil, such as might be spilled on the road or on water, are the result of interference from light rays reflected from the top and bottom of the film. If one wavelength of light (colour) is removed by destructive interference an observer no longer sees white light but colours. Remember that white light results from the presence of all the wavelengths of visible light.

Polarisation

Unpolarised light contains waves which oscillate in all possible planes perpendicular to the direction of propagation. This is illustrated in Figure 11.7. In contrast, polarised light consists of waves which oscillate in one plane only. This is also illustrated in Figure 11.7. The fact that light can be polarised is further evidence that it exists as waves.

How do we make use of this?

You probably already know that polarising filters may be used in

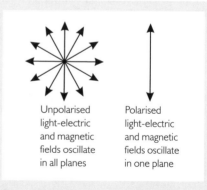

Unpolarised light-electric and magnetic fields oscillate in all planes

Polarised light-electric and magnetic fields oscillate in one plane

Figure 11.7
Unpolarised and polarised light.

sunglasses. In this case the filter transmits only vertically polarised light. Since sunlight is unpolarised, polarising filters transmit only half of the direct sunlight incident upon them. Furthermore, sunlight reflected from a flat surface, which is responsible for glare, is partially polarised in a plane parallel to the reflecting surface. Thus light reflected from a lake is partially polarised in the horizontal plane and therefore much less of it is transmitted by the filter. Consequently, polarising sunglasses do not simply cut down the transmission of light oscillating in all planes, as do ordinary tinted sunglasses, but instead they eliminate light which is not oscillating in a vertical plane, and this has the greatest effect upon reflected light (glare).

Lenses

A *lens* is a transparent object with a curved surface. The *cornea* and the lens of the eye are obvious examples of naturally occurring lenses, while artificial lenses are found in car headlamps, microscopes and spectacles. If a lens curves outwards so that its middle is thicker than its ends it is described as a *convex* lens. *Concave* lenses are those for which the reverse is true.

> *What happens to light when it enters a lens?*

Convex lenses

Figure 11.8 shows an example of a symmetrical double convex lens – that is, both sides of the lens are equally convex. Such lenses represent the simplest of cases to consider. A segment of the lens is enlarged so as to show what happens when a beam of light (i) enters the lens. Note the position of the normal to the surface of the lens which the light strikes (n_1) and the direction of the refracted ray as it passes through the lens (ii). Now note the position of the normal to the surface of the lens from which the light exits (n_2) and the direction of the ray as it re-enters the air (iii). Actually, since the lens concerned is symmetrical, we can simplify the diagram and imagine the light ray being refracted only once from a vertical line, referred to as the *principal axis*, which is drawn through the centre of the lens. This is also shown in Figure 11.8.

> *What is the effect of convex lenses on parallel rays of light?*

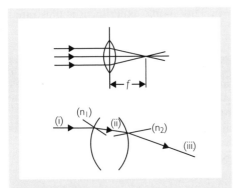

Figure 11.8
The effect of a convex lens on parallel rays of light.

As you can see, the light rays *converge*. Consequently, convex lenses may also be referred to as converging

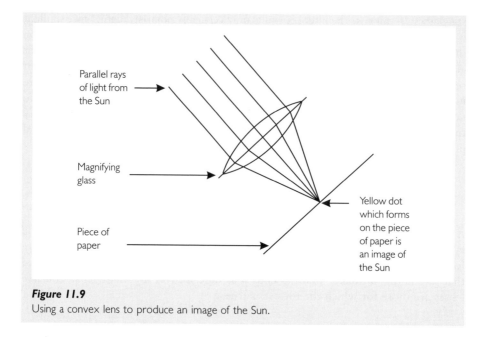

Parallel rays of light from the Sun

Magnifying glass

Piece of paper

Yellow dot which forms on the piece of paper is an image of the Sun

Figure 11.9
Using a convex lens to produce an image of the Sun.

lenses. The shortest distance between the lens and the point at which light rays converge is called the *focal length* and the point of convergence is the *focal point* or *principal focus*. Note that in the case of a convex lens the focal point is behind the lens. If a piece of paper is held at this point, parallel rays of light from a distant object pass through the lens and are brought to a focus on the paper. It is a simple experiment which every child who has owned a magnifying glass has performed on a sunny day. When the magnifying glass is held facing the Sun and the paper is held at the focal point a yellow dot comes into focus on the paper. This is actually an image of the Sun. Such an image is described as being *real* since it can be projected onto a surface and viewed. In addition, the image in this case is also diminished (smaller than the object) and inverted (upside down). This simple experiment is illustrated in Figure 11.9.

Practice point 11.1: convex lenses and the eye

Figure 11.10 is a cross-sectional diagram of the eye. The eye is a hollow spherical structure the outer part of which is formed from three coats. The outermost is the tough *sclera* and interior to this is the vascular *choroid*. The innermost layer is the light sensitive *retina*. At the front of the eye the sclera becomes transparent, and this is known as the *cornea*. The cornea is covered by a thin, transparent membrane called the *conjunctiva*, and this also forms the inner surface of the eyelids.

When we look at the eye from the front we can see through the cornea to a pigmented doughnut-shaped muscle which is part of the choroid.

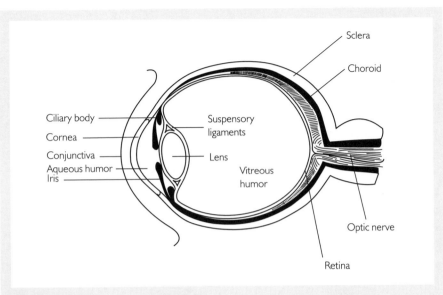

Figure 11.10
Cross-sectional diagram of the eye.

What is this pigmented muscle called?

This is the *iris*. It is the iris to which we refer when ask about the colour of some-one's eyes. The hole at the centre of the iris is called the *pupil*. The pupil becomes constricted (decreased in size) with the contraction of muscle cells in the iris which are arranged in a circular manner. In contrast the contraction of muscle cells which radiate out from the pupil cause it to dilate (enlarge). By this means the amount of light entering the eye under conditions of varying light intensity can be regulated.

Behind the iris the choroid forms a ring of muscle called the *ciliary muscle* (ciliary body) and to this the lens is attached by inelastic *suspensory ligaments*. The space behind the cornea but in front of the lens (*anterior chamber*) is filled with a watery fluid called the *aqueous humor*, while the space behind the lens (*posterior chamber*) is filled with a more viscous substance called the *vitreous humor*.

Figure 11.11 demonstrates how a real but inverted and diminished image of an object is formed on the retina. It is important to note here that although the eye contains a double convex structure called the lens, the cornea also acts like a lens too. In fact the cornea refracts light to a greater degree than does the lens itself.

So why have a lens at all?

The answer to this question is that the lens performs the important role of *accommodation*. When parallel rays from a distant object enter the eye they are focused on the retina by a relatively flat lens. This is shown in Figure 11.12. The lens is made

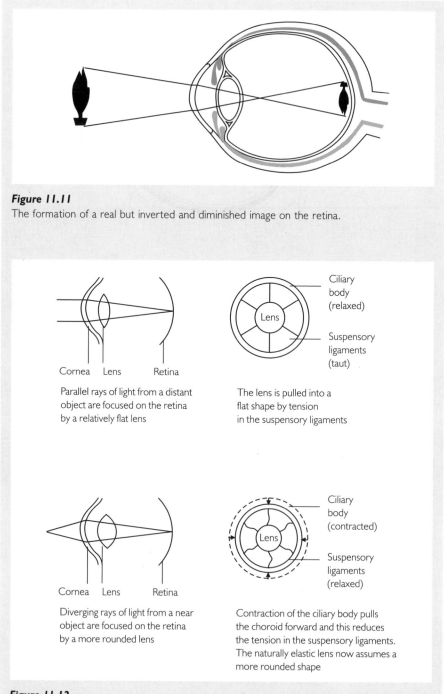

Figure 11.11
The formation of a real but inverted and diminished image on the retina.

Cornea   Lens   Retina

Parallel rays of light from a distant
object are focused on the retina
by a relatively flat lens

Ciliary
body
(relaxed)

Lens

Suspensory
ligaments
(taut)

The lens is pulled into a
flat shape by tension
in the suspensory ligaments

Cornea   Lens   Retina

Diverging rays of light from a near
object are focused on the retina
by a more rounded lens

Ciliary
body
(contracted)

Lens

Suspensory
ligaments
(relaxed)

Contraction of the ciliary body pulls
the choroid forward and this reduces
the tension in the suspensory ligaments.
The naturally elastic lens now assumes a
more rounded shape

Figure 11.12
Accommodation.

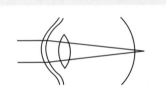

Figure 11.13
Hyperopia.

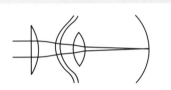

Figure 11.14
Correction of hyperopia.

flat by the relaxation of ciliary muscles. In contrast, diverging rays from a near object require a stronger lens and contraction of the ciliary muscles results in a more rounded lens. This is also shown in Figure 11.12. By this means the eye is able to accommodate both near and distant objects.

Convex lenses are also used in some spectacles. Consider Figure 11.13 for a moment.

What is wrong with this eye?

Light from the image is focused beyond the retina. This condition is called *hyperopia* (long-sightedness). What is required is an additional convex lens to correct the defect, and this is shown in Figure 11.14.

Concave lenses

Figure 11.15 illustrates the effect of a double symmetrical concave lens on parallel rays of light. Note that, once again, refraction can be regarded as having taken place at the principal axis.

What happens to the rays of light in this case?

They *diverge*. Consequently, concave lenses are referred to as diverging.

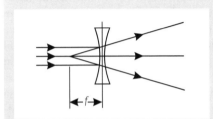

Figure 11.15
The effect of a concave lens on parallel rays of light.

Where is the focal point in this case?

A good question. To work out where the focal point is we have to trace back the diverging rays to discover that it is in front of the lens. This is also shown in Figure 11.15.

Can the image formed by a concave lens be seen?

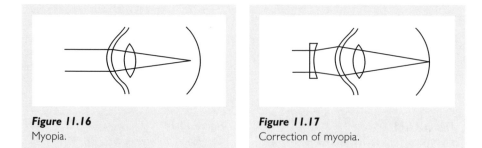

Figure 11.16
Myopia.

Figure 11.17
Correction of myopia.

No. Since the focal point is actually in front of the lens a screen placed there would obviously block the entry of light into the lens. In this case the image is described as being *virtual*. Concave lenses always produce virtual images.

> **Practice point 11.2: concave lenses and vision**
>
> Consider Figure 11.16.
>
> *What is the problem here?*
>
> In this case light from the image is actually focused in front of the retina – a condition called *myopia* (short-sightedness).
>
> *What could be done about this?*
>
> From what has been said already, you will realise that this problem can be corrected using a concave lens, as shown in Figure 11.17.

Optical instruments

The slide projector

When introducing convex lenses we explained how a real but diminished image of the Sun could be formed. Now we show how a convex lens may be used to form an enlarged image of an object in the manner shown in Figure 11.18. In order to illustrate how this is achieved, four rays of light labelled (i)–(iv) are shown. Rays (i) and (iv) are parallel and originate from the top and bottom of the object respectively. In this case the object is a slide of a tree. Note how these rays are diffracted and pass through the focal point. Unlike previous illustrations their continuing path is also shown. Rays (ii) and (iii) both pass through the centre of the lens and are not diffracted. The point of intersection of (i) and (ii) gives the top of the image, while the point of intersection of (iii) and (iv) gives the bottom of the image. Note that the image is both enlarged and inverted – perhaps you now understand why slides are inserted upside down into the projector. Such real, enlarged and inverted

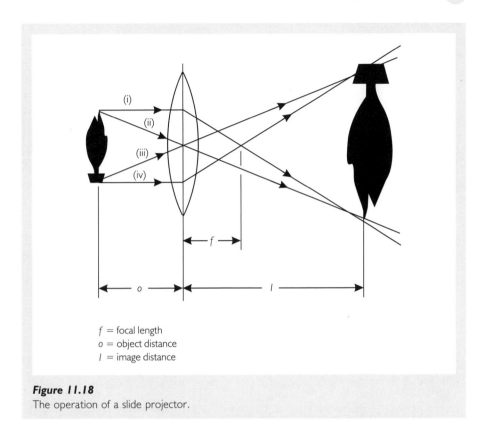

f = focal length
o = object distance
l = image distance

Figure 11.18
The operation of a slide projector.

images are produced by convex lenses when the object is placed a distance from the lens which is close to its focal length.

The magnifying glass

Suppose that we wanted to examine an object closely – to take a detailed look at it.

What would you do to achieve this?

This is not a trick question! You would obviously hold the object close to the eye. Fine: have a go at this – choose a small object and try to examine the detail of it.

Any problems?

Not initially. The closer the object is held to the eye the bigger the image formed on the retina and the easier it is to see the detail. However, if the object is brought very close it becomes out of focus and you can feel the eye strain. In the young, with healthy eyesight, an object may be brought as close as 10 cm and remain in focus.

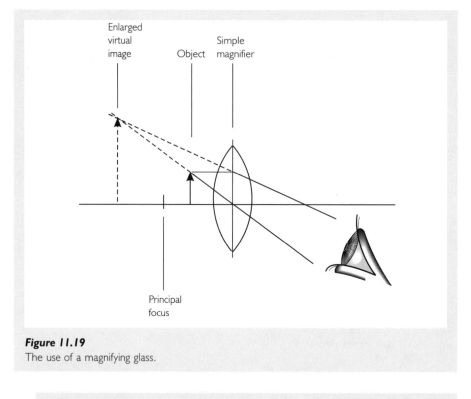

Figure 11.19
The use of a magnifying glass.

So how could we examine the object more closely?

The answer is obviously by using a magnifying glass, which is the simplest of all optical instruments and consists of a single convex lens. It is held close to the eye and the object positioned within its focal length so that a virtual image is formed in the manner shown in Figure 11.19. Note that this is the first time that we have encountered a virtual image produced by a convex lens, so it is probably worthwhile emphasising the arrangement once again. Note that the object is positioned within the focal length on one side of the lens while the observer looks through the other side. It is also worthwhile stressing here that virtual images can be seen, but they are referred to as virtual because they cannot be projected onto a screen. One clinical use of the magnifying glass is in the examination of skin lesions, such as might be undertaken in the dermatology (skin) clinic.

The microscope

The *microscope* is a more complex optical instrument than the magnifying glass since it contains two convex lenses. Its magnifying power is such that it can be used to examine objects which are not visible with the naked eye – cells for example. Once these are prepared and presented on a glass slide they form the *object*. The lens nearest the object is called the *objective*, and this always has a short focal length. It is positioned so that the object distance is just a little longer than the focal length of the

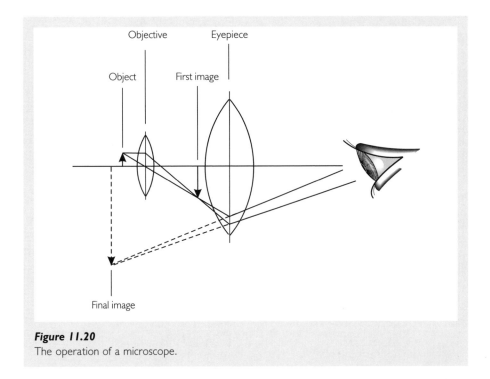

Figure 11.20
The operation of a microscope.

objective. In this way a real and magnified image is produced within the body of the microscope. This is shown in Figure 11.20. The lens through which the observer looks is called, unsurprisingly, the *eyepiece lens*, and this operates like a simple magnifying glass using the real image generated by the objective to produce a very much magnified virtual image.

> *How do we work out the magnification produced by the microscope?*

Usually a microscope has three objectives mounted on a block, which rotates so that they are readily interchangeable. A single eyepiece is used, although this may be removed from the barrel of the microscope and changed if desired. To calculate the magnification produced simply multiply the magnification produced by the eyepiece by that produced by the particular objective in use. A typical eyepiece magnification might be 10× and if used with a 20× objective the resultant magnification is obviously 200×.

Endoscopes

An **endoscope** is simply a tubular device which is inserted into a body orifice or surgical wound for the purpose of examining the appearance of an internal structure – an ulcer of the stomach perhaps. Samples of *exfoliated* (shed) cells may be taken with a brush inserted through the endoscope, or the device may be used to obtain a tissue sample (*biopsy*). The simplest endoscopes are rigid metal tubes, but their

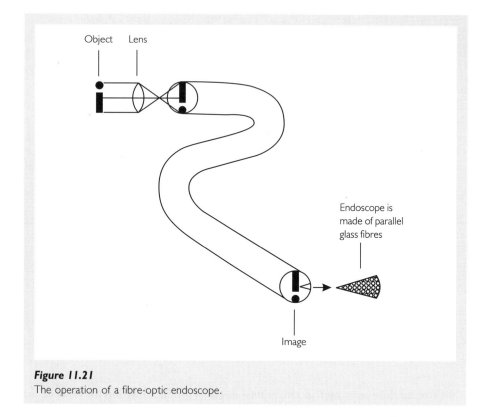

Object   Lens

Endoscope is
made of parallel
glass fibres

Image

Figure 11.21
The operation of a fibre-optic endoscope.

diameter and inflexibility limits their use. Flexible fibre-optic endoscopes have more widespread application. These consist of parallel bundles of flexible glass fibres. Light is passed down the outer fibres so that an object within the body is illuminated. Light reflected from the structure being observed then passes up each fibre by being reflected many times within it. Finally it emerges at the end of the fibre still outside the body. Provided the relative positions of the fibres remains fixed the image presented to the observer will consist of many pieces of visual information equal in number to the number of glass fibres. We might describe the image as a kind of mosaic – not one of tiles, of course, but of light. The operation of an endoscope is illustrated in Figure 11.21.

Endoscopes are available for different purposes and are named after the structure they are designed to visualise. The procedure of examination is described in a similar way. For example, *gastroscopes* are used to examine the stomach – a procedure called *gastroscopy*.

What about bronchoscopy, cystoscopy and colonoscopy?

We are sure that you will have no difficulty in working out what these procedures involve. You might also like to read about the nursing care required before, during and after endoscopy.

Lasers

The word **laser** is an acronym for light amplification by the stimulated emission of radiation. The creation of laser light is not described in this book, but it is helpful to know something about its nature. First of all, recall that white light consists of waves of different wavelengths – for this reason it is sometimes referred to as *incoherent*. Now imagine that light of a single wavelength is produced. Such light may be described as *monochromatic*, but it remains incoherent since the waves are not in phase. In contrast, laser light consists of monochromatic light in which all the waves are in phase, and the expression *coherent light* is used. The light produced may form part of the visible spectrum or it may be infrared or ultraviolet.

What medical use can be made of laser light?

The medical use of lasers is based upon a number of effects of laser light, including *photoblation* and *photocoagulation*. Photoblation is the destruction of tissue by a beam of laser light. The beam may be used for cutting tissue in a similar manner to a scalpel or for more widespread destruction, for example, of tumours. Photocoagulation is the coagulation of blood which occurs as a consequence of the heat produced by the laser light. Consequently, laser surgery is relatively bloodless.

Ultraviolet radiation

Ultraviolet radiation comprises electromagnetic waves within a range of wavelengths of 100–400 nm. It is generally the case that, as far as most of the electromagnetic spectrum goes, the shorter the wavelength the greater the penetrating ability of the waves. However, in the case of ultraviolet, the longer wavelengths penetrate deeper into the body. Three bands of ultraviolet radiation are described; and these are as follows:

ultraviolet A (UVA):   315–400 nm
ultraviolet B (UVB):   280–315 nm
ultraviolet C (UVC):   100–280 nm

UVC

This band has the shortest wavelength and is least penetrating. However, it has a high frequency and carries greater energy than do UVA or UVB.

Are considerations of penetrating power and energy important?

Yes. Since UVC penetrates no deeper than the skin yet carries greatest energy it has the greatest capacity to cause burns. Fortunately, UVC in sunlight is filtered out by the atmosphere's *ozone* layer.

> *Do you know how the ozone layer is currently being affected by environmental pollutants?*

You have probably heard about ozone depletion, and this clearly has important implications in terms of the amount of ultraviolet radiation which is incident upon our bodies.

Despite the potentially damaging effects of UVC we make good use of this band when it is produced artificially by germicidal lamps. As their name implies, such lamps are used to kill microorganisms (bacteria and fungi) and are employed as a means of sterilisation – for example, of the inside of inoculation cabinets used in the microbiology laboratory.

UVA

UVA is the band with the longest wavelength and greatest penetrating power. It is able to penetrate the *epidermis* to reach the *dermis*. However, it has a low frequency and therefore carries the least energy. UVA is a poor initiator of tumour development, but it does damage the enzyme systems responsible for repairing the genetic material (DNA/deoxyribonucleic acid) and so increases the carcinogenic effects of UVB. Because of the deeper penetration of UVA it damages the extracellular matrix of the dermis, causing the skin to become inelastic and wrinkled – an effect sometimes described as premature ageing.

UVA is present in sunlight, even in winter, and it is also produced by sunbeds, therapeutic lamps and even fluorescent tubes. UVA does not trigger tanning as strongly as does UVB, and this fact should be born in mind by those who use sun beds excessively, especially in view of the potential harm caused by UVA. However, UVA is used therapeutically in the treatment of some skin conditions. PUVA (psoralen UVA) is a form of photochemotherapy in which the patient takes an oral dose of the light-sensitising drug psoralen prior to exposure to UVA. It is used in the treatment of psoriasis.

UVB

UVB is sometimes described as the most active component of sunlight.

> *What does this term actually refer to?*

It is a reference to the fact that UVB in sunlight has the greatest impact on our bodies.

> *Why is this?*

Remember that UVC carries the most energy, so it is potentially the most damaging band. However, it is filtered out by the ozone layer. On the other hand, UVA is highly penetrating but carries less energy. In between these bands is UVB, which is of sufficient energy and penetrating power to cause harm. Figure 11.22 shows the penetration of UVA and UVB.

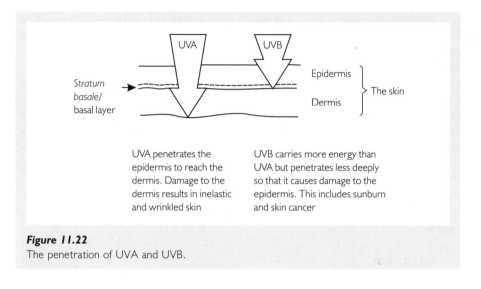

Figure 11.22
The penetration of UVA and UVB.

The effects of ultraviolet light on the body

Ultraviolet light has a number of effects on the body. Several of these are detrimental, but in addition there are positive effects too.

Vitamin D synthesis

UVB is responsible for the conversion in the dermis of the skin of the substance 7-dehydrocholecalciferol to cholecalciferol – one form of vitamin D (D_3). Vitamin D synthesis by the body also involves stages in the liver and kidney. The important point to note here is that we are not entirely dependent upon a dietary intake of vitamin D and that vitamin D synthesis is one positive benefit of UVB. You can find out more about vitamin D and the effects of deficiency in Chapter 6 (Biological molecules and food).

Tanning and sunburn

The colour of the skin of an individual depends upon a number of factors, including the presence of pigments. The most important of these is *melanin*. Melanin is produced in the basal layer of the epidermis by cells called *melanocytes*.

> *Does the number of melanocytes differ in different races?*

You might think so, but in fact this is not the case. The number of melanocytes does not vary with race, age or sex. However, the amount of melanin produced does vary with race and also with exposure to sunlight. In addition, the distribution of melanocytes is not uniform. There are relatively few on the trunk, but more in the penis, scrotum and areola of the nipple.

Melanocytes possess processes which project upwards between the superficial *keratinocytes* (skin cells). Melanin is contained in membrane-bound structures called *melanosomes*. Under the stimulation of ultraviolet, melanosomes pass along these processes and are taken up by the keratinocytes.

The rate of production of melanin increases with increasing exposure to ultraviolet light and it is the accumulation of melanin in the skin which is responsible for tanning. Melanin serves to protect cells from damage caused by ultraviolet light. DNA contained in the **nucleus** is particularly susceptible. Perhaps it is not surprising that melanosomes are not randomly distributed within the keratinocytes but instead are concentrated above the nucleus, thus affording the greatest protection. Despite the protective effects of melanin the development of a tan takes a considerable number of days, and many individuals who do not carefully regulate their exposure to the sun experience the inflammation and *erythema* (redness) which characterises sunburn.

Skin cancer

The incidence of skin cancer in Britain has risen rapidly in the past two decades, so that there are currently over 40 000 new cases and 2000 deaths each year, making it the second most common form of cancer.

Who gets skin cancer and is it difficult to treat?

Skin cancer is interesting since it is the only disease which is more common in the affluent than the poor and reflects the extent of exposure to sunlight. Actually, skin cancer is not a single disease – there are three forms. The most dangerous is *malignant melanoma*, which accounts for 10% of cases. Malignant melanomas spread rapidly, and this is one of the reasons why they are so dangerous. In addition they are more common in younger people and can be linked to periods of sunburn and overexposure. This means that health promotion advice about exposure to sunlight for young people is currently very important.

Basal cell carcinoma and *squamous cell carcinoma* are more common and account for 90% of cases of skin cancer. They are rarely fatal if treated early and it is thought that they result from cumulative exposure to the sun rather than to acute episodes of overexposure. Consequently, they tend to appear on the more exposed parts of the body in later life, especially in individuals who have worked out-of-doors.

Who provides health promotion advice about the effects of the sun?

Advice may be sought from general practitioners, practice nurses or pharmacists, or any nurse may have the opportunity to pass on information about avoiding the damaging effects of the Sun. In addition, most bottles of sunscreen provide the user with some information – perhaps in the form of a leaflet. Occasionally there are differences in the information given by different sources. For example, some classify skin types into four groups, while others use six groups. However, there is general agreement about the advice which is important, and some of this appears in the practice point below.

Practice point 11.3: precautions to take in the sun

1 Select a waterproof sunscreen with an adequate *sun protection factor* (SPF) for your skin. The SPF is a way of denoting the extent of protection afforded by different sunscreens. For example, an SPF of 15 multiplies the period of time it takes to burn by 15. Everyone should use a sunscreen with an SPF of at least 15, while those with paler skin should use one with a higher rating. The sunscreen chosen should also be *hypoallergenic* and perfume-free, and it should screen out UVA as well as UVB.

2 Apply the sunscreen 15–30 minutes before going out in the Sun and rub it in well. Repeat the application approximately every two hours throughout the day and additionally after swimming. Remember to use an SPF 15 lip balm too.

3 Avoid excessive exposure of the skin to the Sun. Wear light-coloured closely woven but loose-fitting clothing. Light colours reflect heat while dark colours absorb it, so light-coloured loose-fitting clothing will help you to feel cooler. However, lightweight clothing provides little protection against ultraviolet light, which will simply pass through it. Consequently, garments should be closely woven. The head and neck should be protected by wearing a wide brimmed hat.

4 Avoid the Sun between 11:00 and 15:00, especially in countries which are near the equator. At other times limit the period time spent in the Sun. Weather forecasts may indicate how long it is likely to take to burn, but clearly this is only a guide and does not take individual factors into account.

5 Wear sunglasses which conform to BS2724:1987, since prolonged exposure to ultraviolet may cause the lens to become opaque – a condition known as a *cataract*. Cataracts are the commonest cause of blindness, although sight may be restored by removal of the lens.

6 Avoid cosmetics, since these may increase sensitivity to the Sun.

7 Babies under six months of age should be kept out of the Sun altogether, while an SPF 50 preparation should be applied to children. In addition, children should always wear T-shirts and hats.

8 Ultraviolet light can be reflected by water, snow and buildings, so it is important to apply sunscreen even if you are sitting in the shade. Ultraviolet also penetrates cloud, so it is still possible to burn on an overcast day.

Summary

1 Transverse waves are those in which particles oscillate in a direction which is perpendicular to the direction of propagation of the wave.

2 Light is an example of an electromagnetic wave and it forms part of the electromagnetic spectrum.

3 Visible light has a wavelength of between 400–700 nm.

4 Different wavelengths of light are seen as different colours.

5 Refraction is the bending of light as it travels through transparent materials of different densities.

6 Lenses are transparent objects with curved surfaces which may be described as being either concave or convex.

7 Hyperopia is the condition in which an image is brought to a focus beyond the retina. It is corrected with a convex lens.

8 Myopia is the condition in which an image is brought to focus in front of the retina. It is corrected with a concave lens.

9 Ultraviolet light is made up of electromagnetic waves within the wavelength range 100–400 nm.

10 UVB is responsible for tanning and the synthesis of vitamin D, but it is also implicated in the development of skin cancer.

Self-test questions

11.1 Which of the following statements are true?

    a Frequency is measured in hertz (Hz)

    b Wavelength is the maximum displacement of a particle from its rest position.

    c The displacement of a particle in a wave is equal to twice its amplitude.

    d The period is the number of complete waves which pass a particular point in one second.

11.2 Which of the following statements are true?

    a Ultraviolet waves have a shorter wavelength than visible light.

    b Electromagnetic waves are mechanical waves.

    c Mechanical waves do not require a material medium through which to pass.

    d Infrared waves have a shorter wavelength than ultraviolet.

11.3 Which of the following statements are true?

    a When light enters a more dense medium it speeds up.

    b When light enters a more dense medium it bends towards the normal.

    c When light enters a less dense medium it speeds up.

    d When light enters a less dense medium it bends away from the normal.

11.4 Which of the following statements are true?

    a Destructive interference results from the meeting of two waves which are 180° out of phase.

    b Constructive interference results from the meeting of two waves which are 180° out of phase.

    c If a particle in the rest position of a wave is compared with one at a peak of an adjacent wave the two are said to be 90° out of phase.

    d If a particle at a peak of a wave is compared to one at a trough of an adjacent wave the two waves are said to be 180° out of phase.

11.5 Which of the following statements are true?

    a Concave lenses always produce real images.

    b A magnifying glass is an example of a convex lens.

    c A convex lens may be used to produce a real and magnified image.

    d In a microscope the objective produces a real, magnified image.

11.6 Which of the following statements are true?

    a In hyperopia the image is focused in front of the retina.

    b In myopia the image is focused in front of the retina.

    c Hyperopia is corrected using a convex lens.

    d Myopia is corrected using a concave lens.

11.7 Which of the following statements are true?

    a Myopia is sometimes referred to as short-sightedness.

    b Hyperopia is sometimes referred to as long-sightedness.

    c Parallel rays of light from a distant object are focused on the retina by a relatively flat lens.

    d Diverging rays of light from a near object are focused on the retina by a rounded lens.

11.8 Which of the following statements are true?

    a UVC carries more energy than UVB.

    b UVB is more penetrating than UVA.

    c UVA is more penetrating than UVC.

    d UVB carries more energy than UVA.

11.9 Which of the following statements are true?

    a UVA is used as a means of sterilisation.

    b UVB does not penetrate the ozone layer.

    c UVB has been implicated in skin cancer.

    d UVA is produced by sunbeds.

11.10 Which of the following statements are true?

    a Everyone should use a sunscreen with an SPF of at least 15.

    b Those with the palest of complexions should use a sunscreen with an SPF of 50.

    c Melanosomes are cells which produce melanin.

    d Increased exposure to sunlight results in increased melanin production.

Answers to self-test questions

11.1 a and c

11.2 a

11.3 b, c and d

11.4 a, c and d

11.5 b, c and d

11.6 b, c and d

11.7 a, b, c and d

11.8 a, c and d

11.9 c and d

11.10 a, b and d

Further study/exercises

11.1 We have described the production of an image upon the retina and we have mentioned visual pigments and vitamin A, but how is this image converted into a series of electrical impulses which are sent to the brain? (*Hint*: describe the action of the retinal cells, rods and cones.)

Tortora G. J. and Grabowski S. R. (1996). *Principles of Anatomy and Physiology*, 8th edn. New York: HarperCollins, pp. 467–471

11.2 Myopia and hyperopia have been described, but there are many vision defects. Find out about the following:

a Astigmatism

b Glaucoma

c Presbyopia

d Cataracts

e Night-blindness

Tortora G. J. and Grabowski S. R. (1996). *Principles of Anatomy and Physiology*, 8th edn. New York: HarperCollins, pp. 463–466

12 Sound and hearing

Learning outcomes

After reading the following chapter and undertaking personal study you should be able to:

1 Identify sound waves as mechanical waves.
2 Describe the characteristics of longitudinal waves.
3 Outline the decibel scale for measuring the perceived loudness of sound.
4 State the importance of wearing hearing protection in noisy environments.
5 Describe the mechanism of hearing.
6 Briefly describe the two main forms of deafness.
7 List some of the important aspects of communicating with someone who has a hearing impairment.
8 Outline the piezoelectric effect and briefly describe the medical use of ultrasound.
9 Describe the Doppler effect and outline the medical use of Doppler ultrasound.

Introduction

Unlike light waves, sound waves are *mechanical* in nature.

> *Do you remember what characterises mechanical waves?*

They require a material medium through which to pass, and they cannot therefore cross the vacuum of space. When we think of sound we usually think of waves passing through air, but sound can be transmitted through different kinds of substance – even our bodies, as we shall see later.

> *How is sound produced and what kind of waves are sound waves?*

Longitudinal waves

Sound waves are produced by vibration – for example, of the tuning fork illustrated in Figure 12.1. When the fork is struck the prongs vibrate – that is, they move in and out. When the prongs move outwards the adjacent air is compressed, but when they

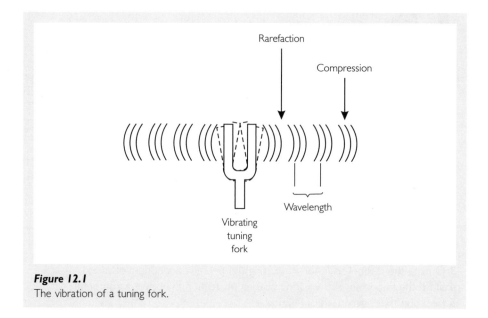

Figure 12.1
The vibration of a tuning fork.

move inwards again the air pressure between the prong and the compression falls; this is referred to as a *rarefaction*. This process is repeated and the series of compressions and rarefactions which results is referred to as a *longitudinal* wave.

> *Why is this term used?*

You should have worked out that it is because the air particles oscillate in the same direction as the direction of propagation of the wave (longitudinally). This is in contrast to transverse waves, where the direction of oscillation is perpendicular to the direction of propagation of the wave. The production of sound by the vibration of the tuning fork is not dissimilar to the way in which the vocal cords vibrate in speech production.

Practice point 12.1: speech

If we run the fingers down the anterior surface of the neck we can feel the rigid structure commonly referred to as the voice box or more correctly the *larynx* (Figure 12.2). The larynx is a hollow cartilaginous structure which is continuous with the *pharynx* above and the *trachea* below. It contains two strong elastic structures, called the *vocal cords*, which are separated by a space called the *glottis*. Sound is produced by the vibration of these cords, which in turn causes the air passing though the glottis to vibrate too.

Sound is normally only generated during expiration, while inspiration occurs silently. Consequently, inspiratory sounds may be important indicators of some form of pathology. For example, *stridor* is a high-pitched inspiratory sound associated

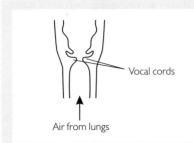

Vocal cords

Air from lungs

Figure 12.2
The larynx and vocal cords.

with spasm of the vocal cords (*laryngospasm*). It may occur following *bronchoscopy*, during an allergic reaction or in the presence of inflammation of the larynx (*laryngitis*). In contrast, the presence of growths on the vocal cords may impair their vibration and produce hoarseness, which also occurs with damage to the *recurrent laryngeal nerve*. Such damage may occur as a consequence of chest surgery or invasion of the nerve by a tumour.

Wavelength

In longitudinal waves the **wavelength** is the distance between two adjacent compressions or two adjacent rarefactions.

Frequency

The *frequency* of longitudinal waves is defined in the same way as for transverse waves – it is the number of complete waves which pass a particular point in one second. In everyday language we often refer to *pitch* rather than frequency. We describe high-frequency sound waves as being high-pitched and vice versa.

> *Is pitch the same as frequency?*

Not quite. Frequency is a measurable quantity, while pitch refers to the individual's perception of the sound. Human ears can detect sound within the frequency range 20–20 000 Hz, but of course it does vary somewhat between individuals.

In addition, although we are familiar with the fact that ageing often results in some degree of reduction of sensitivity to sound, did you know that the range of frequencies which can be heard becomes narrower? There is in fact a selective loss of hearing which affects the higher frequencies. In order to obtain some idea of what it is like to experience this, perform the following simple experiment. Turn on the radio and adjust the volume to a little below a comfortable listening level. This simulates a reduction of overall sensitivity. Now adjust the tone to full bass. This simulates a reduction of sensitivity to higher frequencies. You will still be able to hear what is being transmitted, but you will find yourself having to concentrate quite hard to understand the spoken word.

> *Is there a practical point here? What do we do when trying to communicate with someone who is partially deaf?*

We obviously try to speak more loudly, but in doing so our voice becomes higher pitched, and this may be less easily understood by the patient.

So what determines the frequency at which an object will vibrate?

Perhaps the most obvious is the length of the vibrating medium. The longer the medium the lower the frequency produced. The term *medium* sounds a little vague, but it is used here to encapsulate vibration in all kinds of substances including the metal prongs of a tuning fork, the strings of a stringed instrument or the column of air in a wind instrument. In the case of a tuning fork, the longer the prongs the lower the frequency of the note produced. Some stringed instruments, such as the piano, contain a large number of strings of different lengths, and different frequencies are produced simply by striking different strings. In other stringed instruments, such as the guitar or violin, the frequency of the note produced is changed by compressing the string, and this has the effect of varying the length of the vibrating section. In wind instruments the length of the vibrating column of air may be changed using a system of valves or, in the simple case of a recorder, by covering a different number of holes in the body of the instrument.

The tension in the medium also affects frequency – the greater the tension the higher the frequency. This principle is also used in stringed instruments, which are tuned by adjusting the tension in each of the strings so that each vibrates at a specific frequency. Different frequencies are also produced by the vocal cords through the mechanism of varying the tension within them. The vocal cords are stretched so as to produce higher frequencies and relaxed to produce lower frequencies.

Period

This is the length of time it takes for one wavelength to pass a particular point.

Velocity

The velocity of sound in air increases with increasing air temperature. At room temperature ($20\,^{\circ}$C) the velocity of sound in air is $343\,\mathrm{m\,s^{-1}}$. The velocity of sound also increases with the density of the medium through which it is passing. For example, the velocity of sound in water, also at $20\,^{\circ}$C, is much faster than in air – $1540\,\mathrm{m\,s^{-1}}$.

Amplitude

This is the maximum displacement of a particle from the rest position. The amount of energy carried by the wave and the loudness of the sound are both related to amplitude.

Intensity and loudness

We have already noted in Chapter 11 that intensity is an expression of the amount of energy carried by a wave – it is also proportional to the square of the amplitude.

What is the relationship between intensity and loudness?

Let us imagine a sound of pure frequency – say 5000 Hz. As the intensity increases, then so does the perceived loudness.

So we can equate intensity with loudness?

Not exactly. Intensity is a measurable quantity, while loudness is a subjective perception. The fact is that the human ear is not equally sensitive to all frequencies that are audible, so that two different frequencies each with the same intensity (that is, carrying the same amount of energy) may be perceived as having different degrees of loudness. What is needed is a scale which enables us to compare the perceived loudness of different sounds – the **decibel** scale.

Table 12.1 contains a list of different sounds and for each is identified an approximate rating on the decibel scale along with a figure for the multiple threshold of intensity – that is how many more times the sound is louder that the threshold of hearing. Note that the decibel scale is a *logarithmic* scale. To understand the implications of this, note that rustling leaves are rated at 10 dB and a quiet room at 20 dB.

Does this mean that an empty room is twice as noisy as rustling leaves?

No – look at the multiple of the threshold of intensity. The sound of a quiet room is 100 times (10^2) more intense that the threshold (1), while the sound of rustling leaves is only 10 times louder than the threshold. This means that a quiet room is 10 times as loud as rustling leaves.

Table 12.1 The decibel rating and multiple threshold of intensity for various sounds.

Sound	Decibel level (dB)	Multiple of threshold intensity
Threshold of hearing	0	1
Rustling leaves	10	10
Quiet room	20	10^2
Empty street	30	10^3
Room in house – no-one talking	40	10^4
Quiet conversation	50	10^5
Normal conversation	60	10^6
Loud conversation	70	10^7
Lecture	80	10^8
N.B. Danger of hearing damage with prolonged exposure		
Loudest parts of orchestral music, factory machinery	90–100	$10^9 - 10^{10}$
Loud indoor music	110	10^{11}
Pain threshold	120	10^{12}

Practice point 12.2: hearing protection

Since noise may induce hearing loss (see later), protecting the ears from excessive noise is obviously important. Employers are required to follow strict standards of noise control and provide ear protection for their employees. Many factories have noise levels in excess of 90 dB due to the operation of machinery. Some obvious examples of noisy environments include printing works and bottling plants. Clearly, employees should also be encouraged to take responsibility for their own health and wear the ear protectors provided. The occupational health nurse may have a role in encouraging this. Practices such as using radio headphones beneath ear protectors should be discouraged, since the noise level produced by playing music under such circumstances can be excessive. Clearly, self-inflicted hearing damage caused by the playing of loud music and visiting discos and raves cannot be legislated against, but in Britain local authorities are able to take action against persistent 'noise polluters'.

Resonance

An object may be caused to vibrate at a range of different frequencies, but it will vibrate maximally at a particular frequency. This frequency is called the *resonant* frequency. If a sound wave with the same frequency as the resonant frequency of the object is produced the object will begin to vibrate in time with the wave. This effect is easily demonstrated using two identical tuning forks which have the same resonant frequency. One is struck and the note produced can be heard. In addition the prongs can be seen to be vibrating. When this fork is held near the second fork this too begins to vibrate, and we say that it is resonating.

Resonance is actually an everyday experience. For example, you might be driving a car and note that at a particular speed a part of the dashboard, or an object within the car, starts to vibrate and make a sound.

Why is this?

It is because at a particular speed the vibration produced by the car engine matches the resonant frequency of a component in the dashboard or an object within the car, and these begin to vibrate maximally in time with the engine vibration. Go a little slower or a little faster and the sound disappears, since the engine vibration no longer matches the resonant frequency of the object.

Is resonance important within the body?

Yes. The resonance of air in the mouth, nose, pharynx and nasal sinuses is responsible for the characteristics of the voice produced. In addition, the *basilar membrane* of the inner ear (see later under hearing) contains fibres of different lengths, and these resonate maximally at different frequencies. They thus form the basis of the detection of sounds of different frequency.

The mechanism of hearing

The mechanism of hearing involves the outer ear, middle ear, inner ear and the brain. The brain undertakes the interpretation of nervous impulses conveyed to it via the cochlear nerve, and this is not discussed here. Instead, we shall describe the structures involved in the detection of sound.

The outer ear

The outer ear consists of the *pinna*, the *external auditory canal* and the *tympanic membrane* (eardrum). Locate these structures on the diagram in Figure 12.3. The pinna is the external structure which we commonly refer to as the ear in everyday language. It is not as prominent in humans as in some other animals, but its role is to direct sound waves into the external auditory canal. This is a tubular structure which is open to the atmosphere at its external end, but the internal end is closed by the tympanic membrane.

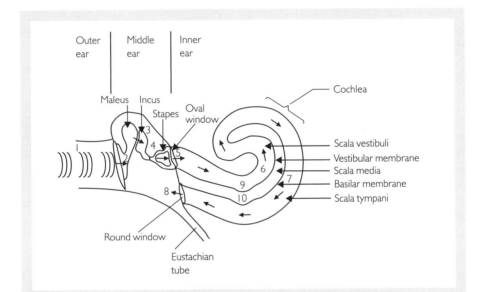

Figure 12.3
The outer ear, middle ear and inner ear. (1) Sound wave passes along the external auditory canal. (2) A compression causes the tympanic membrane to bulge inwards, and as a consequence it pushes against the malleus. (3) The malleus pushes against the incus. (4) The incus pushes against the stapes. (5) The stapes pushes against the oval window and causes it to bulge inwards. (6) The wave is transmitted through the perilymph of the scala vestibuli. (7) The wave continues through the perilymph of the scala tympani. (8) A compression causes the round window to bulge outwards (into the middle ear). (9) The vestibular membrane vibrates in time with the wave. The point of maximal vibration is determined by the frequency of the wave. (10) The basilar membrane vibrates in time with the vestibular membrane.

> *The external auditory canal contains ceruminous glands. What do these glands secrete?*

They secrete *cerumen* (wax) which, together with hairs, discourage insects from entering and help to trap dust too. The end of the external auditory canal inside the skull is covered by the tympanic membrane.

> *What is the effect of sound waves upon the tympanic membrane?*

They make it vibrate.

The middle ear

The middle ear is a small chamber inside the skull which contains three small bones referred to as the *ossicles*. These bones are named the *malleus*, the *incus* and the *stapes*. The literal meaning of these names is the hammer, the anvil and the stirrup and they do indeed look a little like the objects after which they are named. The malleus is in contact with the tympanic membrane and, as a consequence, the two vibrate in time with each other. The malleus in turn moves the incus which in turn moves the stapes. Finally, the stapes pushes against a covered opening called the *oval window*.

> *What exactly do these bones achieve?*

They may be likened to a set of levers which serve to increase the amplitude of vibration. This may be as much as by 20 times. This is important because the wave which passes down the air-filled tympanic membrane will ultimately be transmitted in the fluid of the inner ear, and a greater force is required to set fluid in motion than air. Before we move on to consider the inner ear we should also note that a second membrane-covered opening exists below the oval window, and this is the *round window*. Its role will become clear a little later. Note too that the middle ear is not a closed space – it is vented by a tube called the *eustachian tube*.

> *What is the role of the eustachian tube?*

It connects the middle ear with the pharynx and allows pressure differences on either side of the tympanic membrane to be equalised. You can find out more about this in Chapter 5 (Pressure, fluids, gases and breathing).

The inner ear

The inner ear is formed by a number of structures, some of which are concerned with balance rather than with hearing. The part of the inner ear which is concerned with hearing is the *cochlea*, which is illustrated in Figure 12.3. The cochlea is a spiral-shaped chamber which contains three fluid-filled channels – the *scala vestibuli*, the *scala media* and the *scala tympani*. The scala vestibuli begins on the opposite side of the oval window to the stapes, spirals up to the apex of the cochlea and

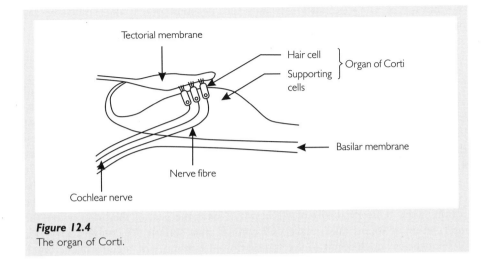

Figure 12.4
The organ of Corti.

leads to the scala tympani, which spirals down the cochlea to the round window. The fluid which fills these two channels is called *perilymph*. Note that the movement of the stapes back and forth causes the oval window to vibrate too, and this sets up a longitudinal wave in the perilymph which is eventually transmitted to the round window. As a consequence, the oval and round windows move synchronously but in opposite directions.

Now let us consider the remaining chamber – the scala media. It too is filled with fluid, but in this case the fluid is called *endolymph*. Note that the scala media is, in effect, sandwiched between the two other membranes. Because of this the floor of the scala vestibuli (*vestibular membrane/Reissner's membrane*) forms the roof of the scala media and the floor of the scala media (*basilar membrane*) forms the roof of the scala tympani. Make sure that you can locate these structures on the diagram in Figure 12.3 before we proceed.

The apparatus of sound detection is called the *organ of Corti*, and it is located in the scala media. It is illustrated in Figure 12.4. The organ of Corti contains *auditory receptor cells* which have microscopic hair-like projections called *cilia*. For this reason they are often referred to as hair cells. The cilia are attached to the *tectorial membrane*, which floats in the endolymph above the hair cells. In the basilar membrane the hair cells are connected to sensory nerve cells (neurons) which collectively form the cochlear nerve.

> How does this arrangement enable us to hear?

We have already noted how vibrations are set up in the perilymph by the movement of the oval window. These vibrations are readily transmitted across the vestibular membrane, which is very thin. This means that the endolymph also vibrates in sympathy with the oval window. Vibrations of the endolymph then cause the basilar membrane to vibrate. Although the entire length of the basilar membrane is capable of vibration, it vibrates maximally at a specific point that is determined by the frequency of vibration of the endolymph.

Have you recognised that this is an example of resonance?

The basilar membrane contains fibres, called *stiff fibres*, of different lengths, all of which begin at the base of the cochlea. Many pass part-way along the basilar membrane, while the longest pass all the way up to the apex of the cochlea. In the case of low-frequency sounds the basilar membrane vibrates maximally at its apex, while in the case of high-frequency sounds the basilar membrane vibrates maximally at its base.

At the point of maximum movement of the basilar membrane the motion of the cilia of the hair cells initiates nervous impulses which are conveyed to a specific region of the *auditory cortex* of the brain. (This is the part of the brain where the interpreting of sound takes place.) Of course, most sounds are complex mixtures of frequencies which result in the stimulation of a number of regions of the organ of Corti and finally to patterns of nervous impulses which the brain has to interpret. More frequent impulses result from sound waves with greater amplitude (loud) than from those with lower amplitude (quiet). In this way the brain receives information about both the intensity of sound and its frequency.

Practice point 12.3: deafness and its forms

A loss of hearing may be partial or complete. Some forms of deafness occur in children, while others are associated with ageing.

When deafness is suspected, how is the hearing assessed?

The simplest form of hearing test is conducted on babies by Health Visitors as part of the developmental assessment. An attempt is made at distraction using a rattle in order to confirm that the infant can indeed hear. In older children and adults an audiometric test is used. This involves listening to sounds of different frequency and intensity through headphones and reporting when a sound can be heard; for example, by pressing a signalling button.

Are there different forms of deafness?

Yes. The causes of deafness can be placed in one of two groups – *conductive* deafness and *sensorineural* deafness.

Conductive deafness

This form results from an impaired conduction of sound to the cochlea and it therefore results from a problem affecting the outer ear or middle ear. Possible causes include the occlusion of the external auditory canal by wax, the accumulation of tenacious secretions in a middle ear infection (*otitis media* – sometimes called glue ear) and damage to the ossicles. As a general rule, conductive deafness is easier to treat than sensorineural deafness. Some treatments are simple indeed. For example, accumulated wax is softened by the daily instillation of olive oil into the external auditory canal over a period of about five days followed by ear syringing. Although this is a simple procedure commonly performed by practice nurses, and others too, there is a risk of causing damage to the tympanic membrane, so only properly trained staff should carry out the procedure. Otitis media is not at all uncommon in children and it may initially be treated with antibiotics. However, if persistent, the tympanic membrane may be perforated surgically and a grommet inserted to allow the fluid in the middle ear to be discharged. Another mechanical problem is the age-related degeneration of the ossicles, which results in their failure to magnify the amplitude of vibration. This form of deafness is usually managed by electronic amplification – that is, the affected individual wears a hearing aid.

Sensorineural deafness

Damage to the cochlea or its associated nervous pathways results in sensorineural deafness. In this form of deafness sounds are not only perceived as being quieter but may also be distorted, and *tinnitus* may be present. (Tinnitus is the subjective experience of noise.) As a consequence a hearing aid may be of little value. Sensorineural deafness may occur as part of the ageing process (*presbycusis*), may be induced by drugs (such as the antibiotic gentamicin), may follow some infections (such as measles, mumps, chickenpox and rubella) or may be induced by noise.

Practice point 12.4: communicating with someone who has a hearing impairment

When communicating with someone who has a hearing impairment it is useful to bear in mind the following points:
1 Do not speak until you have the person's attention.
2 Position yourself directly in front of the person.
3 Do not turn around while still talking to the person.
4 Find out if the person can lip read or if a hearing aid is normally used. This information should be recorded in the nursing notes.
5 If a hearing aid is normally used it should be fitted and working.
6 Do not use exaggerated lip movements.
7 Speak slowly, raising the voice only a little.
8 Confirm that the patient has understood what has been said before continuing a conversation.
9 Rephrase sentences if a difficulty is experienced.
10 Be patient – do not give up and abandon any attempt to communicate.
11 If verbal communication is impossible, try other means such as gestures. You may have to learn some sign language in order to communicate with people who use it.

Ultrasound and the piezoelectric effect

When certain crystals, such as quartz, are deformed by a force, a small voltage is created across them. This effect is termed the *piezoelectric effect*, and crystals which demonstrate it are called piezoelectric crystals. In contrast, if a small voltage is applied across such crystals they then demonstrate a deformation. This is called, unsurprisingly, the *reverse piezoelectric effect*. Alternating current produces intermittent deformations – that is, the crystal vibrates. These vibrations generate sound, the frequency of which is higher than that which can be heard by human ears – we call it **ultrasound**. Piezoelectric crystals are not recent discoveries – the piezoelectric effect was first observed in the late 1800s and in the First World War ultrasound detection of submarines was developed (sonar – sound navigation and ranging).

Practice point 12.5: ultrasound and the body

Ultrasound may be used to produce an image of the internal structures of the body. This involves placing an ultrasound transducer against the body surface and directing the ultrasound waves towards the structure to be imaged. You may recall that a transducer is a device for converting one form of energy into another. The ultrasound transducer uses the reverse piezoelectric effect to generate ultrasound and the piezoelectric effect to detect ultrasound echoes (reflections) bounced back from internal structures. Such reflections occur at any interface between tissues of different densities – for example, between healthy tissue and a tumour. These reflections are then used to produce a visual image of the structure which has been scanned.

Does ultrasound have an advantage over other scanning techniques?

Yes, it has some. Ultrasound scanners are portable and less expensive than many other imaging devices. An ultrasound scan is not invasive and involves little discomfort for the patient. In addition, it is without the risks associated with ionising radiation (see Chapter 13: Radiation in diagnosis and treatment).

Do you know of some of the uses of ultrasound?

Perhaps you have realised that, since ultrasound is believed to be a safe method of scanning, it finds use in obstetrics – in the imaging of the foetus. Ultrasound scans can be used to determine the age of the foetus and to detect foetal abnormalities such as *spina bifida*. Abdominal organs such as the liver may also be scanned using ultrasound, and the detection of *neoplasms* (new growths such as a cancer) is one obvious use.

Finally, short duration high-intensity ultrasound waves may be used to break up kidney stones – a procedure called *lithotripsy*. The alternative would be the surgical removal of the stones.

The Doppler effect

Have you noticed that the pitch of the sound of a speeding motorcycle increases as it gets nearer but decreases as it moves away? This effect is called the Doppler effect.

Would you notice the same change in pitch if you were riding the motorcycle?

No. Provided that the motorcycle was travelling at constant speed, the engine note would remain the same.

Why is it that an observer notices a change in pitch?

In order to answer this question, remember that pitch is related to frequency – we perceive high-frequency sounds as being high-pitched. Remember too, that frequency is the number of waves which pass a particular point in one second. The motorcycle engine emits a sound wave of a certain frequency (*true frequency*) but, as it travels towards us, we receive a greater number of waves per second – that is the *received frequency* is greater. In contrast, as the motor cycle travels away from us, the received frequency is less than the true frequency. The difference between the true frequency and the received frequency is described as a Doppler shift.

So does the Doppler effect have any medical uses?

It does, as the following practice point makes clear.

Practice point 12.6: the Doppler effect and the body

First of all, let us note that the Doppler effect can be demonstrated with all frequencies of sound, including ultrasound.

What would happen if we were to direct an ultrasound wave at flowing blood?

We would be able to detect a Doppler shift in the frequency of the reflected wave. If the blood were flowing towards the transducer the reflected wave would have a higher frequency than that of the original wave, but if the blood were flowing away from the transducer the reflected wave would have a lower frequency than that of the original wave.

What if we directed the ultrasound wave to a narrowed part of the vessel?

Remember that the velocity of blood in narrowed vessels is increased? If you are unsure about this, check out Chapter 5 (Pressure, fluids, gases and breathing). For now it is sufficient to note that the Doppler effect can be used to detect changes in blood flow such as might occur in disease processes such as atherosclerosis.

Summary

1 Sound waves are mechanical longitudinal waves.

2 The perceived intensity of sound is measured on the decibel scale.

3 The mechanism of hearing involves the conduction of sound waves through air in the outer ear, through the bones of the middle ear and through the fluid of the inner ear.

4 Deafness may be classified as either conductive or sensorineural.

5 Sensorineural deafness may result from noise exposure, and wearing hearing protectors in noisy environments is clearly important.

6 Internal structures may be scanned using ultrasound waves produced by the reverse piezoelectric effect and detected using the piezoelectric effect.

7 Short duration high-intensity ultrasound waves may be used to break up kidney stones (lithotripsy).

8 The Doppler effect may be used to detect the narrowing blood vessels in disease processes such as atherosclerosis.

Self-test questions

12.1 Which of the following statements are true?

  a Sound waves are electromagnetic.

  b Sound waves are longitudinal.

  c Sound waves are mechanical.

  d Sound waves are transverse.

12.2 Which of the following statements are true?

  a One wavelength is the distance between adjacent rarefactions.

  b One wavelength is the distance between adjacent compressions.

  c One wavelength is the distance between a compression and an adjacent rarefaction.

  d One wavelength is the number of waves which pass a point in one second.

12.3 Which of the following statements are true?

  a The longer a vibrating medium the lower the frequency of sound produced.

  b The shorter a vibrating medium the higher the frequency of sound produced.

  c The greater the tension in the vibrating medium the lower the frequency of sound produced.

  d The greater the tension in the vibrating medium the higher the frequency of sound produced.

12.4 Which of the following statements are true?

  a Sound is conducted through the outer ear in air.

  b Sound is conducted through the middle ear in fluid.

  c Sound is conducted through the inner ear in fluid.

  d The inner ear is vented by the eustachian tube.

12.5 Which of the following statements are true?

  a The period is the length of time it takes one wavelength to pass a particular point.

    b The velocity of sound in water is greater than the velocity of sound in air.

    c The velocity of sound in a medium decreases with increasing density.

    d Two waves with different frequencies but with the same intensity may be perceived as having different degrees of loudness.

12.6 If an empty street is rated at 30 dB and normal conversation at 60 dB, how much louder is normal conversation than an empty street?

    a Twice as loud.

    b Half as loud.

    c One hundred times as loud.

    d One thousand times as loud.

12.7 Match the structures on the left with the appropriate locations on the right.

a Organ of Corti	i	Outer ear
b Ossicles	ii	Brain
c Cerumen	iii	Middle ear
d Auditory cortex	iv	Inner ear

12.8 Match the structures on the left with the appropriate description on the right.

a Basilar membrane	i	'Roof' of scala tympani
b Vestibular membrane	ii	Located within the scala media
c Tympanic membrane	iii	Outside the cochlea
d Tectorial membrane	iv	'Roof' of scala media

12.9 Which one of the following structures does not form part of the middle ear?

    a Malleus

    b Eustachian tube

    c Stapes

    d Hair cell

12.10 Which one of the following is a form of sensorineural deafness?

    a Accumulated wax in the external auditory canal.

    b Presbycusis.

    c Tenacious fluid in the middle ear.

    d Degeneration of the ossicles.

Answers to self-test questions

12.1 b and c	**12.8** a i
12.2 a and b	b iv
12.3 a, b and d	c iii
12.4 a and c	d ii
12.5 a, b and d	**12.9** d
12.6 d	**12.10** b
12.7 a iv	
b iii	
c i	
d ii	

Further study/exercises

12.1 There has recently been a good deal of attention given to cochlear implants. What are they and who benefits from them?

Bysshe J. (1995). Deafness in childhood: 3. Cochlear implants: who, why and what are the results? *Professional Care of the Mother and Child*, **5**(5), 135–137

Bray M. A., Neault M. W. and Kenna M. (1997). Cochlear implants in children. *Nursing Clinics of North America*, **32**(1), 97–107

12.2 It is well known that hospitals can be noisy places to be in. The next time you are in a care setting where patients need to rest, identify possible sources of noise and suggest ways in which the noise level could be reduced. This exercise might be particularly important at night. Can you find any nursing research on this subject?

Biley F. (1994). Effects of noise in hospitals. *British Journal of Nursing*, **3**(3), 110–113

Hilton B. A. (1986). Noise: who says hospitals are quiet places? *Canadian Nurse*, **82**(5), 24–28

13 Radiation in diagnosis and treatment

Learning outcomes

After reading the following Chapter and undertaking personal study you should be able to:

1 Distinguish between ionising radiation and non-ionising radiation.
2 Describe four forms of ionising radiation and compare their penetrating ability.
3 Explain what is meant by the term *radioactive decay* and describe alpha decay and beta decay.
4 Explain the concept of half-life.
5 Explain how X-rays are produced.
6 Outline the biological effects of ionising radiation.
7 Describe important aspects of radiation protection for health care workers.
8 Briefly describe the following imaging techniques: simple X-ray, cine radiography, computerised tomography and radionuclide scans.
9 Describe the use of ionising radiation in the treatment of cancer.

Introduction

In Chapter 11 (Light and vision) we described visible light as an *electromagnetic* phenomenon and noted that it forms only a small part of the *electromagnetic spectrum*. We also noted that electromagnetic waves radiate from their point of origin and that the word *radiation* may be used to describe them. When used in this way the term *radiation* simply refers to something being emitted, and although electromagnetic waves are often referred to as radiation the term may also be applied to particles too. Some of these particles are described a little later. Furthermore, although mention of the term radiation conjures thoughts of danger, no such meaning is implied in the word itself. When we express a fear of radiation we are, in fact, referring to ionising radiation. Ionising radiation consists of electromagnetic waves or particles, which, when striking matter, cause the production of ions. It is ionising radiation which is considered in this chapter.

The nature of ionising radiation

Ionising radiation includes high-frequency electromagnetic rays such as X-rays and gamma rays, as well as particles such as alpha particles and beta particles.

Alpha particles (helium nuclei)

Alpha particles are actually *helium nuclei* – that is, helium atoms from which two electrons have been lost. Helium nuclei may be represented in the following way:

$$^{4}_{2}He$$

To what does the lower figure refer?

Remember that the lower figure is the *atomic number*; that is, the number of *protons* in the nucleus.

What then is the upper figure?

You will recall that this is the *mass number* – the sum of the number of protons and *neutrons*. In the case of a helium nucleus there are 2 protons and 2 neutrons and the mass number is therefore 4.

Alpha particles possess a positive charge (2+) and because of this they have a relatively poor penetrating power, as Figure 13.1 shows. Alpha particles may in fact be stopped by a sheet of paper and they certainly do not penetrate deeper than the skin.

Beta particles (electrons)

Beta particles are actually *electrons*.

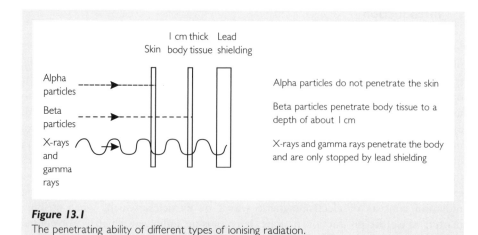

Figure 13.1
The penetrating ability of different types of ionising radiation.

Do you remember what electrons are?

No doubt you recall that they are sub-atomic particles which carry a single negative charge. Once again the possession of charge means that electrons have a poor penetrating power, although this is greater than that of alpha particles. Beta particles penetrate living tissue to a depth of about 1 cm and may be stopped by a thin sheet of metal foil.

Gamma rays and X-rays

These are both highly penetrating forms of electromagnetic radiation. Both readily penetrate the body and are stopped only by lead shielding several centimetres thick. Although gamma rays and X-rays share the same nature they are produced differently. Gamma rays are emitted by the unstable nuclei of certain elements. A few of these are naturally occurring, but most are produced in nuclear reactors. In contrast, X-rays are produced electrically in an *X-ray tube* or a *linear accelerator*.

Radioactivity and radioactive decay

The nuclei of the **isotopes** of certain elements are unstable and break-up.

Remember that isotopes are atoms which have the same atomic number but different mass numbers.

That is, they are atoms of the same element, since they have the same atomic number – the same number of protons in the nucleus – but different numbers of neutrons.

In the process of breaking up, such isotopes spontaneously emit gamma rays or particles, and as a consequence they are referred to as **radioisotopes**. The spontaneous particulate or electromagnetic emissions are referred to as **radioactivity** and the break-up of the nuclei which leads to these emissions is called *radioactive decay*. From what we have noted above about the nature of radiation, you will no doubt have realised that there are three forms of radioactive decay.

Alpha decay

One example of an alpha emitter is uranium-238 (^{238}U). This is chosen as an example because you may very well have heard of uranium even if you have not studied science before.

To what does the number after the name of this element refer?

It is not the atomic number of uranium, since all isotopes of uranium have an atomic number of 92. It is in fact the mass number of a particular isotope of uranium. When thinking about radioactive isotopes, the mass number is always given along with the name of the element, so that different isotopes are readily distinguished.

> *What happens when an atom of uranium-238 emits an alpha particle?*

To answer this question examine the incomplete equation below:

$$^{238}_{92}U \longrightarrow {}^{4}_{2}He + \text{?}$$

When a helium nucleus is emitted from an atom the mass number is reduced by 4 and the atomic number by 2. The resultant element therefore has a mass of 234 and an atomic number of 90. Note that, since the atomic number is no longer 92, this is not simply another isotope of uranium – an atom of a new element has been formed.

> *Which element has an atomic number of 90?*

You could find the answer by looking at the periodic table in Appendix B at the back of this book.

> *Take a moment to do this now.*

You will have found that the name of the element with the atomic number 90 is thorium. So now we are in a position to complete the equation for the radioactive decay of uranium-238:

$$^{238}_{92}U \longrightarrow {}^{4}_{2}He + {}^{234}_{90}Th$$

Actually, ^{234}Th is also an unstable isotope and it too decays. The decay which begins with ^{238}U continues for many years and involves a number of unstable isotopes.

> *When does the process of radioactive decay come to an end?*

You may have realised that the radioactive decay of an atom continues until a stable isotope is formed – in this case it is an isotope of lead.

Beta decay

Beta rays consist of streams of electrons and the key to understanding beta decay is to note that these electrons do not originate from the electron shells surrounding the nucleus of an atom but from the nucleus itself.

> *How can this be if the nucleus is comprised only of protons and neutrons?*

The answer is that, in some isotopes, the neutron is capable of breaking up to form a proton and an electron. The proton is retained in the nucleus and the electron is emitted.

> *What happens to the mass number and the atomic number of an isotope which decays in this manner?*

First of all, note that the mass number does not change.

This is because, although a neutron has broken up, a proton has been produced and protons and neutrons have identical masses. Secondly, note that the atomic number actually increases by one. Remember that the atomic number is the number of protons within the nucleus and, as we have already noted, an additional proton has been created by the break-up of a neutron. So let us look at an example. We shall consider iodine-131 (^{131}I) – an isotope with medical significance, as we shall see later.

$$^{131}_{53}\text{I} \longrightarrow {}^{131}_{54}\text{Xe} + e^-$$

^{131}I decays by beta emission to an element which also has an atomic mass of 131 – remember that the atomic masses do not change. However, the atomic number increases to 54 and the element with an atomic number of 54 is xenon.

Gamma decay

The emission of high-frequency electromagnetic rays referred to as gamma rays does not involve the loss of subatomic particles, so there is no change in either the atomic number or atomic mass. One gamma emitter of medical importance is cobalt-60 (^{60}Co). We shall say more about this isotope a little later.

The concept of half-life

Each radioisotope has a specific rate of decay. Some isotopes decay very quickly, while others decay very slowly indeed. The important thing to note is that the fraction of the total number of nuclei of a particular isotope which decay in a given period of time is constant. Consequently, it makes little sense to attempt to define the total life of an isotope. Instead the concept of *half-life* is used. The physical half-life of an isotope is the length of time it takes for half of the existing nuclei to decay. Table 13.1 contains a list of selected isotopes along with their mode of decay, *physical half-life* and, for some, the *biological half-life* too. This latter

Table 13.1 Half-lives of some important isotopes.

Isotope	Type of decay	Physical half-life	Biological half-life
$^{14}_{6}\text{C}$	Beta	5730 years	35 days
$^{24}_{11}\text{Na}$	Beta	15.02 hours	29 days
$^{42}_{19}\text{K}$	Beta	12.36 hours	43 days
$^{60}_{27}\text{Co}$	Beta	5.27 years	–
$^{131}_{53}\text{I}$	Beta	8.04 days	180 days
$^{235}_{92}\text{U}$	Alpha	7.04×10^8 years	–
$^{238}_{92}\text{U}$	Alpha	4.47×10^9 years	–

concept is explained shortly, but for now you will be able to see that the range of values is very wide indeed.

> *If we took a 1 g sample of $^{60}_{27}$Co, which has a half-life of approximately five years, how much of it would remain after the lapse of five years?*

Only half – 0.5 g. It is worth pointing out that the mass of matter in our sample would not have fallen to 0.5 g. Instead, our sample would consist of 0.5 g of the original isotope plus its decay products.

> *Will our sample be only half as radioactive after five years?*

The answer is no.

> *Can you work out why?*

Perhaps you have realised that the radioactivity will not decrease by half, since many of the decay products are themselves radioactive.

Half-life, as we have described it above, is more correctly referred to as physical half-life in order to distinguish it from biological half-life. The biological half-life is the time taken for the amount of a specific radioisotope which has entered the body to be reduced by half as a consequence of natural, biological processes, whether the isotope has decayed or not. For example, the isotope may be eliminated in the urine. Clearly in clinical practice where isotopes have been administered to patients both the physical half-life and the biological half-life are important, so a further concept, that of *effective half-life* is used. Effective half-life refers to the time required for an amount of a specific isotope to fall to half its original value as a result of both its radioactive decay and its biological elimination.

Practice point 13.1: isotopes and body fluids

In view of what has been noted above about the difference between the physical and biological half-lives you will no doubt have realised that if an isotope which has been administered to a patient is eliminated before it has decayed then the body fluids in which it is present will be radioactive and will require careful handling. If you work in an environment where such patients are cared for there will be a hospital policy which will describe the safe disposal of urine, faeces and vomitus, and you should of course follow it.

X-rays

We have described X-rays as electrically produced electromagnetic waves, and one use to which they are put is the treatment of cancer. We shall look at this a little later. However, the commonest use of X-rays is in the production of a visual

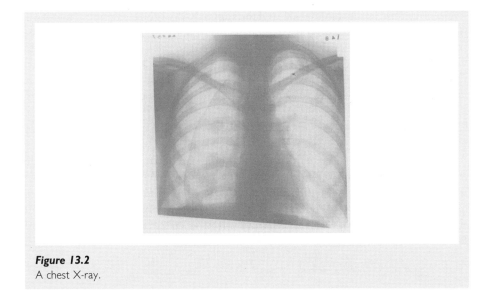

Figure 13.2
A chest X-ray.

image of the internal structures of the body. It is such a common procedure that the term X-ray is used in reference to it as well as to describe the rays themselves, although you may also see the terms *radiograph* and *roentgenogram* used. During the procedure, X-rays are directed towards the patient who lies above, or stands in front of, an X-ray sensitive film. This film is analogous to the light-sensitive film in the back of a conventional camera. Those rays which meet less dense areas of the body are able to pass through it. They subsequently strike the film and expose it, while other rays are stopped by denser tissue of the body. The resultant image which is formed may be likened to a shadow. It is, however, a shadow cast by X-rays and not by light, and because of the penetrating power of the rays it is a shadow of internal structures. Note too that, once the film has been developed, the unexposed areas appear white while the exposed areas appear black. Figure 13.2 shows what an X-ray actually looks like.

How are X-rays produced?

X-rays are produced when a stream of electrons strike a metal target. In a simple X-ray tube the electrons are generated by a heated element and they are then attracted to the target by an electrical potential difference (see Chapter 10: Electricity, magnetism and medical equipment) of up to 200 000 volts. Such apparatus is suitable for generating X-rays with sufficient energy to form an image of the body (that is, to take an X-ray). However, the high-energy X-rays required to treat cancers require a *linear accelerator*. This piece of equipment uses a different method to generate a stream of very high-energy electrons. Nonetheless, the X-rays themselves are still produced when these electrons strike a metal target.

Are there limitations to the usefulness of an X-ray?

Every procedure has its limitations, and the simple X-ray is no exception. First of all, note that the image produced is two-dimensional, and one body structure may be partially obscured by another which is in front of it in the same way that one individual may be obscured by another in a group photograph. This problem can be partially overcome by taking X-rays from different angles. For example, you may see the expressions *A–P* and *lateral* written on X-ray request cards.

> *What do these expressions mean?*

A–P stands for anterior–posterior, meaning an image taken with the X-ray tube in front of the patient and the film behind. The term *lateral* is self-explanatory – it means a view taken from the side, and of course the left or right should be specified.

A second limitation of the simple X-ray is that the density of all the body's soft tissues is approximately the same as that of water, so the boundary between one soft tissue structure and another is indistinct.

> *Can this be overcome in some way?*

You may have guessed that it can in some cases. Take the stomach as an example.

> *How could the structure of this hollow organ be made more visible?*

The description of the stomach as hollow is a clue to the answer to the above question. We could of course ask the patient to drink a *radiopaque* medium. Radiopaque means opaque to X-rays – that is, X-rays do not readily penetrate it. The medium may or may not be opaque to light. Having swallowed such a medium the patient is tilted so that it coats the interior of the stomach, and when an X-ray is taken the wall of the stomach is contrasted against the medium. For this reason such a radiopaque medium is also referred to as a *contrast medium*. In the above example the contrast medium consists of an emulsion containing barium sulphate and the procedure is termed a *barium meal*. This procedure may be used in order to confirm the presence of a lesion within the stomach.

> *What kind of lesion can be detected?*

If a gastric ulcer (stomach ulcer) is present a crater filled with barium can be seen. A small quantity of barium within the crater may even be observed for some time after the stomach has emptied. However, if a neoplasm (new growth/tumour) is present the outline of the stomach will be irregular. You might like to find out more about contrast radiography, for which a variety of contrast media are available. Some of these are suitable for injection into the bloodstream.

The effects of ionising radiation

We have described ionising radiation as that which is capable of producing ions. A number of mechanisms are involved in the process of ionisation, including the ejection of electrons from their orbit when an atom is struck by the radiation. Some of the ions produced are *oxidising agents* and some are *reducing agents*, or in some other way they may be toxic to the cell. Sometimes the cell is able to repair the damage produced, but if the intensity of the radiation is too great or the period of repair between exposures is too short then cells may die. Even if a cell does survive, damage to its genetic material (deoxyribonucleic acid/DNA) may impair its function. Such a genetically damaged cell may be described as a *mutant*. When such a cell reproduces, the altered genetic material will be replicated and subsequently a malignant tumour (cancer) may develop. If the mutated cell is a gamete (sex cell – oocyte or spermatocyte) then the embryo which results from the fusion of this sex cell with another may possess a genetic defect. Without a fuller description of the effects of ionising radiation it becomes obvious that unnecessary radiation exposure should be avoided.

Practice point 13.2: radiation protection for health care workers

In an introductory science text such as this, any advice given about radiation protection must be of a general nature. However, when any procedure involving ionising radiation is performed, you should follow the relevant hospital policy. In addition, in the United Kingdom, all X-ray and radiotherapy departments are required to have a radiation protection supervisor who can be contacted for advice.

Having noted the above, probably the most important thing to understand about radiation is that the intensity varies inversely with the square of the distance from the source. This is referred to as the *inverse square law*.

What does this actually mean?

Firstly, note that the intensity varies inversely with distance – the greater the distance from the source, the lower the intensity. Secondly, note that intensity is not reduced in direct proportion to the distance but as a square of the distance.

If you were to double the distance between yourself and a source of radiation, by how much would the intensity be reduced?

You should have worked out that the intensity would be reduced by a factor of four (2^2). That is when the distance to the source is doubled the intensity is reduced to a quarter. This is illustrated in Figure 13.3.

Therefore, if you are present in a ward when portable X-ray equipment is being used, stand away from the patient and certainly do not stand directly in front of an X-ray beam. It obviously makes more sense to stand behind the X-ray machine rather than in front of it!

Of course, what we have noted about the inverse square law is true when there is no barrier between the individual and the source. Another method of reducing radiation exposure is to use some form of *shielding*. In the X-ray department the

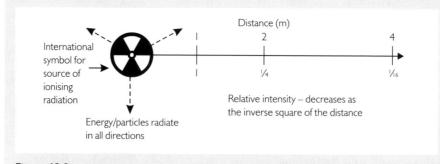

Figure 13.3
The inverse square law.

radiographer will stand behind a partition while taking the X-ray and the patient is safely observed through a window of lead glass.

What if the patient is anxious about the procedure or in danger of falling?

It is rarely necessary for a nurse to remain with a patient during an X-ray. If the patient is anxious give reassurance before the procedure then step behind the partition while the X-ray is being taken. If you are concerned about the patient's physical safety then pillows and bean bags should be used to help the patient maintain position and side rails should be used to ensure safety.

Nonetheless, some procedures do require staff to be present don't they?

Indeed they do – in angiography for example (see later). In these cases the staff exposed to radiation should wear lead rubber aprons. These need to be cared for carefully. They are always hung from strong hangers when not in use and they should never be folded over the back of a chair or (worse still) crumpled on the floor. If this is allowed to happen cracks may appear in the apron, and these will act like a funnel through which radiation will pass.

Finally, some patients have radioisotopes implanted in their bodies as part of the treatment of cancer. In such cases you will need to restrict the time spent in the room with such individuals.

We could summarise radiation protection under three headings – distance, shielding and time.

Is there no way of measuring the exposure of staff to radiation?

Yes there is. Staff who are frequently exposed to radiation, such as radiographers and nurses working in radiotherapy wards, should wear badges which contain radiographic film. Different regions of the badge have a different thicknesses of shielding, so the film is exposed to different degrees according to the amount of radiation exposure.

The measurement of radiation

You may see radioactivity expressed in a number of different units. Those which are currently in use include the becquerel (Bq), the gray (Gy) and the sievert (Sv).

To what do these units refer?

The becquerel measures the radioactivity of a source of radiation in terms of the number of disintegrations per unit of time. One becquerel equals one nuclear disintegration per second. Although this unit is an important one for scientists, it is not so useful for health care workers. What is needed is a unit which measures the amount of energy absorbed by the body. Such a unit is the gray. One gray is equivalent to the absorption of one joule of energy per kilogram of body tissue.

However, even this unit has limited use. This is because different forms of radiation produce different effects even when the absorbed dose in grays is the same – that is, even when the same amount of energy has been absorbed. A third unit is required which takes into account the different biological effects of the different forms of radiation. This unit is the sievert, and it measures *absorbed dose equivalent*. It is derived by multiplying the absorbed dose in grays by a numerical value which is given to each form of radiation. This value is called the *relative biological effectiveness* (RBE). The higher the RBE the greater the biological effect for the same amount of absorbed energy. So then:

absorbed dose equivalent (Sv) = absorbed dose (Gy) × RBE

The RBE for gamma rays, X-rays and beta particles is 1, but for alpha particles it is 15.

Other important radiological diagnostic procedures

In addition to the simple X-ray there are a number of other important radiological diagnostic procedures, and some of these are outlined in this section.

Cine radiography

When a simple X-ray is taken the X-ray film serves the same purpose as a photographic film, and we might describe the result as an X-ray still. Cine radiography involves the use of cine X-ray film and a moving X-ray image is obtained. It is therefore analogous to the cine camera.

When is cine radiography used?

No doubt you have worked out that it is when a dynamic process is to be imaged – blood flow for example.

> *Suppose a patient was suspected of having a narrowed blood vessel. How could this be investigated?*

Perhaps you have realised that an investigation involving a contrast medium is required. This is injected into the bloodstream at a suitable point and the medium then circulates through the vessel under investigation. Clearly a still X-ray image would be of no value in the observation of circulating blood, and therefore a cine image is required. The procedure described above is referred to as *angiography* and one common form is *coronary angiography* – imaging the coronary arteries. In this case a catheter (tube) is inserted through the aorta (main artery of the body) via the brachial artery of the arm or via the femoral artery of the leg. As the catheter approaches the heart its tip is directed towards one of the coronary ostia (opening of a main coronary artery). The contrast medium is injected through the catheter under force and at the same time the cine X-ray image is filmed. By this means an image of the coronary arteries is obtained as the contrast medium flows through them. This enables medical staff to locate a narrowed or occluded portion and to plan surgery to bypass it.

Fluoroscopy

When certain substances are struck by ionising radiation they emit rays of greater wavelength than those received – that is they emit visible light when they are struck by invisible radiation. This effect is called *fluorescence* and it can be used to produce an image of the body by positioning the patient between an X-ray tube and a fluorescent screen. The procedure is called *fluoroscopy*. Note that an image is present only so long as X-rays strike the fluorescent screen. If a permanent record is required a radiograph or cine X-ray has to be taken.

> *What use does fluoroscopy have?*

X-ray fluoroscopy is often combined with other investigations, especially those in which a device has to be positioned within the body. For example, in coronary angiography, the doctor views a fluoroscopic image of the chest while a radiopaque catheter is directed through the aorta to the heart.

Computerised tomography (CT scan)

In recent years the number of scanning techniques available to doctors has increased greatly.

> *Does it help to know something about these techniques?*

Yes it does. Even if you are not closely involved in the procedures themselves it is helpful to patients if you can explain what to expect. The size of some scanners can be intimidating, especially when heavy motorised parts move close to the body. However, you will have no difficulty understanding what computerised tomography is if you remember that the Greek word *tomos* means slice!

You will remember that, in the case of a simple X-ray, one structure may be obscured by another which is in front of it.

> *Would it help if doctors could view an image of the body just as though it had been 'sliced up'?*

Yes it would, and that is exactly what CT images look like – a series of transverse (horizontal) slices through the body.

> *How is such an image produced?*

When a CT scan is undertaken the patient lies very still on a table while a thin beam of X-rays is passed through them to a detector. As the CT scanner rotates the X-rays are directed through the patient from every angle. A computer is used to build up the slice-by-slice image of the patient's body based upon the intensity of the X-rays generated and the intensity of the X-rays that reach the detector.

Radioisotopes in diagnosis

The administration of radioisotopes to patients as part of a diagnostic test is not at all uncommon. Such tests depend upon the metabolisation of radioisotopes by the body in the same way as non-radioisotopes of the same element. The use of radio-isotopes in diagnosis may be illustrated using the example of the thyroid gland. This endocrine gland is located in the neck and it produces the hormone thyroxine, which is important in carbohydrate metabolism. The mineral iodine is essential for the manufacture of thyroxine and a radioisotope of iodine (^{131}I) can be used to investigate thyroid dysfunction in two ways:

Urinary excretion of ^{131}I: in this test the patient is given a measured dose of ^{131}I after which the urine is collected for 48 hours and the amount of isotope excreted is then measured. In thyrotoxicosis (overactive gland) a greater than normal amount of ^{131}I is concentrated in the thyroid gland and less is excreted, while in myxoedema (underactive gland) the reverse is true. The amount of isotope administered to the patient is small, and consequently, it is perfectly safe to be with the patient during this test.

Thyroid scan: if the urinary excretion of ^{131}I is abnormal the patient may have a thyroid scan. In this test ^{131}I, ^{125}I or ^{123}I is administered to the patient. The radiation emitted by the isotope is then detected by a gamma camera which builds up an image from the emissions from the gland. The radioactive iodine is most concentrated in overactive regions of the gland, and emissions from these regions are correspondingly greater, giving rise to so-called 'hot' areas. By this means Graves's disease, in which there is an overall increase in the activity of the gland, can be distinguished from toxic nodular goitre, in which there are isolated hot areas. In contrast, underactive regions take up less iodine than normal and give rise to correspondingly reduced emissions and 'cold' areas. Such an area may be suggestive of thyroid cancer.

Radiation in the treatment of cancer

The damaging effects of ionising radiation upon cells have already been described, but in this section we turn to the use of radiation to kill cancerous cells – a treatment referred to as *radiotherapy*.

Are all the cells of the body equally sensitive to ionising radiation?

No. Rapidly dividing cells are commonly more sensitive to radiation than cells which divide at a slow rate. Since cancer cells usually divide rapidly, radiation has a greater effect upon them than upon normal tissue. Nonetheless, the sensitivity of malignant tumours (cancers) to radiation does vary. Some are highly *radiosensitive* and others much less so. The degree of radiosensitivity is obviously an important factor in determining the suitability of radiotherapy as a treatment in a particular case.

It is also important to note that some of the normal cells of the body, particularly epithelial cells, also divide rapidly, and these include cells of the bone marrow, intestinal epithelium and hair follicles. Consequently, radiotherapy does produce some unpleasant side-effects through its action upon these tissues, including anaemia (erythrocytes are produced in the bone marrow), diarrhoea and hair loss. There are other effects too, and you might like to find out what they are.

Can the effects of radiotherapy upon normal healthy tissue be limited?

Yes they can. Healthy tissue normally recovers more quickly from radiation exposure than does cancerous tissue. Consequently, the radiation dose may be divided between a number of treatments, each treatment providing a fraction of the dose required. The time period between treatments enables the healthy tissue to recover. In addition, beams of radiation may be directed at a malignant tumour at different angles through the skin in a technique called *multiple field therapy*. This means that the healthy tissue receives only a part of the dose which reaches the cancer.

Are there different methods of administering radiotherapy?

Yes there are. In *external beam radiotherapy* a beam of X-rays or gamma rays is directed at the body from an external source. This technique is also called *teleradiotherapy* – a term derived from a Greek word meaning *far*. X-rays of sufficient energy for use in radiotherapy are produced from the rapid deceleration of high-energy electrons in a linear accelerator, while gamma rays are emitted by the nuclei of the atoms of certain radioisotopes such as ^{60}Co. The ability of the linear accelerator to produce a beam of deeply penetrating high-energy X-rays which can be focused precisely upon a tumour means that X-ray radiotherapy is much more common than gamma ray radiotherapy.

Does the patient become radioactive following treatment?

Fortunately the answer is no. Although high doses are administered during radio-therapy, the dose is focused upon the tumour and the exposure localised. Once the treatment is complete it is perfectly safe to be with and touch the patient. Of course it is important to stay out of the treatment room when radiotherapy is being administered, although staff can observe patients via closed-circuit television and communicate via an intercom.

External beam radiotherapy is not the only form of radiotherapy. In internal radio-therapy a radioisotope is implanted within the body. For this reason it is sometimes referred to as *brachytherapy*, from a Greek word meaning *near*. There are three forms of internal radiotherapy. In intracavity implantation the isotope is contained in some form of receptacle which is implanted in the body cavity in order to irradiate adjacent structures. For example, ^{137}Cs (Caesium-137) may be used to treat carcinoma of the cervix or vagina. In interstitial implantation the isotope is surgically implanted directly into a malignant tumour in the form of beads, needles or wires. This method ensures that the tumour receives the greatest dose of radiation. One example of inter-stitial radiotherapy is the insertion of ^{192}Ir (Iridium) needles into a tumour of the tongue. Finally, radioisotopes may also be administered systemically. Perhaps the most notable is the oral administration of ^{131}I in the treatment of cancer of the thyroid gland. The use of this isotope in the diagnosis of thyroid dysfunction has already been described, but in higher doses the beta radiation emitted will actually kill malignant cells.

> *Is the rest of the body not irradiated as well?*

Remember that the body metabolises radioisotopes in the same way as non-radioisotopes and that the thyroid gland concentrates iodine within it. Conse-quently, although the thyroid tumour cells are killed, the rest of the body receives only a very small radiation dose.

> *How do staff protect themselves from radiation emitted by patients who have radioisotopes implanted within their bodies?*

One important consideration is the length of time spent with the patient. Patients receiving internal radiotherapy are cared for in single rooms and the length of time spent with each patient is strictly controlled. The exposure of staff can be limited still further by the technique of *afterloading*. In this form of treatment an empty tube which will eventually contain the isotope is implanted into the relevant body cavity. Subsequently the isotope is remotely loaded into the tube, but it may be removed again when staff need to enter the room.

Summary
1 Ionising radiation includes streams of particles, such as alpha particles (helium nuclei) and beta particles (electrons), as well as electromagnetic waves such as X-rays and gamma rays.
2 Alpha particles and beta particles result from the decay of unstable nuclei.
3 X-rays are produced electrically, while gamma rays are emitted by some radioisotopes.

4 Alpha particles are stopped by the skin, beta particles penetrate soft tissue to a depth of about 1 cm and X-rays and gamma rays are only stopped by thick lead shielding.

5 The length of time it takes for half of the original number of nuclei of an unstable isotope to decay is referred to as the half-life.

6 The production of ions in biological tissue, some of which may be toxic, accounts for some of the damaging effects of ionising radiation.

7 Distance from source, time of exposure and shielding are the factors which should be considered in connection with radiation exposure.

8 Important imaging techniques in which ionising radiation is employed include the simple X-ray, cine radiography, CT scans and radionuclide scans.

9 Ionising radiation may also be used to destroy malignant cells, and it therefore plays an important role in the treatment of cancer.

Self-test questions

13.1 Which one of the following is an accurate definition of isotopes?

    a Atoms with the same number of electrons but with a different number of protons.

    b Atoms with the same number of neutrons and electrons.

    c Atoms with the same number of neutrons but with a different number of protons.

    d Atoms with the same number of protons but with a different number of neutrons.

13.2 Which of the following statements are true?

    a Alpha particles are more penetrating than beta particles.

    b Beta particles are more penetrating than X-rays.

    c X-rays are more penetrating than alpha particles.

    d Gamma rays are more penetrating than alpha particles.

13.3 Match the forms of radiation on the left with the descriptions on the right.

    a Alpha particle   i   Electromagnetic radiation emitted by an isotope

    b Gamma ray   ii Helium nucleus

    c Beta particle   iii Electromagnetic radiation produced electrically

    d X-ray   iv Electron

13.4 Suppose that a radioisotope has a physical half-life of 5 years. How much of an original mass of 16 g will remain after 20 years?

    a 8 g

    b 4 g

    c 2 g

    d 1 g

13.5 Suppose that you are standing 2 m away from a source of radiation and you then move to a position 4 m away. By how much will the intensity of the radiation have fallen?

    a 1/2

    b 1/4

    c 1/8

    d 1/16

13.6 Which form of ionising radiation is emitted in the following reaction?

$$^{60}_{27}Co \longrightarrow \ ^{60}_{27}Co + ?$$

a Alpha particles
b X-rays
c Gamma rays
d Beta particles

13.7 When a $^{238}_{92}U$ atom emits an alpha particle it decays to which of the following?

a $^{59}_{27}Co$
b $^{239}_{94}Pu$
c $^{235}_{92}U$
d $^{234}_{90}Th$

13.8 Identify the missing element in the following equation:

$$^{131}_{53}I \longrightarrow \ ? + e^-$$

a $^{131}_{54}Xe$
b $^{131}_{53}Xe$
c $^{130}_{53}Xe$
d $^{126}_{51}Xe$

13.9 Match the investigations on the left with the descriptions on the right.

a Radionuclide scan	i	A technique which results in images of transverse slices through the body
b CT scan	ii	An image of a contrast medium passing through a blood vessel
c Angiography	iii	A technique which involves the administration of a radioisotope
d Simple X-ray	iv	A still two-dimensional image of the internal structures of the body

13.10 Which one of the following currently describes the reason why it is possible to trace the metabolic pathways of the body using isotopes of the naturally occurring elements of the body?

a They are used in such small amounts that they do no harm.
b They have very short half-lives and so do no harm.
c They emit only gamma rays.
d They are metabolised in the same way as non-radioactive isotopes.

Answers to self-test questions

13.1 d	**13.6** c
13.2 c and d	**13.7** d
13.3 a ii	**13.8** a
b i	**13.9** a iii
c iv	b i
d iii	c ii
13.4 d	d iv
13.5 b	**13.10** d

Further study/exercises

13.1 Patients who are undergoing radiotherapy often experience unpleasant side-effects. What are the important aspects of nursing care when these effects are present?

Corner J. and Bailey C. eds. (1999). *Cancer Nursing Care in Context*. Oxford: Blackwell Science

Lochhead J. N. M. (1983). *Care of the Patient in Radiotherapy*. Oxford: Blackwell Science

Souhami R. L. and Tobias J. S. (1998). *Cancer and its Management*, 3rd edn. Oxford: Blackwell Science

13.2 Other diagnostic procedures include positron emission tomography (PET scan) and nuclear magnetic resonance imaging (NMR scan). What do these procedures involve?

Armstrong P. and Wastie M. L. (1998). *Diagnostic Imaging*, 4th edn. Oxford: Blackwell Science

Ball J. L. and Price T. (1995). *Chesney's Radiographic Imaging*, 6th edn. Oxford: Blackwell Science

Culmer P. (1995). *Chesney's Care of the Patient in Diagnostic Radiography*, 7th edn. Oxford: Blackwell Science

14 Genetics in health and illness

Introduction

In this chapter we concentrate on genetic inheritance and how genetic diseases are passed from parents to children. In order to understand this we must first consider the structure and function of the nucleus and explain two forms of cell division – **mitosis** and **meiosis**. In addition, any study of genetics would be incomplete without an explanation of protein synthesis, since the function of genes is to control the production of protein.

The nucleus

The **cell** (Figure 14.1) is the smallest living structure. The entire contents of a cell are collectively referred to as **protoplasm** and are separated from the outside by the **cell membrane**. Most cells have located within the protoplasm an oval or spherical structure called the **nucleus**. The contents of the nucleus (**nucleoplasm**) is separated from the rest of the cell contents (**cytoplasm**) by the *nuclear membrane*. However, this membrane does contain *pores* (*channels*) through which molecules may enter or leave the nucleus.

> *Name the structures that are found within the nucleus?*

You may have already heard of chromosomes. These structures can be seen with a light microscope, but they are only visible when the cell is actively dividing.

> *How many chromosomes do human cells possess?*

To answer this question it is first of all important to explain that the body's cells form two groups – **gametes** and *somatic cells*. Gametes are the sex cells – *oocytes*

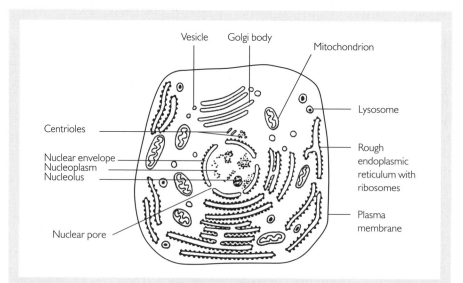

Figure 14.1
A typical animal cell. (Redrawn from Cornwell A. and Miller, R. (1997). *Longman Study Guides. A-Level and AS-Level Biology*. Harlow: Longman.)

in females and *spermatozoa* in males. All other cells are described as somatic. In somatic cells there are 23 pairs of chromosomes – the so called **diploid** number.

> *Why do we say 23 pairs and not simply 46?*

Chromosomes are arranged in pairs according to their size and shape. Such pairs are referred to as *homologous* pairs. One member of each homologous pair is derived from the mother and one from the father. One pair of chromosomes carries information about sex, and these are the sex chromosomes. The remaining 22 pairs are called *autosomes.*

> *How many pairs of chromosomes do gametes possess?*

Gametes possess only 23 chromosomes – the so called **haploid** number. Twenty-two of these are autosomes and one is a sex chromosome. Sex chromosomes are given the designation X and Y. The female possesses two X chromosomes, while the male possess one X chromosome and one Y chromosome. Consequently, all oocytes possess one X chromosome, as do half of all spermatozoa, while the other half have one Y chromosome. For this reason when *fertilisation* (fusion of gametes) takes place the sex of the offspring is determined by which spermatozoon fuses with the oocyte.

> *What are chromosomes made of?*

Chromosomes are made of *deoxyribonucleic acid* (DNA) and its associated proteins. Prior to cell division DNA forms a mass of threads called **chromatin**. Electron micrographs (photographs taken with an electron microscope) show that chromatin resembles beads on a string. Each bead-like structure is called a *nucleosome* and consists of DNA wrapped twice around a core of eight proteins called *histones*. During the early stages of cell division DNA is replicated (copied) and the chromatin coils to form the discrete structures that we call chromosomes. At this stage it can be seen that each chromosome consists of two *chromatids* joined by a structure called a *centromere*. Each pair of chromatids carries the same genetic information.

> *What is DNA made of?*

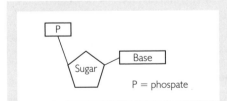

Figure 14.2
The structure of a nucleotide. (Redrawn from Millican C. and Barber, M. (1997). *Longman Study Guides. GCSE Biology.* Harlow: Longman.)

The structure of DNA was first described by Crick and Watson in 1956. It is made of sub-units called **nucleotides**. Each nucleotide has three components – a five carbon atom sugar molecule, a phosphate group and an organic base. The structure of a nucleotide is shown in Figure 14.2. Nucleotides that form DNA have the sugar *deoxyribose* and one of four bases – *adenine, thymine, guanine* or *cytosine.*

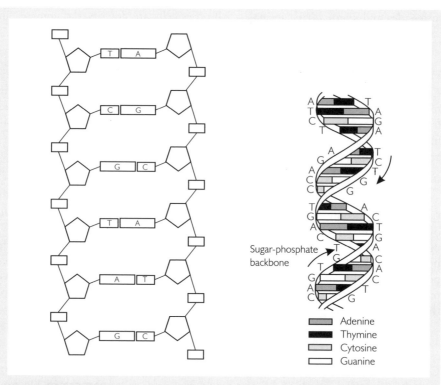

Figure 14.3
The structure of DNA. (Redrawn from Millican C. and Barber, M. (1997). *Longman Study Guides. GCSE Biology*. Harlow: Longman.)

DNA possesses two chains that are formed from the bonding of the sugar of one nucleotide with the phosphate group of another. Consequently, these chains are sometimes referred to as sugar-phosphate 'backbones'. These two chains are then joined by bonds formed between the bases. Adenine always bonds with thymine and guanine always bonds with cytosine. On first examination this arrangement looks rather like a ladder, but actually the two sugar–phosphate chains spiral around each other. In fact, that is how the structure of DNA is described – as a *double helix* (Figure 14.3).

Is DNA the only nucleic acid?

No. Another important nucleic acid is **ribonucleic acid** (RNA).

How does RNA differ from DNA?

In three important ways. Firstly, RNA consists of only a *single strand* of nucleotides. Secondly, in the nucleotides that form RNA the sugar is *ribose*, not deoxyribose, and thirdly, the base *uracil* occurs instead of thymine.

DNA – the genetic code

DNA is often described as the body's genetic code.

In what way is DNA a code?

The bases of DNA are arranged in triplets (groups of three) that are referred to as **codons**. Each codon acts as a code for a specific amino acid. The are 20 amino acids in nature.

How many codons would you expect?

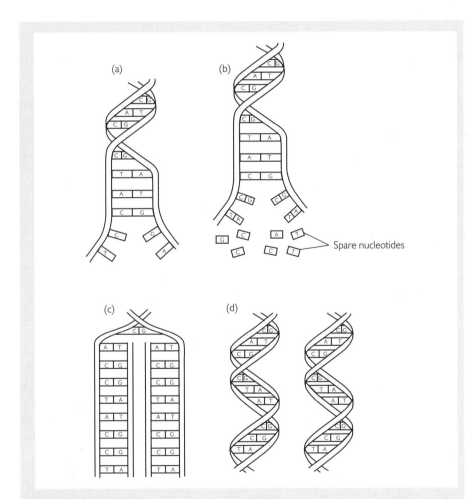

Figure 14.4
DNA replication. (Redrawn from Millican C. and Barber, M. (1997). *Longman Study Guides. GCSE Biology.* Harlow: Longman.)

You might think that there should be 20, but in fact there are 64. Some amino acids are specified by more than one codon – a condition referred to as *redundancy*. In addition, some codons have functions other than coding for specific amino acids. Some indicate the point at which a code for a particular protein begins and these are referred to as *start codons*. Others indicate the point at which the code for a particular protein ends, and are referred to as *stop codons*.

> What is a gene?

Genes are segments of DNA that code for a specific protein. They form a section of a strand of DNA located between a start codon and a stop codon. There are thousands of genes in human DNA.

DNA replication

If a cell is to divide (reproduce) it must *replicate* its DNA so that 23 pairs of chromosomes can be passed on to each of the daughter cells. The first process in DNA replication is the unwinding of the double helix. Next newly synthesised bases pair with the bases of the two original strands. In this way two DNA molecules result (Figure 14.4). Each one consists of a strand of the original molecule and a new strand. Consequently, DNA replication is said to be *semi-conservative*.

Mitosis and the cell cycle

Cells need to reproduce so that damaged cells can be replaced and growth take place. Some cells, such as those of the basal layer of the skin, divide rapidly throughout life, while others, such as brain cells, do not divide at all after birth. Figure 14.5 is a diagrammatic representation of the life cycle of a somatic cell. The cell cycle can be divided into two parts – *interphase* and *cell division*.

Interphase

When a cell is not actively dividing it is said to be in interphase. Interphase can itself be divided into distinct phases, which are represented on the diagram in Figure 14.5. S stands for synthesis and refers to DNA synthesis (DNA replication), G_1 and G_2 are periods of growth, while G_0 is a period of rest. In G_0 the cell is actively metabolising but not growing or synthesising DNA.

Cell division

Cell division in somatic cells is referred to as **mitosis** although strictly speaking this term refers only to the manner in which chromosomes are distributed between two newly forming cells (daughter cells). The division of the parent cell into two

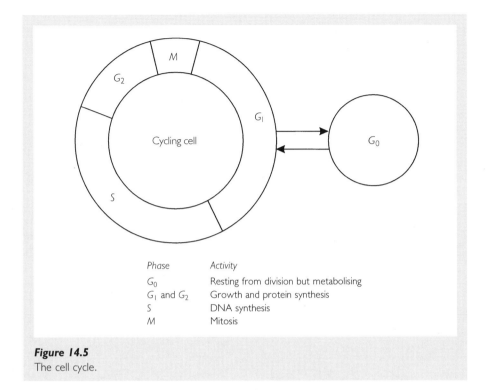

Phase	Activity
G_0	Resting from division but metabolising
G_1 and G_2	Growth and protein synthesis
S	DNA synthesis
M	Mitosis

Figure 14.5
The cell cycle.

daughter cells is called *cytokinesis* (cytoplasmic division). Mitosis is itself divided into four stages (Figure 14.6).

Prophase: in this first stage of mitosis, chromosomes become visible and it can be seen that each is made of two chromatids. At the end of prophase the nuclear membrane breaks down and fibres form between structures called *centrioles*. Centrioles are normally found adjacent to the nucleus in a region called the *centrosome*, but in prophase they move to opposite poles of the cell. The fibres that form between them are referred to as *spindle* fibres.

Metaphase: during this stage chromosomes become attached along the centre of the mitotic spindle (*equatorial plate*).

Anaphase: this stage is characterised by splitting of the centrosomes so that sister chromatids separate and begin to move to opposite ends of the parent cell. From this point each chromatid is regarded as a separate chromosome.

Telophase: in this stage the DNA of both sets of chromosomes begins to uncoil and form thread-like chromatin once again. The mitotic spindle disappears and a nuclear membrane forms around both sets of chromosomes.

Cytokinesis: cell division is completed when the cytoplasm of the parent cell is divided between the two newly forming cells.

In view of the above sequence of events we can summarise mitosis as the process of cell division that occurs in somatic cells and that leads to the production of two diploid daughter cells.

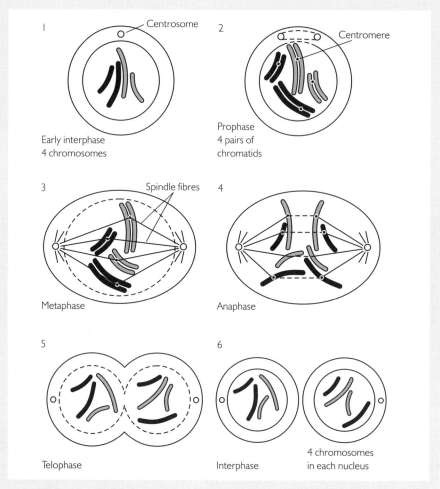

Figure 14.6
Mitosis. (Redrawn from Sheeler P. (1996). *Essentials of Human Physiology*, 2nd edn. Dubuque: Wm. C. Brown.)

Practice point 14.1: cancer

Cell division is normally a well-regulated process, so that the rate of production of new cells is adequate for the purposes of growth and repair. However, sometimes cells escape the normal controlling mechanisms and a **neoplasm** (new growth) results. Neoplasms can produce unpleasant effects through a number of mechanisms, including the compression of vital structures such as nerves and blood vessels. Such effects are referred to as *space-occupying lesion effects*. Neoplasms are sometimes referred to as *tumours* and they are of two kinds.

Benign neoplasms consist of cells that resemble those from which they arise, while the cells of *malignant* neoplasms are highly abnormal. As well as space-occupying lesion effects, malignant tumours spread out and invade adjacent structures in a way that was thought to resemble the spreading of a crab's legs. For this reason malignant tumours are often referred to as *cancers*. You may wish to find out more about the difference between benign and malignant neoplasms.

Meiosis

Cell division in the *gonads* (reproductive organs) takes place by **meiosis** – a different process from mitosis in somatic cells.

What is the reason for this difference?

To answer this question, recall that gametes differ from somatic cells in having only 23 chromosomes rather than 23 pairs of chromosomes. Meiosis is a process whereby only one of each homologous pair is distributed to each gamete. For this reason meiosis is said to be a *reduction division*. In fact, meiosis involves two divisions. Prior to the first division DNA replication takes place, as it does in mitosis. The first meiotic division therefore results in two diploid cells. The second meiotic division takes place without additional DNA replication, and so the final result is four haploid cells. Meiosis is clearly a more complicated process than mitosis, but each stage is given the same name as in mitosis. However, a Roman numeral is added so as to indicate whether the first or second division is being referred to (prophase I, prophase II etc.). The details of a meiotic division are not given here, but key events that take place in prophase I are explained.

Key events in prophase I

Certain events occur in prophase I that do not occur in prophase of mitosis. In prophase I chromosomes thicken and shorten. They then arrange themselves in homologous pairs, each individual chromosome consisting of two chromatids. For this reason, when a homologous pair join together the structure which is formed is referred to as a *tetrad* (meaning four). It is worth remembering that one member of a homologous pair is derived from the mother and the other from the father. Following tetrad formation pieces of DNA are exchanged between homologous pairs in a process that is referred to as *crossing-over* (Figure 14.7).

What is the significance of crossing-over?

Remember that gametes possess only 23 chromosomes rather than 23 pairs. Crossing-over ensures genetic variability by ensuring that each gamete receives chromosomes that contain a combination of maternal and paternal genetic information.

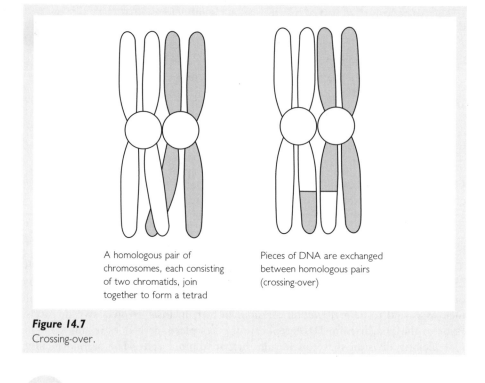

A homologous pair of chromosomes, each consisting of two chromatids, join together to form a tetrad

Pieces of DNA are exchanged between homologous pairs (crossing-over)

Figure 14.7
Crossing-over.

Protein synthesis

We have already noted that DNA is a code for protein synthesis.

Why is this so important?

Proteins form essential components of our body. Some, such as collagen, are of structural importance; actin and myosin are the contractile proteins of muscle; and others are referred to as functional and perform the roles of enzymes, hormones and antibodies. In essence, the way that our bodies function is due to proteins of one kind or another. Consequently, whatever controls protein synthesis ultimately determines our physical characteristics and body function.

The code for protein synthesis is located within the nucleus. How is it employed to make protein?

This is what we consider next, but first we must return to RNA – in fact to three forms of this nucleic acid.

Ribosomal RNA (rRNA)

Proteins are synthesised in cytoplasmic organelles called **ribosomes**. Ribosomes are actually made of rRNA that is synthesised in the nucleus and that enters the cyto-plasm via nuclear pores. Ribosomes are made of two sub-units – a large sub-unit and a small sub-unit – and these need to join together before protein synthesis can take place.

Messenger RNA (mRNA)

In essence, mRNA is a copy of a DNA gene that is made in the nucleus in a process referred to as **transcription** (literally, copying). mRNA then leaves the nucleus via a nuclear pore and enters the cytoplasm. This copy of the genetic code forms the instruction for protein synthesis that is 'read' by a ribosome. No doubt you now understand why this form of RNA is described as a messenger. In effect, it acts as a 'messenger' from the nucleus to the ribosome!

Transfer RNA (tRNA)

tRNA molecules consist of three nucleotides, the base sequence of which is referred to as an **anticodon**. Each tRNA molecules picks up a specific amino acid and func-tions as a means of transferring it to a ribosome, where protein synthesis actually takes place in a process called **translation**.

Transcription

Remember that transcription is the process by which mRNA is synthesised in the nucleus. The first stage of this process is the separation of DNA base pairs so that the double helix unwinds just as it does prior to replication. However, only one of the strands of DNA is involved in mRNA synthesis, and this is referred to as the *sense strand*. The other strand is referred to as the *antisense strand*. As the DNA double helix unwinds, the bases of mRNA nucleotides pair with the corre-sponding DNA nucleotides in the manner shown in Figure 14.8. In this way a single strand of mRNA is assembled alongside the sense strand.

> *Does this process begin and end at random?*

No. Remember that certain DNA codons perform the role of identifying the point at which a gene begins (start codons) and others the point at which it ends (stop codons). Consequently, mRNA synthesis begins with a start codon and ends with a stop codon. In addition, recall that while the bases adenine, guanine and cytosine are common to both DNA and RNA, DNA alone has thymine and RNA alone has uracil.

> *If a DNA codon contained the bases ATA what would be the corresponding mRNA codon?*

The correct answer is of course UAU.

> *What now happens to the mRNA strand?*

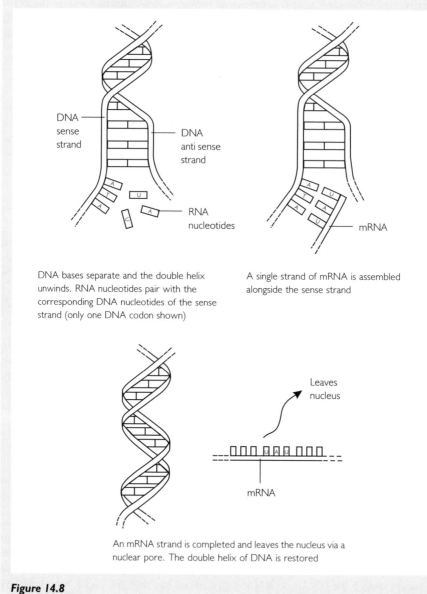

DNA bases separate and the double helix unwinds. RNA nucleotides pair with the corresponding DNA nucleotides of the sense strand (only one DNA codon shown)

A single strand of mRNA is assembled alongside the sense strand

An mRNA strand is completed and leaves the nucleus via a nuclear pore. The double helix of DNA is restored

Figure 14.8
Transcription.

Once an mRNA strand has been completed it leaves the nucleus via a nuclear pore and enters the cytoplasm.

Translation

In the cytoplasm a ribosome becomes attached to a mRNA strand at a start codon

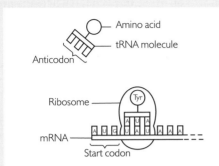

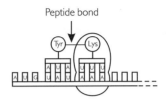

The ribosome has reached the first codon after the start codon (UAU). A tRNA molecule with a corresponding anticodon (AUA) becomes attached. This tRNA molecule carries the amino acid tyrosine (Tyr)

The ribosome moves along to the next codon (AAA). A tRNA molecule with a corresponding anticodon (UUU) becomes attached. This tRNA molecule carries the amino acid lysine (Lys). Tyrosine and lysine become joined by a peptide bond

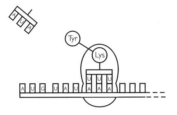

The first tRNA molecule leaves the ribosome and becomes free to pick up another tyrosine molecule. The ribosome will now move on to the next codon.

Figure 14.9
Translation.

and the process of translation begins. We might summarise this process by saying that the code in the form of a sequence of bases in mRNA is 'translated' into a sequence of amino acids in a protein chain. During translation the ribosome moves along the length of the mRNA molecule from the start codon to the stop codon. This process is shown in Figure 14.9.

Let us imagine that the ribosome has reached the first codon after the start codon – that is UAU in Figure 14.9. At this point a tRNA molecule with the corresponding anticodon (AUA) becomes attached to the mRNA molecule. Remember that each tRNA molecule carries a specific amino acid. The tRNA molecule with the anticodon AUA carries the amino acid tyrosine. The ribosome now moves along the mRNA strand and a second tRNA molecule becomes attached. At this point the amino acids of the first and second tRNA molecules are joined by a peptide bond. The first tRNA molecule then leaves the ribosome and becomes free to pick up another amino acid, while the ribosome moves on to the third codon where the process is repeated. In this way the code for the synthesis of a specific amino acid is 'read' by the ribosome until a stop codon is reached.

Genetic inheritance

We now move on to consider the inheritance of genes using examples unrelated to illness before proceeding to consider some common genetic diseases. The position that a gene occupies on a chromosome is referred to as a *locus* (plural *loci*). Genes that occupy the same loci on homologous chromosomes are said to be **alleles** or allelic genes.

What is the significance of alleles?

Alleles code for the same trait such as eye colour. However, they do not necessarily code for the same expression of that trait. For example, one allele may code for blue eyes and the second for brown eyes. When both alleles code for the same expression of a trait the individual is said to be **homozygous** for that trait. If alleles code for different expressions of a trait the individual is said to be **heterozygous**. The genetic make-up of an individual, that is which genes are possessed, is termed the **genotype**, while the expression of the genes is termed the **phenotype**.

In the case of the heterozygous individual, which of the two alleles is expressed?

When one allele is fully expressed over another it is said to be *dominant*, while the allele that is not expressed at all is said to be *recessive*. Characteristics that are expressed by dominant genes can be observed in both the homozygous dominant and heterozygous states, while characteristics that are expressed by recessive genes can only be observed when an individual is homozygous recessive. When identifying genotypes it is customary to use a single letter that is given as a capital in the case of a dominant gene and as a lower-case letter for a recessive gene. The simplest cases are those in which a characteristic is determined by only two alleles of a single gene present on autosomes – *simple autosomal inheritance*.

Simple autosomal inheritance

Consider the example of dimples in cheeks, the presence of which is determined by a dominant gene. Let D stand for the dominant (dimples) gene and d the recessive (absence of dimples) gene. Table 14.1 gives the possible genotypes and phenotypes in this case.

Some traits are of course expressed by recessive genes – flat feet is an example. Let D stand for the dominant (normal arches) gene and d stand for the recessive (flat feet) gene.

Table 14.1. **The simple autosomal inheritance of the dimples in cheeks gene.**

Genotype	Phenotype
DD (homozygous dominant)	Dimples present
Dd (heterozygous)	Dimples present
dd (homozygous recessive)	Dimples absent

Table 14.2. **The simple autosomal inheritance of flat feet.**

Genotype	Phenotype
DD (homozygous dominant)	Normal arches
Dd (heterozygous)	Normal arches
dd (homozygous recessive)	Flat feet

> *What are the possible genotypes and phenotypes in this case?*

Check your answer against Table 14.2.

In Table 14.3 a number of normal human traits and genetic diseases are listed according to whether the gene concerned is dominant or recessive.

The Punnet square

The *Punnet square* is a simple device that is used to demonstrate the possible combination of genes derived from both parents. It consists of a square that is divided into four boxes. The possible alleles derived from the father are placed to the left of the square and the possible alleles derived from the mother are placed above it. In order to work out the possible genotypes of children the alleles of parents are combined in turn. Suppose we take the case of parents both of whom are heterozygous for the dimples gene (Dd). Remember that this gene is dominant, so both

Table 14.3. **Phenotypes expressed by dominant and recessive genes.**

Phenotype due to the expression of dominant genes (DD or Dd)	Phenotype due to the expression of recessive genes (dd)
Normal skin pigment	Albinism
Freckles	Absence of freckles
Ability to roll tongue into a U shape	Inability to roll tongue into a U shape
Normal arches	Flat feet
Dimples in cheeks	Absence of dimples
Unattached earlobes	Attached earlobes

parents have dimples. The use of the Punnet square in this case is demonstrated below.

	D	d
D	DD	Dd
d	Dd	dd

What are the possible genotypes and phenotypes of the children?

They are DD (dimples present), Dd (dimples present) and dd (dimples absent).

What is the probability of the parents in the above example having a child with dimples?

It is three out of four, or 75%. The probability of having a child without dimples is of course 25%.

Incomplete dominance

Incomplete dominance is the condition that exists when both alleles are expressed in the heterozygous state. The inheritance of hair texture is a good example. In this case the individual who is homozygous for the curly hair gene has curly hair and the individual who is homozygous for the straight hair gene has straight hair.

What is the phenotype of the heterozygous individual?

The heterozygous individual does indeed have an intermediate hair texture – that is, wavy hair.

Multiple allele inheritance

Although only two alleles are inherited for each gene, that does not necessarily mean that only two alleles exist. In some cases there are *multiple alleles*. The ABO blood group system is a good example. In this case there are three alleles; I^A codes for blood group A, I^B codes for blood group B and i codes for blood group O. I^A and I^B are co-dominant genes, while i is a recessive gene.

What are the possible genotypes and phenotypes in this case?

Check your answer against Table 14.4.

Let us now use a Punnett square to answer a question.

What is the probability of a child of blood group O being born to heterozygous parents of blood group A?

To answer this question, first note the genotypes of the parents. Since they are both heterozygous for the blood group A they must have the genotype I^Ai. We can now

Table 14.4. **The possible genotypes and phenotypes in the ABO blood group system.**

Possible genotypes	Possible phenotypes (blood groups)
$I^A I^A$	A
$I^A i$	A
$I^B I^B$	B
$I^B i$	B
$I^A I^B$	AB
ii	O

use a Punnett square to work out the possible genotypes of the children, as shown below.

	I^A	i
I^A	$I^A I^A$	$I^A i$
i	$I^A i$	ii

You should now be able to see that there is a probability of one in four children born to the above parents having blood group O (25%).

Polygenic inheritance

When a trait is controlled by several genes the term *polygenic inheritance* is used. Examples of polygenic inheritance traits include eye colour, hair colour and skin colour. In the case of skin colour there are three separate genes involved and each has two alleles; Aa, Bb and Cc. The genotype AA, BB, CC leads to a very dark-skinned phenotype and the genotype aa, bb, cc leads to a very light-skinned phenotype. Suppose two such individuals were to have children.

What would the possible phenotypes be?

Once again we need to draw a Punnet square.

	abc	abc
ABC	Aa, Bb, Cc	Aa, Bb, Cc
ABC	Aa, Bb, Cc	Aa, Bb, Cc

In this case all the children would have a skin tone of intermediate darkness.

Inheritance of sex

We have already seen that all oocytes, and 50% of spermatozoa, carry an X chromosome, while the remaining 50% of spermatozoa carry a Y chromosome. We can illustrate the inheritance of sex using a Punnett square.

	X	X
X	XX	XX
Y	XY	XY

On this basis 50% of children should be male and 50% female. However, it is important to point out that chromosomes determine genetic sex. Initially, both male and female embryos develop identically until about seven weeks after fertilisation. After this time genes present on the sex-determining region of the Y chromosome (SRY) are responsible for initiating male development. When the SRY is absent the embryo continues to develop physiologically as a female despite possessing a Y chromosome.

Genetic disease

We now turn to consider a number of diseases that have a genetic origin, beginning with simple autosomal examples.

Cystic fibrosis (simple autosomal recessive)

Cystic fibrosis is often used as an example of an autosomal recessive disorder since it is relatively common, affecting about one in 2000 live births. It is not necessary to know a great deal about a genetic disorder in order to understand how it is transmitted. However, you may be interested to learn that the gene of importance in this case encodes a protein that regulates the movement of chloride ions across epithelial membranes that line airways, ducts and hollow organs. On initial examination it appears to be a rather insignificant function, but when the defective gene is present the cystic fibrosis sufferer produces abnormally tenacious mucus. This is difficult to cough up, and cystic fibrosis sufferers require daily physiotherapy. Even when this is performed they remain predisposed to serious lung infections. They also experience many other health problems, and you may like to find out more about what these are. For the time being let us consider the case of a child with cystic fibrosis being born to parents neither of whom has the disease. Let D stand for the (dominant) normal gene and d for the (recessive) cystic fibrosis gene.

What are the genotype of the parents?

To answer this question, remember that since the defective gene is recessive the disease is only present in the homozygous recessive individual (dd) and the child in the above example must have inherited defective genes from both parents. However, since neither parent has the disease they must both be heterozygous (Dd).

What term is used to describe individuals who possess a single recessive gene for a disease?

They are referred to as *carriers*; they possess the defective gene, but since it is recessive the disease is not expressed in the heterozygous state. Carriers can of course pass the gene on to their children, as the above case shows. The Punnet square helps us to explain the above example more fully.

	D	d
D	DD	Dd
d	Dd	dd

Parents who have a child with cystic fibrosis often ask the question 'what is the chance of having another child with the disease?'.

> *What answer would you give?*

If you have checked the Punnett square you will no doubt realise that it is 25%. There is the same probability of having a normal child, but a 50% probability that a child will be a carrier.

Huntington's disease (simple autosomal dominant)

Huntington's disease involves a progressive degeneration of the nervous system, the symptoms of which include depression, chorea (involuntary dance-like movement) and dementia. However, these do not become evident until the third or fourth decade. The disease occurs even when a single defective gene is inherited – that is, it is an example of autosomal dominant inheritance.

Sickle-cell anaemia (co-dominance)

Remember that co-dominance (incomplete dominance) is the condition that exists when both alleles are expressed in the heterozygous state. Sickle cell anaemia (sickle-cell disease), which affects individuals of African and Caribbean descent, is a good example. This condition results from a gene defect that causes an abnormality of the red blood cell pigment haemoglobin. This molecule is important in oxygen transport, since oxygen binds to it to form oxyhaemoglobin. When the defective gene is present in the homozygous state the abnormal haemoglobin produced precipitates within red blood cells, which, as a consequence, become sickle-shaped. Abnormally shaped red blood cells have a shorter lifespan than normal red blood cells and anaemia (reduced number of red blood cells) results. In the heterozygous state normal and abnormal haemoglobin is produced, and therefore the tendency for red blood cells to sickle is less. This leads to a milder form of the condition that is referred to as sickle-cell trait.

The normal gene is usually designated Hb^A and the sickling gene as Hb^S. Let us now consider the case of parents both of whom have sickle-cell trait and who are therefore heterozygous ($Hb^A Hb^S$).

> *What are the probabilities of the following children being born?*
> *a Unaffected child*
> *b Child with sickle-cell trait*
> *c Child with sickle-cell disease*

In order to answer this question you will need to draw a Punnet square. If you have done this correctly you will see that the answers are (a) 25%, (b) 50% and (c) 25%. If you are unsure about this check your Punnet square against the one given below.

	Hb^A	Hb^S
Hb^A	$Hb^A Hb^A$	$Hb^A Hb^S$
Hb^S	$Hb^A Hb^S$	$Hb^S Hb^S$

Sex-linked disorders

Thus far the traits and disorders that we have considered have been the result of genes carried on autosomes. In addition, they have affected males and females equally. We now turn to sex-linked disorders that affect males rather than females and in which a defective recessive gene is carried on an X chromosome.

Sex-linked disorders are probably best demonstrated by haemophilia A. The term haemophilia literally means 'blood-love' and it is applied to conditions in which blood coagulation is impaired and there is a tendency to bleed severely following injury (or even spontaneously). Haemophilia is caused by the absence of one of a number of chemical substances called *clotting factors*. In the case of haemophilia A clotting factor VIII is missing. The defective gene responsible is carried on the X chromosome and this is usually designated X' in order to distinguish it from a normal X chromosome. The genotype of haemophiliac males is therefore $X'Y$.

> *Why, when the defective gene is recessive, do males with the genotype $X'Y$ develop haemophilia?*

The answer is that X and Y chromosomes are not homologous in the true sense. The Y chromosome is only about a third of the size of the X chromosome and lacks some of the genes that are present on the X chromosome, including the gene for the production of factor VIII. Consequently, in the male with the $X'Y$ genotype the defective gene is not masked. In contrast, females with the genotype $X'X$ possess a normal (dominant) gene on the second X chromosome, so the defective (recessive) gene is not expressed, such females are, however, able to pass on the defective gene, and so are referred to as carriers.

> *What about females with the genotype $X'X'$*

The $X'X'$ genotype usually results in spontaneous abortion early in pregnancy, so haemophiliac females are extremely rare.

Use a Punnett square to work out the possible genotypes and phenotypes of children born to parents of the genotypes XY and X'X.

	X'	X
X	XX'	XX
Y	X'Y	XY

The possible genotypes and phenotypes are X'X (carrier female), XX (normal female), X'Y (haemophiliac male) and XY (normal male).

It should also be noted that some segments of the Y chromosome have no counterpart on the X chromosome and this type of inheritance is referred to as Y-linked.

Chromosomal abnormalities

Thus far we have considered abnormalities of specific genes. In contrast, *chromosomal* abnormalities (aberrations) affect the whole of a chromosome or a significant part of it. The most familiar is probably Down's syndrome in which there are three copies of chromosome number 21 (trisomy 21). The defect results from either a failure of homologous chromosomes to separate in the first meiotic division or of a failure of sister chromatids to separate in the second meiotic division. Such a failure is referred to as *non-disjunction*, and it results in a gamete that possesses both copies of chromosome number 21. The oocyte is most commonly affected. When fertilisation by a normal spermatozoon takes place the child inherits three copies of the 21st chromosome. The characteristics of Down's syndrome include low IQ, poor muscle tone, short stature, protruding tongue and epicanthal folds.

Summary

1 Chromosomes are structures that are present in the nucleus and formed from tightly coiled deoxyribonucleic acid (DNA).
2 The bases in DNA are arranged in triplets (codons) that form a code for protein synthesis.
3 There are 23 pairs of chromosomes in human somatic cells and 23 in human gametes.
4 Somatic cells divide by mitosis, a process that leads to the formation of two diploid daughter cells.
5 Gametes are the result of a meiotic division; a process that results in the formation of four haploid daughter cells.
6 The synthesis of mRNA from the sense strand of DNA is referred to as transcription.
7 The synthesis of protein that occurs when the anticodons of tRNA molecules pair with the codons of mRNA is referred to as translation.
8 The patterns of genetic inheritance may be described as simple autosomal (dominant and recessive), incomplete dominance (co-dominance), multiple allele inheritance, polygenic inheritance and sex-linked inheritance.

9 The possible combinations of alleles inherited from parents of known genotype can be demonstrated using a Punnett square.

Self-test questions

14.1 Which of the following statements are true?
  a Chromosomes are visible throughout the life cycle of the cell.
  b Chromatin is present within the cytoplasm of a cell.
  c Chromosomes are formed from tightly coiled DNA.
  d Histones are formed from DNA.
14.2 Which of the following statements are true?
  a Gametes possess 23 pairs of chromosomes.
  b Somatic cells possess 22 pairs of autosomes.
  c Somatic cells possess one pair of sex chromosomes.
  d Gametes possess one sex chromosome.
14.3 Which of the following statements are true?
  a DNA consists of a single strand of nucleotides.
  b RNA contains the base uracil.
  c RNA consists of a single strand of nucleotides.
  d DNA contains the base uracil.
14.4 Which of the following statements are true?
  a Mitosis occurs in somatic cells.
  b Meiosis leads to the formation of four haploid cells.
  c Mitosis leads to the formation of two haploid cells.
  d Tetrad formation occurs in prophase I of meiosis.
14.5 Which of the following statements are true?
  a A triplet of DNA bases is referred to as a codon.
  b A tRNA molecule contains three nucleotides.
  c A triplet of mRNA bases is referred to as an anticodon.
  d An mRNA molecule consists of three nucleotides.
14.6 Which of the following statements are true?
  a Transcription is the process by which mRNA is synthesised.
  b The DNA strand involved in mRNA synthesis is referred to as the antisense strand.
  c The process that leads to the synthesis of protein is referred to as transformation.
  d The nucleus is the site of protein synthesis.
14.7 Match the term on the left with the appropriate description on the right.
  a alleles   i   non-sex chromosomes
  b autosomes   ii   genes that code for the same trait
  c ribosomes   iii   proteins associated with DNA
  d histones   iv   the sites of protein synthesis
14.8 What is the probability in simple autosomal recessive inheritance of the birth of a child with a trait when the parents are heterozygous for this trait?

a 25%

b 50%

c 75%

d 100%

14.9 What are the possible combinations of alleles of children born to parents with the blood groups AB and O? (*Hint:* use a Punnett square.)

a $I^A i$, $I^A i$, $I^B I^B$, $I^B I^B$

b $I^A i$, $I^A i$, $I^B i$, $I^B i$

c $I^A i$, $I^A i$, $I^A i$, $I^A i$

d $I^A I^A$, $I^B I^B$, $I^A I^B$, ii

14.10 Match the genetic diseases on the left with the appropriate descriptions on the right.

a cystic fibrosis   i   simple autosomal dominant

b haemophilia A   ii   simple autosomal recessive

c Huntington's disease   iii   incomplete dominance

d sickle-cell anaemia   iv   sex-linked

Answers to self-test questions

14.1 c

14.2 b, c and d

14.3 b and c

14.4 a, b and d

14.5 a and b

14.6 a

14.7 a ii

     b i

     c iv

     d iii

14.8 a

14.9 b

14.10 a ii

     b iv

     c i

     d iii

Further study/exercises

14.1 Describe the health problems experienced by individuals with cystic fibrosis and identify their nursing needs.

Bramwell E. (1998). Cystic fibrosis: care of cystic fibrosis in the community. *Community Nurse*, **3**(12), 16–17

Dyer J. and Morais A. (1996). Supporting children with cystic fibrosis in school. *Professional Nurse*, **11**(8), 518–520

Duncan-Skingle F. and Foster F. (1991). The management of cystic fibrosis. *Nursing Standard*, **5**(21), 32–34

Glew J. (1993). One of the family . . . cystic fibrosis . . . specialist CF nurses. *Nursing Times*, **89**(15), 46–48

14.2 Outline the role of the genetic counsellor

Harrison S. (1995). Viewpoint. Practice tips: in the genes shop. *Nursing Standard*, **10**(8), 50

Visser A. and Bleiker E. (1997). Genetic education and counselling. *Patient Education and Counselling*, **32**(1–2), 1–7

Appendix A
Expressing numbers and SI units

Scientists have to deal with a very wide range of numerical values. For example, the mass of the Earth is about 5 980 000 000 000 000 000 000 000 kg, while that of a hydrogen atom is about 0.000 000 000 000 000 000 000 000 001 674 kg.[1] The range of values which health care professionals meet with is not quite so wide, but sometimes they do find themselves writing out a great many zeros!

> *Could we write very large or very small numbers in a more convenient way?*

Yes we could – we could use *scientific notation.*

Scientific notation

To change a large number to scientific notation move the decimal point to the left and place it between the first and second digits. Count the number of spaces moved and express this as a positive power of 10. In this way the mass of the Earth becomes 5.98×10^{24} kg.

> *So what about very small numbers?*

To change a small number to scientific notation, move the decimal point to the right and stop immediately after the first digit which is not a zero. In this way the mass of a hydrogen atom becomes 1.674×10^{-27} kg.

> *Is there not an even more convenient method of expressing very large and very small numbers?*

Indeed there is – we could use *numerical prefixes.*

Numerical prefixes

Numerical prefixes can be used to replace powers of 10. For example, instead of writing the diameter of a red blood cell (erythrocyte) as 0.000 008 m or 8×10^{-6} m

---

[1] Coleman G. J. and Dewar D. (1997). *The Addison-Wesley Science Handbook for Students, Writers and Science Buffs*. Don Mills, Ontario: Addison-Wesley

Table A.1 **Numerical prefixes.**

Powers of ten		
Power	Prefix	Symbol
10^3	kilo	k
10^6	mega	M
Sub-powers of ten		
Power	Prefix	Symbol
10^{-1}	deci	d
10^{-2}	centi	c
10^{-3}	milli	m
10^{-6}	micro	μ
10^{-9}	nano	n

we could write 8 μm (8 micrometres or 8 microns). Some important prefixes are given in Table A.1.

SI units

The Système Internationale (SI) is an agreed system of standard units which is used by most scientists throughout the world. It is also the system normally used in health care, although, for convenience, hospitals sometimes use non-SI units too. The system is comprised of a number of *base units* (Table A.2) and a much larger number of *derived units* (Table A.3), which, as their name implies, are derived from the base units.

Length

The SI unit of length is the *metre*, and this unit is certainly used by health care professionals. For example, the height of adults is recorded in metres. It may be more convenient to use *centimetres* to measure the height of infants and children. You will also see *millimetres* used too – for example, in expressing the length of a suture needle. Of course much smaller units are also used. For example, the bacteria

Table A.2 **Selected SI base units.**

Quantity	Symbol	Name	Abbreviation
Amount of a substance	n	mole	mol
Electrical current	I	ampere	A
Length	L	metre	m
Mass	m	kilogram	kg
Temperature	T	kelvin	K
Time	t	second	s

Table A.3 **Selected SI derived units.**

Quantity	Symbol	Unit
Acceleration	*a*	m/s^2 or m s^{-2}
Area	*A*	m^2
Electrical potential	*V*	volt (v)
Electrical resistance	*R*	ohm (Ω)
Energy	*E*	joule (J)
Force	*F*	newton (N)
Frequency	*f*	hertz(Hz)
Molality	*m*	mol/kg
Molarity	*M*	mol/l
Period	*T*	s
Power	*P*	watt (W)
Pressure	*P*	pascal (Pa)
Radiation absorbed dose	*D*	gray (Gy) – J/kg
Radioactivity	*A*	becquerel (Bq)
Solubility	*s*	g/l
Speed	*s*	m/s or m s^{-1}
Volume	*V*	m^3
Wavelength	λ	m
Weight	*W*	newton (N)
Work	*W*	joule(J)

which causes anthrax (*Bacillus anthracis*) is about 6 μm in diameter, while the smallpox virus is only about 300 nm in diameter.

Mass

The SI unit of mass is the *kilogram*. Adult body weight is normally expressed in kilograms. In contrast, the *gram* is used to express some drug doses. For example, a normal adult dose of aspirin is 1.2 g. However, many other drugs are given in smaller doses, expressed in *milligrams*. For example, the initial adult dose of the drug propranolol, which is used to treat hypertension (high blood pressure), is 40 mg. Occasionally drug doses are expressed in *micrograms*. For example, the adult daily maintenance dose of the heart drug digoxin may be as low as 125 mcg.

> *What's that again? mcg?*

Yes that's right. Although the normal numerical prefix for *micro* is the Greek letter μ, when μg is written by hand it may be mistaken for mg. This would clearly be unhelpful, since it would appear that 1000 times the required dose should be given! Consequently, doctors use mcg as the abbreviation for micrograms.

Volume

The most convenient unit of volume for health care professionals is not the cubic metre (m^3) but the *litre* (1 l = 10^{-3} m^3). Of course this unit is widely used elsewhere, including the measurement of volume of soft drinks and petrol. Bags of intravenous fluid are commonly of 1 l volume, but the *millilitre* (ml) is an even more commonly used unit. For example, the volume of drug solutions for injection is usually measured in millilitres. Incidentally, 1 ml is equal to $1\,cm^3$.

Appendix B
The periodic table[1]

[1] Sackheim G. I. (1996). *An Introduction to Chemistry for Biology Students*, 5th Edn. California: Benjamin/Cummings

Atomic weights are based on carbon-12.
Numbers in parentheses are the mass numbers of the most stable isotopes.

Key:
Atomic number	1
Name	Hydrogen
Symbol	H
Atomic mass	1.01

Transition elements

Main Table

I	II											III	IV	V	VI	VII	VIII or 0
1 Hydrogen H 1.01																	2 Helium He 4.00
3 Lithium Li 6.94	4 Beryllium Be 9.01											5 Boron B 10.81	6 Carbon C 12.01	7 Nitrogen N 14.01	8 Oxygen O 16.00	9 Fluorine F 19.00	10 Neon Ne 20.18
11 Sodium Na 22.99	12 Magnesium Mg 24.305											13 Aluminium Al 26.98	14 Silicon Si 28.09	15 Phosphorus P 30.97	16 Sulphur S 32.06	17 Chlorine Cl 35.45	18 Argon Ar 39.95
19 Potassium K 39.10	20 Calcium Ca 40.08	21 Scandium Sc 44.96	22 Titanium Ti 47.90	23 Vanadium V 50.94	24 Chromium Cr 52.00	25 Manganese Mn 54.94	26 Iron Fe 55.85	27 Cobalt Co 58.93	28 Nickel Ni 58.70	29 Copper Cu 63.546	30 Zinc Zn 65.38	31 Gallium Ga 69.72	32 Germanium Ge 72.59	33 Arsenic As 74.92	34 Selenium Se 78.96	35 Bromine Br 79.90	36 Krypton Kr 83.80
37 Rubidium Rb 85.47	38 Strontium Sr 87.62	39 Yttrium Y 88.91	40 Zirconium Zr 91.22	41 Niobium Nb 92.91	42 Molybdenum Mo 95.94	43 Technetium Tc 98.91	44 Ruthenium Ru 101.07	45 Rhodium Rh 102.91	46 Palladium Pd 106.42	47 Silver Ag 107.87	48 Cadmium Cd 112.41	49 Indium In 114.82	50 Tin Sn 118.69	51 Antimony Sb 121.75	52 Tellurium Te 127.60	53 Iodine I 126.90	54 Xenon Xe 131.30
55 Cesium Cs 132.91	56 Barium Ba 137.33	57 * Lanthanum La 138.91	72 Hafnium Hf 178.49	73 Tantalum Ta 180.95	74 Tungsten W 183.85	75 Rhenium Re 186.21	76 Osmium Os 190.20	77 Iridium Ir 192.22	78 Platinum Pt 195.09	79 Gold Au 196.97	80 Mercury Hg 200.59	81 Thallium Tl 204.37	82 Lead Pb 207.20	83 Bismuth Bi 208.98	84 Polonium Po (209)	85 Astatine At (210)	86 Radon Rn (222)
87 Francium Fr (223)	88 Radium Ra 226.03	89 † Actinium Ac (227)															

Lanthanides *

58 Cerium Ce 140.12	59 Praseodymium Pr 140.91	60 Neodymium Nd 144.24	61 Promethium Pm (145)	62 Samarium Sm 150.40	63 Europium Eu 151.96	64 Gadolinium Gd 157.25	65 Terbium Tb 158.93	66 Dysprosium Dy 162.50	67 Holmium Ho 164.93	68 Erbium Er 167.26	69 Thulium Tm 168.93	70 Ytterbium Yb 173.04	71 Lutetium Lu 174.97

Actinides †

90 Thorium Th 232.04	91 Protactinium Pa 231.04	92 Uranium U 238.03	93 Neptunium Np 237.05	94 Plutonium Pu (244)	95 Americium Am (243)	96 Curium Cm (247)	97 Berkelium Bk (247)	98 Californium Cf (251)	99 Einsteinium Es (252)	100 Fermium Fm (257)	101 Mendelevium Md (258)	102 Nobelium No (259)	103 Lawrencium Lr (260)

Metals

Non-metals

Noble gases

Glossary

absolute zero The temperature at which the kinetic energy of matter is at minimum, that is, $0\,K$ or $-273\,°C$.

acceleration The rate of change of velocity in metres per second per second (m/s^2 or ms^{-2}).

acetyl co-enzyme A (acetyl Co A) A substance that is formed when an acetyl group combines with a co-enzyme molecule. Acetyl Co A then enters the Krebs cycle.

acid A substance that donates hydrogen ions (H^+) during a chemical reaction.

acidity The state of an excess of free hydrogen ions (H^+).

acidosis The state that exists when blood is more acidic than normal (that is, pH is less than 7.35).

action potential A wave of current in nerve and muscle cells that results from the movement of ions across the plasma membrane.

activation energy The energy that must be supplied in order for a chemical reaction to proceed.

adenosine diphosphate (ADP) A substance that consists of the organic base adenine, a five carbon atom sugar and two inorganic phosphate groups. The compound that results when a phosphate group is removed from adenosine triphosphate.

adenosine triphosphate (ATP) The body's chemical form of energy; a molecule that consists of the organic base adenine, a five carbon atom sugar and three inorganic phosphate groups. ATP is produced when a third phosphate group is added to an adenosine diphosphate molecule.

adhesive forces Forces of attraction between unlike particles, for example, between water and the vessel in which it is contained.

aerobic A process that requires the presence of oxygen.

aerosol A colloidal mixture in which a solid or a liquid is dispersed in a gas.

agar A gel manufactured from seaweed and used as a medium on which to grow microorganisms.

albumin A group of water-soluble proteins, certain of which are present in blood.

albumin solution A colloidal solution derived from blood that contains 95% albumin and which is used as a plasma volume expander.

alcohols Compounds that contain a hydroxyl group (OH) attached to a carbon atom, the simplest of which conform to the general formula RCH_2OH.

aldehydes Compounds that contain a carbonyl group (C=O) at the end of a carbon atom chain and conform to the general formula RCHO.

aliphatic Organic compounds in which the ends of the carbon atom chain are not joined together; that is, a ring is not formed.

alkalosis The state that exists when blood is more alkaline than normal (that is, pH is greater than 7.45).

alkanes A group of saturated aliphatic hydrocarbons that conform to the general formula C_nH_{2n+2}.

alkenes A group of unsaturated aliphatic hydrocarbons that contain one double bond and conform to the general formula C_nH_{2n}.

alkynes A group of unsaturated aliphatic hydrocarbons that contain one triple bond and conform to the general formula C_nH_{2n-2}.

alleles Genes that occupy identical loci on homologous chromosomes.

alpha particle A helium nucleus ($^4_2$He) consisting of two protons and two neutrons.

amines A group of compounds that are related to ammonia (NH_3) in which one of the hydrogen atoms is replaced by a hydrocarbon chain. Amines contain an amino group (NH_2) and conform to the general formula RCH_2NH_2.

amino acids The component molecules of proteins – organic acids that contain a basic amino group (NH_2) and a carboxyl group (COOH).

Essential Those that are required in the diet since they cannot be synthesised in the body.

Non-essential Those that are not essential in the diet since they can be synthesised in the body provided the essential amino acids are present.

ammonia A gaseous compound that has the molecular formula NH_3.

ampere (amp/A) The SI unit of electric current, equivalent to the flow of one coulomb of charge per second.

amphetamines A group of amines that act as nervous system stimulants and that are abused due to their ability to produce euphoria.

amplitude The maximum displacement of a particle of a wave from its rest position.

anabolism The synthesis reactions of the body.

anaerobic A process that proceeds in the absence of oxygen.

anion A negatively charged ion.

anode A positively charged electrode.

anti-codon A triplet of nucleotides in tRNA that acts as a code for a particular amino acid.

apoenzyme The protein part of an enzyme; the non-protein part is referred to as a co-factor.

aqueous mixture A mixture of a substance in water.

aqueous solution A solution in which water is the solvent.

aromatic This term literally means having an aroma, but it has come to refer to cyclic organic compounds.

atom The smallest particle of an element.

atomic mass The average mass of all stable isotopes of an element expressed in atomic mass units and reflecting the relative proportions of the different isotopes.

atomic mass unit (amu/u) A unit equal to one twelfth of the mass of the most abundant isotope of carbon (^{12}C).

atomic number The number of protons in the nucleus of an atom.

basal metabolic rate (BMR) The metabolic rate measured in a subject who is fasted, rested and inactive but awake.

base A substance that accepts hydrogen ions (H^+) during a chemical reaction.

beta oxidation The process of forming acetyl CoA by removing acetyl groups from a carboxylic acid chain.

beta particle The name given to electrons that are emitted from an unstable nucleus as a consequence of the break-up of a neutron.

blood pressure The pressure exerted by blood upon the walls of the vessel in which it is circulating.
Diastolic During ventricular diastole (relaxation).
Systolic During ventricular systole (contraction).

Boyle's law The law that states that, provided the temperature of a gas does not change, then the pressure of a fixed amount of gas will increase as the volume decreases and vice versa.

Brownian motion The random movement of particles of a gas or a liquid.

buffer An aqueous solution that resists a change in pH upon the addition of an acid or a base.

capacitor A device used to store electric charge.

carbohydrate An organic compound containing only carbon, hydrogen and oxygen.

carbonic acid The compound that has the molecular formula H_2CO_3 and that is formed from a reaction between carbon dioxide and water.

carbonic anhydrase The enzyme responsible for catalysing the reaction between carbon dioxide and water to produce carbonic acid.

carboxylic acids Organic compounds that contain a carboxyl (COOH) group and conform to the general formula RCOOH; also known as fatty acids since they react with the alcohol glycerol to form neutral fats.

catabolism The chemical reactions of the body that involve the breaking down of substances.

catalyst A substance that increases the speed of a chemical reaction but remains unchanged by the reaction.

catecholamines A group of amines that includes adrenaline (epinephrine), nora-drenaline (norepinephrine) and dopamine.

cathode A negatively charged electrode.

cathode ray A stream of electrons produced in a cathode ray tube.

cation A positively charged ion.

cell The smallest unit of living matter.

cell membrane *See* plasma membrane.

Celsius scale A scale used to measure temperature in which there are 100 divisions between the melting point of ice and the boiling point of water. Each division is referred to as one degree Celsius ($°C$).

centre of gravity The point at which the entire weight of an object can be thought of as acting for the purpose of considering torque.

Charles' law The law that states that if the pressure of a gas is constant, its volume is directly proportional to its absolute temperature.

chromatin Literally 'coloured material'; the thread-like mass of DNA and its associated proteins that is found within the nucleus of a cell.

chromosome Literally 'coloured body'; the structures that are formed within the nucleus of a cell by the super-coiling of DNA and its associated proteins.

chylomicrons Water-soluble lipoprotein droplets present in blood.

codon A triplet of nucleotides in DNA or mRNA that acts as a code for a specific amino acid.

co-enzyme A type of co-factor; a non-protein organic molecule that forms a component of an enzyme and that is essential for its activity.

co-enzyme A A co-enzyme that is derived from the B group vitamin pantothenic acid and that combines with an acetyl group to form acetyl co-enzyme A.

co-factor A non-protein component of an enzyme such as a co-enzyme or an inorganic ion that is essential for the activity of the enzyme.

cohesive forces The forces of attraction that exist between like molecules.

colloid Also referred to as a colloidal suspension or colloidal mixture; an aqueous mixture in which the particles that are dispersed in the fluid are smaller than those of a mechanical suspension but too large to form a true solution.

combination reaction A type of chemical reaction in which two or more reactants combine to form a new substance.

compound A substance that is comprised of atoms of more than one element linked by chemical bonds.

concentration An expression of the relative proportions of solute to solvent in an aqueous mixture commonly given as a percentage or in terms of a number of moles or millimoles of solute per litre of solvent (mol/l or mmol/l).

concentration gradient A difference in concentration between two regions.

condensation reaction A type of chemical reaction in which water is produced.

conductor A substance through which electrical current flows or through which heat readily passes.

control centre A component of a homeostatic control mechanism; a part of the body that regulates a controlled condition.

controlled condition A variable, such as body temperature, that is regulated by a homeostatic control mechanism.

convection A process involved in heat loss whereby air that has been warmed rises and is replaced by cooler, denser air.

covalent bond The chemical linking of atoms through the sharing of electrons.

covalent compound A compound that is formed when atoms are joined by covalent bonds.

crenation The shrinking of a cell so that it develops a scalloped surface due to the outward movement of water by osmosis.

crystal A solid that has a geometric form.

crystalloid solution Solutions in which the solute has a crystalline structure in the solid phase.

cyclic compound An organic compound in which the ends of the carbon atom chain are joined together to form a ring structure.

cytochrome chain *See* electron transport chain.

cytoplasm Literally 'cell substance'; the living material of the cell, including the cellular organelles but excluding the nucleus.

Dalton's law The law that states that in a mixture of gases, the total pressure is a sum of the pressures exerted by each of the gases alone.

deamination The removal of an amino (NH_2) group from a molecule.

decibel (dB) The unit of measurement of the relative loudness of sound.

decomposition A chemical reaction in which a single reactant is broken down to produce one or more products.

defibrillation The application of an electric current to the heart for the purpose of converting ventricular fibrillation to sinus rhythm.

defibrillator The electrical device used to administer an electric current for the purpose of defibrillation.

dehydration The condition that results from a negative water balance.

denaturation A change in the three-dimensional structure of an enzyme that results in its inactivity.

density An expression of the mass of a substance per unit volume.

deoxyribonucleic acid The nucleic acid that consists of nucleotides made up of the sugar deoxyribose, a phosphate group and one of four nitrogenous bases (adenine, thymine, guanine and cytosine). Genetic information is encoded in the sequence of these bases.

depolarisation The loss of the resting membrane potential that accompanies the influx of sodium ions during an action potential.

dialysate A solution used in dialysis.

dialysis The exchange of solutes and water between blood and a dialysate across a man-made or biological membrane.

    haemodialysis In this form an extra-corporeal circulation is established through a dialyser (artificial kidney) and dialysis occurs across an artificial membrane.

    peritoneal In this form the dialysate is infused into the peritoneal cavity and dialysis occurs across the peritoneum.

diastole Relaxation of the heart; the term may be applied to the atria or the ventricles.

diathermy The use of high-frequency alternating current for the purpose of cutting tissue or coagulating blood so as to seal blood vessels.

diatomic The term literally means consisting of two atoms; it is used of the gases oxygen (O_2), nitrogen (N_2) and hydrogen (H_2) – the so-called diatomic gases.

diffusion The movement of solute particles from an area of high solute concentration, through a semi-permeable membrane, to an area of low solute concentration until an equilibrium is reached.

diploid Possessing the full number of chromosomes that is characteristic of somatic cells; that is 46 in the human.

disaccharide A compound that is formed from the joining together of two simple sugars (monosaccharides).

displacement A type of chemical reaction in which a less reactive element is displaced from a molecule by a more reactive element.

dissociation The breakdown of a substance into ions when dissolved in solution.

dissolving The act of forming a true solution.

Doppler effect The apparent change in frequency of sound due to the motion of the source relative to the listener.

earth The conductive connection that forms the route by which current flows to earth.

effector An organ that produces an effect as a result of stimulation by the nervous system or the effect of a hormone. Effectors include muscles and endocrine glands.

electric current The flow of electrons from a region of high electron density to a region of low electron density. The SI unit of current is the ampere (amp).

electrode A solid conductor through which electric current passes in a variety of electrical devices.

electrolyte A substance that dissociates into ions in solution and that accounts for the ability of the solution to conduct electricity.

electromagnet A magnet in which the electromagnetic force is dependent on the flow of electric current.

electromagnetic radiation Energy in the form of transverse waves that do not require a material medium through which to pass but which result from the oscillation of electrical and magnetic fields.

electron A subatomic particle of negligible mass that carries a single negative charge and orbits the nucleus of the atom.

electron cloud The region around the nucleus of an atom that is occupied by electrons.

electron transport chain A sequence of linked chemical reactions in which electrons are passed from one substance to another and ATP is formed. The reactions take place in the mitochondria.

element A substance that is composed of identical atoms; that is, they all have the same atomic number.

elemental particle A sub-atomic particle such as a proton, a neutron or an electron.

emulsion A colloidal solution in which one liquid is dispersed in another.

endergonic A chemical reaction in which the products possess more energy than the reactants and in which energy is therefore required for the reaction to proceed.

endoscope A rigid or fibre-optic instrument that is used to examine hollow organs or body cavities.

energy The ability to do work.

energy levels/shells The orbitals in which the electrons of an atom are located.

enzymes Protein catalysts.

esters Organic compounds formed by a reaction between a carboxylic acid and an alcohol and that conform to the general formula RCOOR.

ethanoic acid The carboxylic acid with the molecular formula C_2H_3COOH; also known as acetic acid and, in dilute solution, as vinegar.

evaporation A change in state from liquid to gas.

exergonic A chemical reaction in which the products possess less energy than the reactants and in which energy is liberated during the reaction.

extracellular fluid Fluid that is located outside cells; the term includes blood and interstitial fluid.

fats *See* lipids.

fatty acids *See* carboxylic acids.

feedback mechanism A homeostatic response to a physiological change.
   negative A type of feedback mechanism in which change is not reinforced but suppressed – most homeostatic mechanisms are of this type.
   positive A type of feedback mechanism in which change is reinforced.

fibrillation A highly disorganised and uncoordinated heart rhythm that may affect either the atria or the ventricles. Ventricular fibrillation is one cause of cardiac arrest.

filtration The process whereby substances are separated by means of a membrane that has pores of a given size.

fluid *See* liquid.

force An influence that is responsible for changing the state of rest or uniform motion of an object; the SI unit of force is the newton (N).

formula An expression of the chemical composition of a compound.
   molecular States the number of each of the atoms present.
   structural Shows the number of each of the atoms present and their structural arrangement too.

friction The force that results from the movement of one object across another.

fulcrum A point of support about which turning force acts.

functional group A group of two or more atoms that are present in a number of different molecules and that confer similar properties upon all the molecules in which they are present.

gamete A sex cell: spermatozoa in males and oocytes in females.

gamma rays Electromagnetic radiation that has a shorter wavelength than X-rays and is emitted by the nuclei of certain radioisotopes.

gas The state of matter in which particles of matter are able to move independently of each other. Gases fill the container in which they are placed.

gaseous *See* gas.

gel A colloidal suspension formed by the dispersal of a liquid in a solid. A gel exists on the borderline of solidity.

gene The smallest unit of heredity corresponding to a section of DNA that codes for a specific mRNA; that is, a section of DNA located between a start codon and a stop codon.

genetic code A code for the synthesis of protein in the form of a sequence of bases that form part of a nucleic acid molecule.

genetics The study of genes and heredity.

genome The complete gene complement of an organism.

genotype The genetic composition of an individual.

gluconeogenesis The manufacture of glucose from non-carbohydrate sources such as amino acids.

glyceride A type of lipid that is formed from the reaction between the alcohol glycerol and one, two or three carboxylic acids. Glycerides are classified as monoglycerides, diglycerides and triglycerides.

glycerol A three-carbon alcohol that reacts with carboxylic acids to form glycerides.

glycogen The body's carbohydrate store of energy; a polysaccharide consisting of branched chains of glucose molecules.

glycogenesis The process whereby glycogen is manufactured from glucose.

glycogenolysis The breakdown of glycogen to release glucose.

glycolysis A process that occurs in the cytoplasm; it is the first step in glucose catabolism and results in the formation of pyruvic acid.

glycoprotein A protein that has a sugar molecule attached.

glycosuria The presence of glucose in the urine.

gravity The force of attraction between objects; especially between an object and the Earth.

group A vertical column in the periodic table of elements.

haemodialysis *See* dialysis.

half-life

biological The time taken for the amount of a specific isotope that has entered the body to be reduced by a half as a consequence of natural biological processes.

effective The time taken for the amount of a specific isotope that has entered the body to be reduced by half as a consequence of its radioactive decay and its biological elimination.

physical The time taken for half of a given number of atoms of a radioisotope to decay.

haploid Possessing half the number of chromosomes that is characteristic of somatic cells; that is 23 in the human.

heat Energy possessed by a substance as a consequence of the vibration of its particles.

Henry's law The law that states that the mass of a gas that will dissolve in water at a given temperature is proportional to the partial pressure of the gas and to its solubility coefficient.

heterozygous The presence of different alleles of the same gene on homologous chromosomes.

homeostasis The concept of a relatively stable internal environment.

homologous chromosomes A pair of morphologically identical chromosomes that carry genes for the same traits.

homozygous The presence of identical alleles on homologous chromosomes.

hormone A chemical substance that is secreted directly into the bloodstream by an endocrine (ductless) gland and that produces an effect in a specific target tissue elsewhere in the body.

hydrocarbon A compound composed solely of hydrogen and carbon atoms.

hydrogen bond Weak forces of attraction that exist between polar molecules or between different regions of the same molecule in which hydrogen atoms with a partial positive charge are attracted to atoms with a partial negative charge.

hydrolysis A type of decomposition reaction that requires a water molecule.

hydrophilic Water-attracting.

hydrophobic Water-repelling.

hydrostatic pressure The pressure that results from the collision of particles of a fluid with the walls of the vessel in which it is contained.

hydroxyl ion The OH^- ion.

hyperglycaemia An elevated blood sugar level.

hyperosmotic A solution that has a higher osmotic pressure than another.

hypertension An elevated blood pressure.

hyperthermia An abnormally elevated body temperature.

hypertonic A solution that is capable of causing crenation of cells.

hypoglycaemia An abnormally low blood sugar level.

hypo-osmotic A solution that has a lower osmotic pressure than another.

hypotension An abnormally low blood pressure.

hypothermia An abnormally low body temperature.

hypotonic A solution that is capable of causing the plasmolysis of cells.

hypoxaemia A low partial pressure of oxygen in arterial blood (P_aO_2).

hypoxia Decreased availability of oxygen to the tissues.

induced fit A model of enzyme action in which the active site is believed to undergo a change in shape in order to accommodate the substrate.

inertia The tendency of an object to resist a force that is applied to it.

infrared Electromagnetic radiation that has a wavelength of 10^{-5}–10^{-4} m.

inorganic chemistry The branch of chemistry that deals with compounds that do not contain the element carbon.

insulator A substance through which electric current and heat do not readily pass.

intercellular fluid *See* interstitial fluid.

interstitial fluid Fluid that surrounds cells and that is outside the vascular compartment.

intravascular fluid Fluid in the vascular compartment; that is, blood.

intravenous infusion The administration of a fluid directly into a vein.

inverse square law The law that states that the intensity of radiation is inversely proportional to the square of the distance from the source.

ion Atoms or groups of atoms that have become charged as a consequence of the loss or gain of electrons.

ionic bond The chemical link between ions that arises as a result of the attraction of ions carrying opposite charges.

ionic compound A compound that results from ionic bonding.

ionisation A type of decomposition reaction in which a compound dissociates into ions.

ionising radiation Radiation in the form of electromagnetic waves or particles that is capable of producing ions.

isomers Compounds that have the same molecular formula but different structural formulae.

iso-osmotic A fluid that has the same osmotic pressure as another.

isotonic A solution that causes neither crenation or plasmolysis of cells.

isotopes Atoms with the same atomic number but with different mass numbers. Isotopes are therefore atoms of the same element but they possess different numbers of neutrons.

joule (J). The SI unit of energy.

karyotype The arrangement of chromosomes based upon their size, shape and position of centromeres.

kelvin (K) The SI unit of temperature.

ketoacidosis/ketosis Metabolic acidosis that results from the production of excess ketoacids.

ketoacids Ketone molecules that also possess an acidic carboxyl (COOH) group.

ketones Organic compounds that contain the carbonyl functional group (C=O) part-way along a carbon atom chain and that conform to the general formula RCOR.

kilogram (kg) The SI unit of mass.

kinetic energy The energy of movement.

Krebs cycle A series of biochemical reactions that take place in the matrix of mitochondria in which electrons are transferred to coenzymes and as a consequence carbon dioxide and ATP are formed.

laser An acronym for light amplification by the stimulated emission of radiation. Laser light consists of monochromatic waves of visible or invisible electromagnetic radiation all of which are in phase.

latent heat The energy required to produce a change in state, for example, to change a solid into a liquid (latent heat of fusion) or to change a liquid into a gas (latent heat of vaporisation).

lever A bar that connects a mechanical force to an object.

light Electromagnetic radiation with a wavelength range of 400–800 nm.

lipids A diverse group of organic compounds, including glycerides (neutral fats) that share the properties of insolubility in water and solubility in organic solvents.

lipoprotein A molecule that has a lipid and a protein component.

liquid A state of matter in which particles are weakly attracted to each other. Liquids conform to the shape of the container into which they are poured.

magnet An object that demonstrates the property of magnetism.

magnetic field The region in which magnetic force can be demonstrated.

magnetism The properties and effects of a magnetic substance.

mass The amount of matter of a substance.

mass number The sum of the number of protons and the number of neutrons in an atom.

matter Anything that occupies space and has mass.

mechanical suspension A mixture in which particles are dispersed but not dissolved in water.

meiosis Cell division characterised by the formation of four daughter cells, each of which possesses the haploid number of chromosomes. Meiosis takes place in the gonads and leads to the production of gametes (sex cells).

meniscus The bending of the surface of a liquid due to the relative strength of the forces of cohesion and adhesion.

metabolic rate An expression of the amount of energy released in the body in a given period of time.

metabolism The sum of all the biochemical reactions of the body.

metre (m) The SI unit of length.

microwaves Electromagnetic radiation with a wavelength of 1 mm–30 cm.

minerals Inorganic substances required in the diet.

mitochondria Cellular organelles that possess an inner membrane and an outer membrane and that are the site of the Krebs cycle and the electron transport chain.

mitosis Cell division characterised by the formation of two daughter cells, each of which possesses the diploid number of chromosomes.

mixture A substance that contains separate elements or compounds that are not bound together.

molality An expression of concentration in terms of the number of moles or millimoles of solute per kilogram of solvent (mol/kg or millimol/kg)

molarity An expression of concentration in terms of the number of moles or millimoles of solute per litre of solvent (mol/l or millimol/l).

mole The relative atomic mass of an element or the molecular weight of a compound in grams; one mole of a substance contains 6×10^{23} particles.

molecular formula *See* formula.

molecular mass The sum of the relative atomic masses of the atoms that make up a compound.

molecule Two or more atoms joined by a covalent bond.

momentum The product of the mass and the velocity of an object.

monosaccharide A simple sugar – one that cannot be broken down into a simpler sugar.

nebuliser A device in which the Bernoulli effect is used in order to produce an aerosol. Nebulisers are used for dispersing liquid medication in a gas so that it may be inhaled.

neoplasm Literally new growth; a swelling due to abnormal and excessive cell growth.

neutral Having a pH of 7.

neutral fats Triglycerides.

neutralisation The process of combining acids and bases to produce a neutral solution.

neutron A sub-atomic particle found in the nucleus that has identical mass to that of a proton but that carries no charge.

nucleoplasm Literally nuclear substance; a colloidal fluid, enclosed by the nuclear membrane in which chromatin is suspended.

nucleotide A molecule formed from a sugar, a nitrogenous base and an inorganic phosphate group. Nucleotides are the component molecules of nucleic acids.

nucleus (a) the central part of an atom that contains protons and neutrons.

(b) a cellular organelle that contains the genetic material.

oedema The condition of an excess of interstitial fluid.

oncotic pressure The fraction of the plasma osmotic pressure for which plasma proteins are responsible.

organelles Literally 'little organs'; membrane-bound structures found within cells that perform specific functions on behalf of the cell.

organic chemistry The branch of chemistry that deals with compounds that contain carbon.

osmol A unit that corresponds to the number of moles of a solute multiplied by the number of particles that the solute dissociates into.

osmolality The number of osmols per kilogram of solution.

osmolarity The number of osmols per litre of solution.

osmosis The movement of solvent particles across a semi-permeable membrane from an area of high solvent concentration to an area of low solvent concentration.

osmotic pressure The pressure required to prevent osmosis.

oxidation A type of chemical reaction in which a substance acquires an oxygen atom, loses a hydrogen atom or loses an electron.

partial pressure The pressure exerted by one of the gases present in a mixture of gases or dissolved in solution.

particle Any unit of matter.

partner exchange reaction A type of chemical reaction in which there is a substitution of elements between two molecules.

pascal (Pa) The SI unit of pressure; a pressure of $1\,kPa$ exists when a force of one newton ($1\,N$) acts over an area of one square metre (m^2).

Pascal's law The law that states that the pressure exerted upon a static liquid is transmitted uniformly throughout the liquid.

peptide bond The bond that is formed between the amino group of one amino acid and the carboxyl group of a second amino acid.

period A horizontal row of the periodic table.

periodic table A table in which the chemical elements are placed in groups (vertical columns) and periods (horizontal rows). The order of the elements is such that atomic number increases from left to right along a period, while elements with similar properties are found in the same group.

peritoneal dialysis *See* dialysis.

phenotype The physical expression of the genotype.

phospholipid A group of lipids that consist of the alcohol glycerol combined with two carboxylic acids and one phosphate group.

phosphorylation A type of combination reaction that involves a phosphate group.

pH scale A numerical scale (0–14) corresponding to the negative logarithm of the hydrogen ion concentration.

pitch The subjective perception of the frequency of a sound wave.

plasma expander A colloidal suspension that is administered intravenously and that increases plasma osmotic pressure, thus assisting in the maintenance of plasma volume.

plasma membrane The membrane that separates the contents of a cell from the environment.

plasma protein A protein such as albumin that is normally present in blood plasma.

plasma protein solution A colloidal suspension derived from blood plasma that contains 85% albumin.

plasmolysis The rupture of the plasma membrane that occurs as a consequence of the movement by osmosis of water into the cell.

polar molecule A covalent molecule in which there is an unequal sharing of electrons so that part of the molecule has a slight negative charge and part has a slight positive charge.

pole The end of a bar magnet.

polyatomic ion/polyatomic radical An ion that is made up of atoms of more than one element but which behaves as though it were an ion of a single element.

polypeptide A chain of many amino acids joined together by peptide bonds.

polysaccharide A carbohydrate that consists of many sugar molecules joined together.

pore A small opening through a membrane.

potential difference The difference in electrical potential, measured in volts, between two points.

potential energy The energy possessed by virtue of the position or chemical composition of matter.

power The rate of doing work measured in watts (W).

pressure The force exerted per unit surface area over which the force acts.

products The substances that are produced as a consequence of a chemical reaction.

proteins Molecules consisting of long chains of at least 100 amino acids.

proton A sub-atomic particle found in the nucleus; it has the same mass as a neutron but it carries a single positive charge.

protoplasm All the living substance of a cell that is contained within the plasma membrane.

pulley A grooved wheel on an axle that is used to change the direction of a force or to produce a mechanical advantage.

pyrexia An abnormally elevated body temperature that results from the action of a pyrogen.

radiation Energy in the from of electromagnetic waves or particle that radiate from a source.

radioactivity The emission of electromagnetic radiation.

radioisotope An isotope that emits radiation.

rarefaction A reduction in pressure that follows a compression in a longitudinal wave.

reactants The substances that take part in a chemical reaction.

receptors Sensory nerve endings that are capable of responding to a stimulus.

reducing agent A substance that is capable of reducing another substance.

reduction A type of chemical reaction in which a substance loses an oxygen atom, gains a hydrogen atom or gains an electron.

reflection A term that is used to describe the manner in which a wave 'bounces off' a surface that it strikes.

refraction The bending of a beam of light as it moves from one transparent medium to another of different density.

relative atomic mass The average mass, in atomic mass units, of a single atom of an element that takes into account the relative proportions of the different isotopes of that element.

repolarisation The restoration of the resting membrane potential following depolarisation.

resistance The property of a material that opposes the flow of electrons. The SI unit of resistance is the ohm (Ω).

resonance The tendency of two objects with identical resonant frequencies to vibrate in time with each other.

resonant frequency The frequency at which an object vibrates maximally.

respiration
  external The act of breathing (ventilation).
  internal The energy liberating reactions of the cell.

reversible reaction A type of chemical reaction in which the reactants can be reformed from the products.

ribonucleic acid Nucleic acid that consists of a singe strand of nuceotides. Each nucleotide consists of the sugar ribose, a phosphate group, and one of four nitrogenous bases (adenine, uracil, guanine and cytosine). Three forms exist: messenger RNA (mRNA), ribosomal RNA (rRNA) and transfer RNA (tRNA).

ribosomes Organelles that are the site of protein synthesis.

salt An ionic compound that does not dissociate into hydrogen ions (H^+) or hydroxyl ions (OH^-).

saturated fat A lipid in which the carbon atoms are linked by single covalent bonds.

scalar A quantity that has size but not direction.

selectively permeable membrane A biological membrane the permeability of which may change due to the opening and closing of channels through it.

semi-permeable membrane A membrane that has pores of a certain diameter. Particles of a smaller diameter can therefore diffuse through it, while those with a larger diameter can not.

set-point The level or value of a controlled condition, such as body temperature, that the body strives to maintain.

sol A colloid composed of a solid dispersed in a liquid.

solid A state of matter in which the particles are not free to move but are held by forces of attraction in relatively fixed positions.

solute The least abundant component of a solution.

solution An homogenous mixture in which the solute particles exist as single ions, atoms or molecules. These particles do not settle out and cannot be separated by simple filters or semi-permeable membranes. The Tyndall effect cannot be demonstrated with true solutions.

solvent The most abundant component of a solution.

specific gravity The ratio of the density of a liquid to that of water. For example, the normal specific gravity of urine is in the range 1.001–1.035.

sphygmomanometer A manometer that is used in the non-invasive measurement of arterial blood pressure.

stimulus An event that provokes a change in a controlled condition.

structural formula *See* formula.

substitution reaction *See* displacement.

substrate The substance on which an enzyme acts.

surface tension The term used to describe the observation that the surface of a liquid behaves as though tense as a consequence of forces of cohesion. Surface tension is responsible for droplet formation.

surfactant A substance that lowers surface tension.

synthesis reaction *See* combination reaction.

torque Turning force.

transcription Formation of a strand of mRNA in which the sequence of bases is determined by the sequences of bases in DNA. In effect the genetic code is transcribed (copied) from DNA to mRNA. Transcription takes place in the nucleus.

transformer An electrical device that is used to produce a change in voltage.

translation The synthesis of new protein on a ribosome; the sequence of amino acids in the protein is determined by the sequence of bases in the mRNA attached to the ribosome. In effect the genetic code carried by mRNA is translated into the sequence of amino acids in protein.

triglycerides Neutral fats formed from the reaction between the alcohol glycerol and three carboxylic acid chains.

triplet Three nucleotides that form part of a nucleic acid molecule and act as a code for a specific amino acid.

tumour *See* neoplasm.

Tyndall effect The scattering of a beam of light that is produced by mechanical suspensions and colloidal suspension but not by true solutions.

ultrasound Sound waves of a frequency that is too high to be heard by the human ear.

ultraviolet Electromagnetic radiation that has a wavelength range of 20–390 nm.

unsaturated fat A lipid in which some of the carbon atoms are linked by double bonds. In monounsaturates there is a single double bond, while in polyunsaturates there are more than one.

valence shell The outermost energy shell of an atom; only the electrons in this shell take part in chemical bonding.

valency A numerical expression of the number of chemical bonds that en element can form.

vector A quantity that possesses both magnitude and direction.

Venturi barrel A device in which the Bernoulli effect is used in order to entrain one gas into another. Venturi barrels are important components of certain oxygen masks.

vitamins Organic molecules that are essential for life and that are required in relatively small amounts.

volt (V) The SI unit of potential difference.

watt (W) The SI unit of power.

wavelength The distance between two identical points on adjacent waves.

zwitterion A covalent compound that possesses charged ionic regions.

Index